Elements of Linear Algebra

Elements of Linear Algebra

P.M. Cohn

Department of Mathematics,
University College London, UK

CHAPMAN & HALL

London · Glasgow · Weinheim · New York · Tokyo · Melbourne · Madras

Published by Chapman & Hall, 2–6 Boundary Row, London SE1 8HN, UK

Chapman & Hall, 2–6 Boundary Row, London SE1 8HN, UK

Blackie Academic & Professional, Wester Cleddens Road, Bishopbriggs, Glasgow G64 2NZ, UK

Chapman & Hall GmbH, Pappelallee 3, 69469 Weinheim, Germany

Chapman & Hall USA, One Penn Plaza, 41st Floor, New York NY 10119, USA

Chapman & Hall Japan, Kyowa Building, 3F, 2-2-1 Hirakawacho, Chiyoda-ku, Tokyo 102, Japan

Chapman & Hall Australia, Thomas Nelson Australia, 102 Dodds Street, South Melbourne, Victoria 3205, Australia

Chapman & Hall India, R. Seshadri, 32 Second Main Road, CIT East, Madras 600 035, India

---

First edition 1994

Typeset in 10/12 pt Times by Thomson Press (India) Ltd., New Delhi

Printed in Great Britain by Clays Ltd, Bungay, Suffolk

ISBN 0 412 55280 9

A catalogue record for this book is available from the British Library

∞ Printed on permanent acid-free text paper, manufactured in accordance with ANSI/NISO Z39.48-1992 and ANSI/NISO Z39.48-1984 (Permanence of Paper).

To the memory of my great^{7} grandparents (1.37%)
Elias Gomperz Cleve (d. 1689)
Sara Miriam Gomperz, née Benedict (d. 1691)

Contents

Preface

In 1958 I wrote a brief account of *Linear Equations*, which was followed by an equally short book on *Solid Geometry*. Both books have been surprisingly popular, so in making the first major revision in over 30 years I have endeavoured to retain the simple style of the original. It seemed practical to combine both in one book, since they essentially deal with the same subject from different points of view. The revision also provided an opportunity to include some basic topics left out of the earlier versions.

The core of the book is formed by Chapters 2 and 4, where the solution of systems of linear equations is discussed, taking the regular case (systems with a unique solution) in Chapter 2, followed by the general case in Chapter 4. To facilitate the discussion, vectors are introduced in Chapter 1 and matrices in Chapter 3. All this can be done quite concisely and, together with the account of determinants in Chapter 5, constitutes the old *Linear Equations*. Of course the whole has been revised and clarified, with additional examples where appropriate.

The next two chapters, 6 and 7, together with parts of Chapters 8 and 9, represent *Solid Geometry*. The short chapter on spheres from that text has been relegated to the exercises, but transformations have been discussed in more detail, and some attention is given to the n-dimensional case (which is of practical importance in statistics, mechanics, etc.). An account of linear mappings between vector spaces has also been included; it follows on naturally and provides a motivation for the description of normal forms in Chapter 8. This concerns mainly the canonical form of matrices under similarity transformations (Jordan normal form), but also the transformation to diagonal form of symmetric matrices by congruence transformations, which may be regarded as a special case of similarity. Some care has been taken to arrange matters so that the more difficult reduction to the Jordan normal form comes at the end of the chapter and so can be omitted without loss of continuity. Throughout the text there are brief historical remarks which situate the subject in a wider context and, it is hoped, add to the reader's interest.

The remaining chapters, 9 and 10, are devoted to applications. Systems of linear equations are used so widely that it was necessary to be very selective; it would have been possible to present amusing examples from many fields, but it seemed more appropriate to confine attention to cases which used methods of mathematical interest. Chapter 9, on algebraic and geometric applications, deals with the simultaneous reduction of two quadratic forms and the resulting classification of quadric surfaces. Section 9.6, on linear programming, describes the basic problem, of solving systems of linear inequalities, and its solution by the simplex algorithm; like many of these later sections, it can only give a taste of the problem, and the interested reader will be able to pursue the topic in the specialized works listed in the Bibliography at the end. Another normal form, the polar form, is described, and the chapter ends with a brief account of the method of least squares, which provides a good illustration of the use of matrices.

Chapter 10 deals with applications to analysis. The algebraic treatment of linear differential equations goes back to the nineteenth century, and in fact some of the algebraic methods, such as the theory of elementary divisors, were originally developed in this context. This theory goes well beyond the framework of the present book, where we only treat the two basic applications – the use of the Jordan normal form to solve linear differential equations with constant coefficients, with illustrations from economics, and the application of the simultaneous reduction of quadratic forms to find the normal modes of vibration of a mechanical system. Much of the theory has an analogue for linear difference equations, which is described and illustrated by the Fibonacci sequence. A more recent development is the use of algebraic methods in the calculus of functions of several variables, which in essence is the recognition of the derivative as a linear mapping of vector spaces. A brief outline explains the role of Jacobians, Hessians and the use of the Morse lemma in the study of critical points. The contraction principle of functional analysis is illustrated by an iterative method to invert a matrix.

The book is not primarily aimed at mathematics students, but addresses itself to users of mathematics, who would follow up its study with a book devoted to their special field. To quote from the original preface to *Linear Equations*,

> the book is not merely intended as a reference book of useful formulae. Rather it is addressed to the student who wants to gain an understanding of the subject. With this end in mind, the exposition has been limited to basic notions, explained as fully as possible, with detailed proofs. Some of the proofs may be omitted at first, but they do form an integral part of the treatment and, it is hoped, will help the reader to grasp the underlying principles.

To assist the reader, exercises have been included at the end of each chapter, with outline solutions at the end of the book. The reader should bear in mind that a proof is often more easily understood by first working through a particular case.

I should like to thank my colleagues Bert Wehrfritz and Mark Roberts for their comments on the manuscript (and the latter also for his help with proof reading), and Nick Rau for providing me with the illustrations from economics in Chapter 10. I am also indebted to the staff of Chapman & Hall for carrying out their tasks with the usual courtesy and efficiency, with a particular word of thanks to their copy editor Mr Richard Leigh who saved me from some mathematical errors. It is perhaps too much to hope that no errors remain, but I shall be glad of any reader's help in diminishing their number in future editions.

P. M. Cohn
London
March 1994

Note to the reader

No specific knowledge is presupposed, although in Chapter 10 some familarity with elementary calculus is expected. The reader is strongly advised to work through the examples in the text as they occur, and to attempt the exercises at the end of each chapter. At a first reading, proofs may be skipped (or better, skimmed), but it is important to go back to them, since studying them will help an understanding of the text.

Introduction

A basic part of linear algebra is concerned with the solution of linear equations. This will occupy us in the first few chapters and will provide an opportunity to introduce vectors, matrices and determinants. It is followed by a brief introduction to coordinate geometry, which illustrates the use of vectors and matrices, and introduces the notion of a linear mapping. A separate chapter explains normal forms of matrices, an important topic with many applications in both algebra and analysis, some of which are outlined in the two final chapters.

To solve a system of linear equations means to find all the values of the unknowns which satisfy the equations (here **linear** means 'of the first degree in the unknowns'). For problems with two or three unknowns this is not difficult, but when there are more unknowns a systematic method is needed. We begin by looking at some examples.

The simplest type of system is that of one equation in one unknown:

$$ax = k \tag{0.1}$$

Everyone will recognize the solution of this equation to be $x = k/a$, but that is true only when $a \neq 0$. To be quite precise, we have to distinguish two cases:

(i) $a \neq 0$. In this case there is exactly one value of x satisfying (0.1), for any value of k, namely $x = k/a$.
(ii) $a = 0$. In this case no value of x satisfies (0.1) if $k \neq 0$; but if $k = 0$, then every value of x satisfies (0.1).

This distinction is fundamental in all that follows. For when we try to solve a system of several linear equations in a number of unknowns (not necessarily the same as the number of equations), the process of solution may be more complicated, but there are again two cases corresponding to (i) and (ii) above, with a subdivision of case (ii) according to the value of the right-hand sides of the equations.

The following example illustrates this in the case of two equations in two unknowns. Consider the equations

$$3x + 2y = 7 \tag{0.2a}$$

$$2x + 5y = 12 \tag{0.2b}$$

If x and y exist to satisfy (0.2), we eliminate y by subtracting twice the second equation from 5 times the first, and we eliminate x by forming 2.(0.2a) − 3.(0.2b):

$$(5.3 - 2.2)x = 5.7 - 2.12 \tag{0.3a}$$

$$(2.2 - 3.5)y = 2.7 - 3.12 \tag{0.3b}$$

which simplifies to $x = 1$, $y = 2$. So this is the only pair of values which can satisfy (0.2), and it does in fact satisfy (0.2), as is easily checked. We see that there is exactly one set of values of x and y satisfying (0.2), or as we shall say, the equations (0.2) have a **unique solution**. We remark that it would be enough to solve one of the equations (0.3), say (0.3b), giving $y = 2$. We can then substitute this value for y in either of the equations (0.2), say (0.2a), and obtain an equation for x:

$$3x = 7 - 2y = 7 - 4 = 3$$

which yields $x = 1$.

The same method applied to the system

$$3x + 2y = 7 \tag{0.4a}$$

$$6x + 4y = 11 \tag{0.4b}$$

gives, on elimination of y and x in turn,

$$(4.3 - 2.6)x = 4.7 - 2.11 \tag{0.5a}$$

$$(6.2 - 3.4)y = 6.7 - 3.11 \tag{0.5b}$$

Here the left-hand sides are zero whatever the values of x and y, but the right-hand sides are not zero. We conclude that (0.4) has no solution, a fact which could have been found directly by observing that the left-hand side of (0.4b) is twice the left-hand side of (0.4a), so that the equations can only have a solution if the same relation holds between the right-hand sides. This is not so in (0.4), but in the system

$$3x + 2y = 7 \tag{0.6a}$$

$$6x + 4y = 14 \tag{0.6b}$$

it is the case: (0.6b) is just twice (0.6a), so any values of x and y satisfying (0.6a) also satisfy (0.6b) and so form a solution of (0.6). There are infinitely many such solutions: if we assign an arbitrary value, say λ, to x, then there is just one solution of (0.6) with this value for x, namely

$$\begin{aligned} x &= \lambda \\ y &= \tfrac{1}{2}(7 - 3\lambda) \end{aligned} \tag{0.7}$$

This solution, involving the parameter λ, is the **complete** solution of the system (0.6) in the sense that we get all the different solutions of (0.6) by assigning all possible numerical values to λ in (0.7).

We note that the solubility in each case depended on the numerical value of a certain expression formed from the coefficients, occurring as the coefficient of x and $-y$ in (0.3) and (0.5). This expression, called the **determinant** of the system, had to be non-zero for a unique solution to exist. We shall meet it again in a more general context in Chapter 5. The case where a unique solution exists will be taken up in Chapter 2, while Chapter 4 deals with general systems. The actual work of finding a solution is greatly facilitated by introducing a convenient notation, and this will be done in Chapters 1 and 3.

If we interpret x and y as rectangular coordinates in the plane, then a linear equation in x and y represents the points lying on a straight line, and the solution of two such equations corresponds to the intersection of the

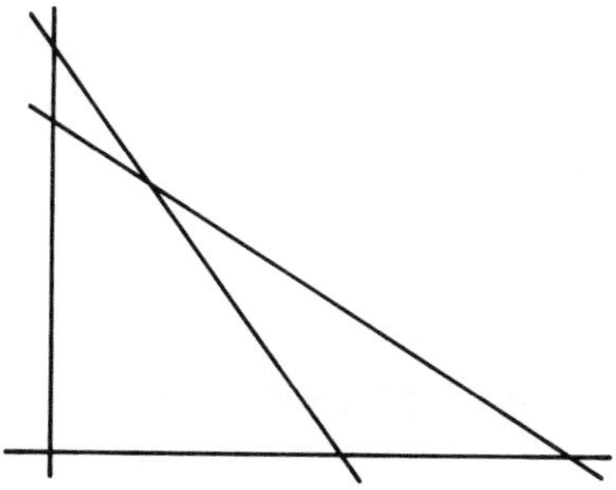

Fig. 0.1

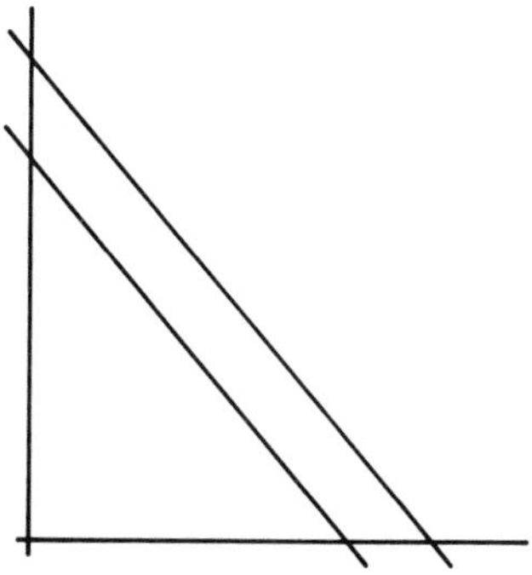

Fig. 0.2

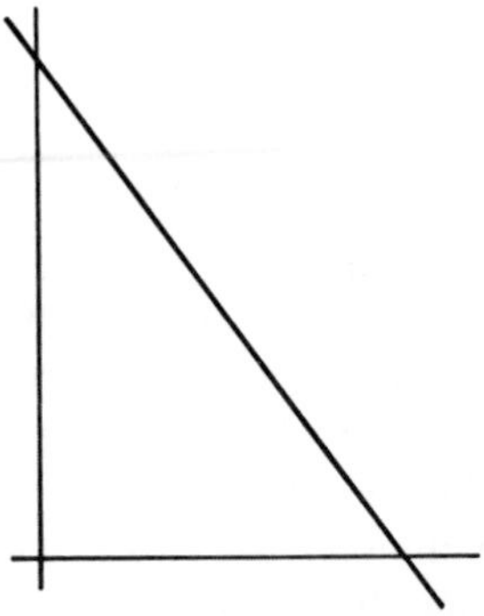

Fig. 0.3

straight lines represented by these equations. Let us draw these lines for each of the systems (0.2), (0.4), (0.6). In the first case we have two lines intersecting in a single point (Fig. 0.1) corresponding to the unique solution of (0.2). The system (0.4), which has no solution, corresponds to two parallel lines (Fig. 0.2), and (0.6), which has an infinity of solutions, corresponds to two coincident lines (Fig. 0.3); here the 'intersection' is the whole line.

Our aim is to give a method of finding the complete solution of a given system of linear equations (when a solution exists), which is sufficiently systematic to be applied to any number of equations in any number of unknowns. While the actual process used is not much more than the familiar elimination of the unknowns, the point of view emphasizes the relation of the general problem to the simplest case (0.1). Corresponding to the types (i) and (ii) of the solution of (0.1) we shall distinguish two cases and deal (in Chapter 2) with the 'regular' system before tackling the general case (Chapter 4). In order to have a concise way of discussing the problem we shall begin by explaining the language of vectors.

1

Vectors

1.1 NOTATION

The system we met in the Introduction consisted of two equations in two unknowns. In the general case we have a system of m linear equations in n unknowns, which we can call an $m \times n$ system:

$$\begin{aligned} a_{11}x_1 + a_{12}x_2 + \cdots + a_{1n}x_n &= k_1 \\ a_{21}x_1 + a_{22}x_2 + \cdots + a_{2n}x_n &= k_2 \\ \cdots \quad \cdots \quad \cdots & \\ a_{m1}x_1 + a_{m2}x_2 + \cdots + a_{mn}x_n &= k_m \end{aligned} \tag{1.1}$$

The unknowns need not all occur in each equation; this is allowed for by including 0 as a possible value for the as. Generally the as and ks are real numbers, though all that is said still applies if complex numbers are used. In practice the coefficients (and hence the solutions) are only known approximately and are given as decimal fractions. However, this makes no difference in principle and in our examples we shall give quite simple values to the coefficients so as not to obscure the argument.

In any given instance of (1.1), where the as and ks have particular numerical values, the equations must of course be written out in full, but as long as we are dealing with the general system, we can abbreviate it in various ways. Thus we can write (1.1) more briefly as

$$a_{i1}x_1 + a_{i2}x_2 + \cdots + a_{in}x_n = k_i \quad (i = 1, \ldots, m) \tag{1.2}$$

When we give i the values $1, 2, \ldots, m$ in succession we just get (1.1). The sum on the left-hand side of (1.2) can be shortened further by writing down the general term (say the jth term) of this sum, $a_{ij}x_j$, and placing a Σ (Greek capital sigma) on the left, to indicate summation, with some indication of the range of j. Thus (1.2) may be written as

$$\sum_{j=1}^{n} a_{ij}x_j = k_i \quad (i = 1, \ldots, m) \tag{1.3}$$

or simply $\Sigma_j a_{ij}x_j = k_i$, when the ranges of i and j are clear from the context.

1.2 DEFINITION OF VECTORS

In (1.3) we had to deal with the m-tuple $(k_1, k_2, \ldots, k_m)$ of constants and the solution was an n-tuple $(x_1, x_2, \ldots, x_n)$. It is convenient to have a consistent terminology and notation. A row of numbers $(x_1, \ldots, x_n)$ is called a **vector**, more precisely an ***n*-vector**, and is usually denoted by a small letter in bold type, thus we may write $\mathbf{x} = (x_1, \ldots, x_n)$. The numbers $x_1, \ldots, x_n$ are called the **components** or **coordinates** of $\mathbf{x}$, while the integer n is called its **dimension**. In this context, numbers are also called **scalars**. In physics entities like displacements, velocities or forces may be represented by vectors; the word 'vector' literally means 'carrier'. Numerical quantities such as temperatures, lengths, speeds or weights are called 'scalars', because they can be read off on a scale.

The dimension n can have any value $1, 2, 3, \ldots$. The 1-vectors are the simplest case of vectors, though perhaps a rather trivial one. The 2-vectors are pairs of numbers: $\mathbf{x} = (x_1, x_2)$; if with the vector (x_1, x_2) we associate the point P of the plane whose coordinates in a given coordinate system are x_1, x_2, we have a correspondence between 2-vectors and points of the plane, as in Fig. 1.1 (and in the illustrations in the Introduction), which will enable us to interpret our results geometrically, at least for dimension 2. In a similar way the 3-vectors can be represented by the points in space; for $n > 3$ there is no such intuitive picture to help us, but there is no difference in principle between 3-vectors and n-vectors for $n > 3$, and many quite concrete problems require n-vectors with $n > 3$ for their solution. For example, a mechanical system of two particles requires $2 \times 3 = 6$ coordinates for its description; to describe it completely, we shall need their velocities as well as coordinates (because the equations of motion are of the second order) and so we need 12 coordinates in all. Thus the state of the system is represented by a point in 12-dimensional space.

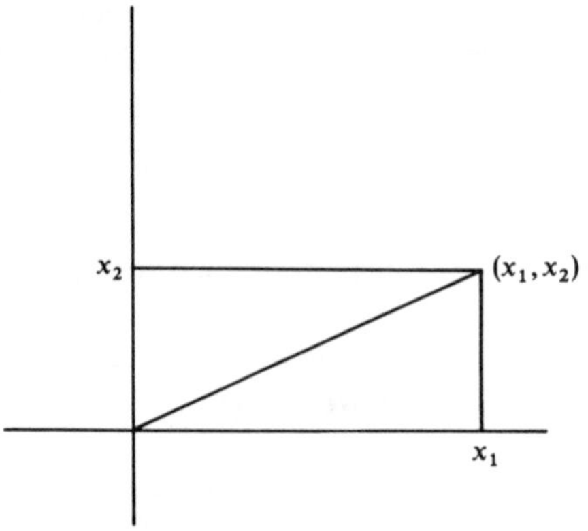

Fig. 1.1

The order of the components in a vector is essential. For example, $(3, 7, -1) \neq (7, 3, -1)$, and generally, two vectors are equal:

$$(x_1, \ldots, x_n) = (y_1, \ldots, y_n) \tag{1.4}$$

if and only if the corresponding components are equal:

$$x_i = y_i \quad \text{for } i = 1, \ldots, n$$

Thus any equation between n-vectors is equivalent to n scalar equations; we express this fact by saying that we may **equate components** in any equation (1.4).

1.3 ADDITION OF VECTORS

If vectors are to serve our purpose of describing the system (1.1), we must know how to add them and multiply them by scalars. In defining these operations we shall suppose that the dimension, though arbitrary, is the same for all vectors occurring; we denote it by n. The reader may find it helpful at first to keep in mind a definite case, such as $n = 3$.

Given two vectors $\mathbf{x} = (x_1, \ldots, x_n)$ and $\mathbf{y} = (y_1, \ldots, y_n)$, we define their **sum** as the vector

$$\mathbf{x} + \mathbf{y} = (x_1 + y_1, \ldots, x_n + y_n)$$

For example, $(3, 7, -1) + (2, -1, 0) = (5, 6, -1)$. The particular vector whose components are all zero is called the **zero vector** and is denoted by $\mathbf{0}$; thus $\mathbf{0} = (0, 0, \ldots, 0)$.

The addition of numbers satisfies certain laws, and it turns out that these laws also hold for the addition of vectors. We list them here, but do not give a proof, as they are all easily deduced from the corresponding laws of numbers by equating components.

V.1 $\mathbf{x} + \mathbf{y} = \mathbf{y} + \mathbf{x}$ *(commutative law of addition)*
V.2 $(\mathbf{x} + \mathbf{y}) + \mathbf{z} = \mathbf{x} + (\mathbf{y} + \mathbf{z})$ *(associative law of addition)*
V.3 $\mathbf{x} + \mathbf{0} = \mathbf{x}$
V.4 *For each vector* $\mathbf{x}$ *there is a vector* $-\mathbf{x}$ *such that* $\mathbf{x} + (-\mathbf{x}) = \mathbf{0}$.

In V.4, if $\mathbf{x} = (x_1, \ldots, x_n)$, then $-\mathbf{x} = (-x_1, \ldots, -x_n)$, which accounts for the choice of the notation $-\mathbf{x}$.

1.4 MULTIPLICATION BY A SCALAR

If $\mathbf{x} = (x_1, \ldots, x_n)$ and λ is a scalar, we define a new vector $\lambda\mathbf{x}$ or $\mathbf{x}\lambda$ by the equation

$$\lambda\mathbf{x} = (\lambda x_1, \ldots, \lambda x_n)$$

For example, $3(2,1,-4)=(6,3,-12)$, $-5(2,1,-4)=(-10,-5,20)$ and of course $0(2,1,-4)=(0,0,0)$. This vector $\lambda\mathbf{x}$, which may be called the **product** of the scalar λ by the vector $\mathbf{x}$, satisfies the rules

$$\left.\begin{array}{ll}\text{V.5} & \lambda(\mathbf{x}+\mathbf{y})=\lambda\mathbf{x}+\lambda\mathbf{y}\\ \text{V.6} & (\lambda+\mu)\mathbf{x}=\lambda\mathbf{x}+\mu\mathbf{x}\end{array}\right\}\quad \textit{(distributive laws)}$$

$$\text{V.7}\quad (\lambda\mu)\mathbf{x}=\lambda(\mu\mathbf{x})$$

$$\text{V.8}\quad 0\mathbf{x}=\mathbf{0}$$

$$\text{V.9}\quad 1\mathbf{x}=\mathbf{x}$$

where $\mathbf{x}$ and $\mathbf{y}$ are any vectors and λ, μ any scalars. These rules are again easily verified by equating components. We note that in V.8 the scalar zero appears on the left, whereas on the right we have the zero vector $\mathbf{0}$.

From the definitions we see that $(-1)\mathbf{x}$ equals $-\mathbf{x}$ as defined in V.4; this means that V.4 is a consequence of V.5–9, for we have $\mathbf{x}+(-1)\mathbf{x}=(1-1)\mathbf{x}=0\mathbf{x}=\mathbf{0}$. We shall simply write $-\lambda\mathbf{x}$ for $(-\lambda)\mathbf{x}$ and $\mathbf{x}-\mathbf{y}$ instead of $\mathbf{x}+(-\mathbf{y})$.

From V.5 we can deduce the more general form

$$\lambda(\mathbf{x}_1+\cdots+\mathbf{x}_r)=\lambda\mathbf{x}_1+\cdots+\lambda\mathbf{x}_r$$

for any vectors $\mathbf{x}_1,\ldots,\mathbf{x}_r$ and any scalar λ. For by repeated application of V.5 we have $\lambda(\mathbf{x}_1+\cdots+\mathbf{x}_r)=\lambda\mathbf{x}_1+\lambda(\mathbf{x}_2+\cdots+\mathbf{x}_r)=\lambda\mathbf{x}_1+\lambda\mathbf{x}_2+\lambda(\mathbf{x}_3+\cdots+\mathbf{x}_r)=\cdots=\lambda\mathbf{x}_1+\lambda\mathbf{x}_2+\cdots+\lambda\mathbf{x}_r$. This is really an instance of the use of mathematical induction, which we shall meet again in section 2.4, where it will be discussed in more detail.

The set of all vectors of dimension n with these rules of addition and multiplication by real scalars is called the **space** of all n-vectors or the **vector space** of dimension n over the real numbers, and is denoted by $\mathbf{R}^n$. Similarly, for complex scalars we have the vector space $\mathbf{C}^n$ over the complex numbers. Other choices for the scalars (such as the rational numbers) are possible; all that is required is that they admit the four operations of arithmetic: addition, subtraction, multiplication and division, satisfying the usual laws. Such a system is called a **field** and we can form vector spaces over any field, but we shall mainly be dealing with the field $\mathbf{R}$ of real numbers or $\mathbf{C}$ of complex numbers. For $n=1$, $\mathbf{R}^1=\mathbf{R}$ is the field of real numbers itself; similarly $\mathbf{C}$ can be regarded as a 1-dimensional space over itself, but we can also regard $\mathbf{C}$ as a 2-dimensional vector space over $\mathbf{R}$, by representing complex numbers as pairs of real numbers.

Gradually we have come to use mathematical language by speaking of the field $\mathbf{R}$ of real numbers rather than individual real numbers, and this is a good opportunity to introduce a useful piece of notation. If we have a collection S of objects, we express the fact that c is a member of S by writing $c\in S$; for example we write $\mathbf{x}\in\mathbf{R}^n$ to indicate that $\mathbf{x}$ is an n-vector with real components.

1.5 GEOMETRICAL INTERPRETATION

All that has been said can be interpreted geometrically when $n = 2$ or 3. Thus the zero vector represents the origin O, and if the non-zero vectors $\mathbf{x}$ and $\mathbf{y}$ correspond to the points P and Q, respectively, then their sum $\mathbf{x}+\mathbf{y}$ corresponds to the point R which is the fourth vertex of the parallelogram with sides OP and OQ (Fig. 1.2). Any scalar multiple $\lambda\mathbf{x}$ of $\mathbf{x}$ represents a point on the line OP, produced in both directions, and all the points of this line are represented by scalar multiples of $\mathbf{x}$. For $\lambda > 1$ the vector $\lambda\mathbf{x}$ represents the vector $\mathbf{x}$ 'stretched' by the factor λ; for $0 < \lambda < 1$, $\mathbf{x}$ becomes compressed to form $\lambda\mathbf{x}$, while for $\lambda < 0$ the vector also reverses direction, stretched or compressed by a factor $|\lambda|$. Here as usual, $|\lambda|$ denotes the absolute value. For real λ,

$$|\lambda| = \begin{cases} \lambda & \text{if } \lambda \geqslant 0 \\ -\lambda & \text{if } \lambda < 0 \end{cases}$$

For a complex number $\lambda = \lambda_1 + \mathrm{i}\lambda_2$ (λ_1, λ_2 real), we have

$$|\lambda| = (\lambda_1^2 + \lambda_2^2)^{1/2}$$

We note that the point R in Fig. 1.2 lies in a plane with the points OPQ, even if $\mathbf{x}$ and $\mathbf{y}$ are vectors in three dimensions. More generally, the vector $\mathbf{z} = \lambda\mathbf{x} + \mu\mathbf{y}$, for any scalars λ and μ, again lies in a plane with $\mathbf{x}$ and $\mathbf{y}$. We call such a vector $\mathbf{z}$ a **linear combination** of $\mathbf{x}$ and $\mathbf{y}$ and we say: $\mathbf{z}$ **depends linearly** on $\mathbf{x}$ and $\mathbf{y}$. Similarly a vector $\mathbf{z}$ depends linearly on $\mathbf{x}$ if $\mathbf{z} = \lambda\mathbf{x}$, so that $\mathbf{z}$ and $\mathbf{x}$ lie along the same line.

Generally, a vector $\mathbf{z}$ is said to **depend linearly** on the vectors $\mathbf{u}_1, \ldots, \mathbf{u}_r$ if there are scalars $\lambda_1, \ldots, \lambda_r$ such that

$$\mathbf{z} = \lambda_1\mathbf{u}_1 + \cdots + \lambda_r\mathbf{u}_r \tag{1.5}$$

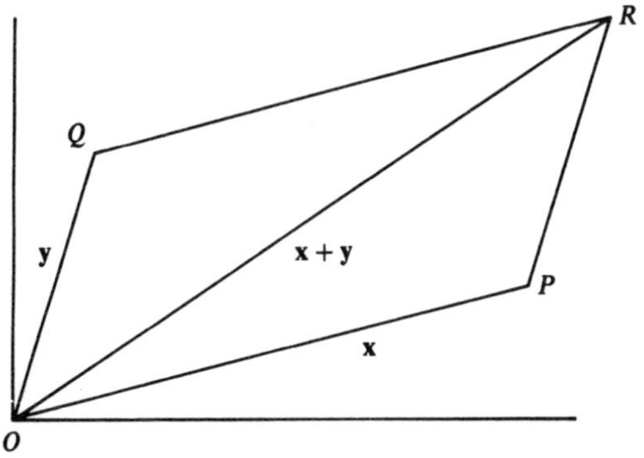

Fig. 1.2

For example, $(4, 6, 2)$ depends linearly on the vectors $(2, 0, -2)$, $(-1, 3, 4)$:

$$(4, 6, 2) = 3(2, 0, -2) + 2(-1, 3, 4)$$

but not on $(2, 0, -2)$, $(1, 3, 4)$, as some trials will show. Later we shall meet systematic tests for linear dependence.

1.6 LINEAR DEPENDENCE OF VECTORS

A set of vectors $\mathbf{u}_1, \ldots, \mathbf{u}_r$ is said to be **linearly dependent** if one of these vectors depends linearly on the rest, say if

$$\mathbf{u}_1 = \lambda_2 \mathbf{u}_2 + \cdots + \lambda_r \mathbf{u}_r \tag{1.6}$$

If the vectors $\mathbf{u}_1, \ldots, \mathbf{u}_r$ are not linearly dependent, i.e. if none of these vectors can be expressed linearly in terms of the remaining ones, we say that the vectors are **linearly independent**.

The definition of linear dependence can be expressed in a slightly different, more symmetric form as follows:

The vectors $\mathbf{u}_1, \ldots, \mathbf{u}_r$ are linearly dependent if and only if there is a relation

$$\alpha_1 \mathbf{u}_1 + \cdots + \alpha_r \mathbf{u}_r = \mathbf{0} \tag{1.7}$$

in which the scalars α_i are not all zero.

For if the vectors are linearly dependent, say if (1.6) holds, then

$$\mathbf{u}_1 - \lambda_2 \mathbf{u}_2 - \cdots - \lambda_r \mathbf{u}_r = \mathbf{0}$$

and this is a non-trivial relation between the **u**s – i.e. one in which the coefficients are not all zero – because the coefficient of $\mathbf{u}_1$ is 1. Conversely, if there is a relation (1.7) in which the α_i are not all zero, say $\alpha_1 \neq 0$, then by rearranging (1.7), we find

$$\mathbf{u}_1 = \beta_2 \mathbf{u}_2 + \cdots + \beta_r \mathbf{u}_r$$

where $\beta_i = -\alpha_i/\alpha_1$ $(i = 2, \ldots, r)$; this shows that the vectors are linearly dependent. For example, the vectors $\mathbf{u} = (2, 3, 1)$, $\mathbf{v} = (2, 0, -2)$, $\mathbf{w} = (-1, 3, 4)$ are linearly dependent, for $2\mathbf{u} - 3\mathbf{v} - 2\mathbf{w} = \mathbf{0}$, hence $\mathbf{u} = (3/2)\mathbf{v} + \mathbf{w}$.

The second form of the dependence condition is often more convenient. It shows that a set of vectors is linearly independent precisely when the only linear relation between the vectors is the trivial one, in which all the coefficients are zero. A set consisting of a single vector $\mathbf{u}$ is linearly dependent if and only if there is a relation $\alpha\mathbf{u} = \mathbf{0}$, with $\alpha \neq 0$, which is true precisely when $\mathbf{u} = \mathbf{0}$. Generally any set which includes the zero vector is linearly dependent, for the set $\mathbf{0}, \mathbf{u}_1, \ldots, \mathbf{u}_r$ satisfies the relation

$$1.\mathbf{0} + 0.\mathbf{u}_1 + 0.\mathbf{u}_2 + \cdots + 0.\mathbf{u}_r = \mathbf{0}$$

in which not all the coefficients are zero.

1.7 SUBSPACES OF A VECTOR SPACE

It may happen that we are using a subcollection of all the vectors in $\mathbf{R}^n$ which again forms a vector space in its own right, for example the vectors in $\mathbf{R}^n$ whose last component is zero again form a vector space. This leads to the notion of a subspace which is defined as follows.

Let V be a vector space. By a **subspace** of V we understand a collection U of vectors in V such that the zero vector is included; for any pair of vectors $\mathbf{u}, \mathbf{u}'$ in U the vector $\mathbf{u} + \mathbf{u}'$ again lies in U; and $\lambda\mathbf{u}$ lies in U for any scalar λ.

The set $\{\mathbf{0}\}$ consisting of the zero vector alone is a subspace, called the **trivial** or **zero space** and usually denoted by 0 (instead of $\{\mathbf{0}\}$). The whole space V is a subspace; any subspace other than V itself is called **proper**. Thus V is the largest and 0 the smallest subspace and any other subspace lies between these two. It is usual to write $U \subseteq V$ or $V \supseteq U$ to indicate that U is a subspace of V and $U \subset V$ or $V \supset U$ to mean that U is a proper subspace, i.e. $U \neq V$.

From the definition we see that given a subspace U of V, for any $\mathbf{u}, \mathbf{u}' \in U$ and any scalars λ, λ' we have $\lambda\mathbf{u}, \lambda'\mathbf{u}' \in U$ and so $\lambda\mathbf{u} + \lambda'\mathbf{u}' \in U$. Conversely, any set U of vectors, including $\mathbf{0}$ and which for $\mathbf{u}, \mathbf{u}' \in U$ also includes $\lambda\mathbf{u} + \lambda'\mathbf{u}'$ for any scalars λ, λ', must be a subspace. For we have $\mathbf{u} + \mathbf{u}' = 1.\mathbf{u} + 1.\mathbf{u}' \in U$ and $\lambda\mathbf{u} = \lambda\mathbf{u} + 0.\mathbf{u}' \in U$, so the definition is satisfied. This shows that we can rephrase the definition as follows:

A subspace of a vector space V is a subset U of V such that

S.1 $\mathbf{0} \in U$,

S.2 *If $\mathbf{u}, \mathbf{u}' \in U$ and λ, λ' are any scalars, then $\lambda\mathbf{u} + \lambda'\mathbf{u}' \in U$.*

It is easily verified that any subspace again satisfies V.1–9 and so is again a vector space. We have already noted that the vectors in $\mathbf{R}^n$ with last component zero form a subspace; another example is the set of vectors in which the first and second components are equal. Suppose now that our vector space is $\mathbf{R}$ itself, i.e. $n = 1$; if U is any non-zero subspace, we can find a non-zero vector $\mathbf{z}$ in U. Then for any $\alpha \in \mathbf{R}$ we have $\alpha = (\alpha/\mathbf{z})\mathbf{z} \in U$ and so $U = \mathbf{R}$. This shows that $\mathbf{R}^1$ has no proper subspaces other than 0.

If V is any vector space and $\mathbf{z}$ is a vector in V, then the set of all vectors $\lambda\mathbf{z}$, where $\lambda \in \mathbf{R}$, is a subspace. More generally, let $\mathbf{u}_1, \ldots, \mathbf{u}_r$ be any vectors in V and consider the set U of all linear combinations

$$\mathbf{z} = \alpha_1\mathbf{u}_1 + \cdots + \alpha_r\mathbf{u}_r \tag{1.8}$$

This set U is a subspace, for it contains $\mathbf{0} = \Sigma 0.\mathbf{u}_i$, and if $\mathbf{z} = \Sigma\alpha_i\mathbf{u}_i$ as in (1.8) and $\mathbf{z}' = \Sigma\alpha_i'\mathbf{u}_i$, then

$$\begin{aligned}\lambda\mathbf{z} + \lambda'\mathbf{z}' &= \lambda(\alpha_1\mathbf{u}_1 + \cdots + \alpha_r\mathbf{u}_r) + \lambda'(\alpha_1'\mathbf{u}_1 + \cdots + \alpha_r'\mathbf{u}_r)\\ &= \lambda\alpha_1\mathbf{u}_1 + \cdots + \lambda\alpha_r\mathbf{u}_r + \lambda'\alpha_1'\mathbf{u}_1 + \cdots + \lambda'\alpha_r'\mathbf{u}_r\\ &= (\lambda\alpha_1 + \lambda'\alpha_1')\mathbf{u}_1 + \cdots + (\lambda\alpha_r + \lambda'\alpha_r')\mathbf{u}_r\end{aligned}$$

This shows that U is indeed a subspace of V. It is called the subspace **spanned** by $\mathbf{u}_1,\ldots,\mathbf{u}_r$ or simply the **linear span** of $\mathbf{u}_1,\ldots,\mathbf{u}_r$; it may also be defined as the least subspace containing $\mathbf{u}_1,\ldots,\mathbf{u}_r$ (cf. Exercise 1.13).

Geometrically, if we interpret vectors as points in space as described in section 1.5, the subspace spanned by a single vector $\mathbf{z}$ is the straight line through the origin in the direction $\mathbf{z}$, provided that $\mathbf{z} \neq \mathbf{0}$. When $\mathbf{z} = \mathbf{0}$, the subspace just reduces to the origin. Similarly the subspace spanned by two vectors is a plane through the origin or a line through the origin when the two vectors are linearly dependent, or the origin when both vectors are zero.

1.8 A BASIS FOR THE SPACE OF n-VECTORS

In general, if a vector $\mathbf{z}$ depends linearly on $\mathbf{u}_1,\ldots,\mathbf{u}_r$, there are different ways of expressing this dependence. For if the dependence is given by (1.8) and if

$$\lambda_1\mathbf{u}_1 + \cdots + \lambda_r\mathbf{u}_r = \mathbf{0} \tag{1.9}$$

is any linear dependence relation between the **u**s, then we also have

$$\mathbf{z} = (\lambda_1 + \alpha_1)\mathbf{u}_1 + \cdots + (\lambda_r + \alpha_r)\mathbf{u}_r$$

and this expression for $\mathbf{z}$ is different from the expression in (1.8) unless all the λ_i are zero. Therefore, if $\mathbf{u}_1,\ldots,\mathbf{u}_r$ are linearly dependent, there are different ways of expressing $\mathbf{z}$ linearly in terms of the **u**s.

On the other hand, if $\mathbf{u}_1,\ldots,\mathbf{u}_r$ are linearly independent, then the expression (1.8) for $\mathbf{z}$ is unique. For if

$$\mathbf{z} = \alpha'_1\mathbf{u}_1 + \cdots + \alpha'_r\mathbf{u}_r \tag{1.10}$$

is another way of expressing $\mathbf{z}$ in terms of the **u**s, then by subtracting (1.10) from (1.8) we obtain

$$(\alpha_1 - \alpha'_1)\mathbf{u}_1 + \cdots + (\alpha_r - \alpha'_r)\mathbf{u}_r = \mathbf{0}$$

Since $\mathbf{u}_1,\ldots,\mathbf{u}_r$ are linearly independent, the coefficients in this relation must all be zero, i.e.

$$\alpha'_i = \alpha_i \quad \text{for } i = 1,\ldots,r$$

and so (1.10) is really the same expression as (1.8). This result may also be expressed as follows:

In the expression (1.8) for the vector $\mathbf{z}$ *the coefficients* α_i *are uniquely determined by* $\mathbf{z}$ *if and only if the vectors* $\mathbf{u}_1,\ldots,\mathbf{u}_r$ *are linearly independent.*

Any linearly independent set of vectors in a space V which spans the whole space V is called a **basis** of V. By what has just been said, every vector of the space can then be uniquely expressed as a linear combination of the basis vectors. For example, let us write $\mathbf{e}_i$ for the n-vector whose ith component is 1 and whose other components are zero; thus $\mathbf{e}_1 = (1, 0, \ldots, 0)$, $\mathbf{e}_2 = (0, 1, 0, \ldots, 0), \ldots,$ $\mathbf{e}_n = (0, 0, \ldots, 0, 1)$. Every vector of dimension n depends linearly on $\mathbf{e}_1, \mathbf{e}_2, \ldots, \mathbf{e}_n$: if $\mathbf{x} = (x_1, x_2, \ldots, x_n)$, then

$$\mathbf{x} = x_1\mathbf{e}_1 + \cdots + x_n\mathbf{e}_n \tag{1.11}$$

The coefficients in (1.11) are uniquely determined as the components of $\mathbf{x}$, hence the $\mathbf{e}_i$ are linearly independent. They also span $\mathbf{R}^n$, as we have just seen, so they form a basis. We shall frequently use the basis $\mathbf{e}_1$, $\mathbf{e}_2, \ldots, \mathbf{e}_n$ introduced above; it is called the **standard basis** of $\mathbf{R}^n$. In the geometric picture for $n = 3$ the $\mathbf{e}_i$ represent points at unit distance along the coordinate axes, and equation (1.11) states then that every vector in 3-space can be written as a linear combination of vectors along the axes.

Of course there are many other bases of the same space; an example is the set $(1, 0, \ldots, 0), (1, 1, 0, \ldots, 0), \ldots, (1, 1, \ldots, 1)$, in which the ith vector consists of i ones followed by $n - i$ zeros. In the next chapter we shall see that every basis of this space has exactly n members.

1.9 THE CONSTRUCTION OF A BASIS

Suppose we are given r linearly independent vectors $\mathbf{u}_1, \ldots, \mathbf{u}_r$. Then every vector $\mathbf{x}$ depending linearly on the $\mathbf{u}_i$ can be uniquely written as

$$\mathbf{x} = \xi_1\mathbf{u}_1 + \cdots + \xi_r\mathbf{u}_r \tag{1.12}$$

Thus $\mathbf{u}_1, \ldots, \mathbf{u}_r$ form a basis for the subspace spanned by these vectors.

As an illustration consider two linearly independent vectors $\mathbf{u}, \mathbf{v}$ in 3-dimensional space, represented by the points P, Q, respectively, referred to a point O as origin. To say that $\mathbf{u}$ and $\mathbf{v}$ are linearly independent means that neither is a scalar multiple of the other. It follows that P and Q are not collinear with O and so determine a plane. Any point R of this plane corresponds to a vector $\lambda\mathbf{u} + \mu\mathbf{v}$, where the scalars λ and μ are uniquely determined by R. Thus the plane may be regarded as a 2-dimensional subspace with λ, μ as coordinates relative to the basis $\mathbf{u}, \mathbf{v}$.

Suppose now that $\mathbf{u}_1, \ldots, \mathbf{u}_m$ is any set of n-vectors, linearly dependent or not. The subspace U spanned by $\mathbf{u}_1, \ldots, \mathbf{u}_m$ consists of all vectors depending linearly on the $\mathbf{u}_i$ and we can proceed as follows to obtain a basis of this subspace: if $\mathbf{u}_1, \ldots, \mathbf{u}_m$ are linearly independent, then they form a basis of U by what has been said. If they are linearly dependent, then one of the vectors, say $\mathbf{u}_m$, depends on the rest and in this case $\mathbf{u}_1, \ldots, \mathbf{u}_{m-1}$ already span U. If these vectors are still dependent, we can again express one of them in terms of the rest and so reduce their number by one. In this way we finally obtain a linearly independent set of vectors spanning U, which by definition is a basis, unless all the $\mathbf{u}_i$ were zero. In that case our subspace U consists of the zero vector alone and so is 0. Formally this may be regarded as a 0-dimensional space, with a basis of no vectors, i.e. an empty basis.

1.10 GENERAL VECTOR SPACES

So far all our vector spaces have been n-tuples of numbers. It is important to realize that, on the one hand, many other types of object may be represented in this way, but, on the other, they can all be regarded as sets of n-tuples. Let us see how this comes about.

We have seen that points in the plane or in space may be represented as vectors. As another example consider polynomials in one variable x. They are linear combinations of powers of x:

$$f = a_0 + a_1 x + \cdots + a_n x^n \tag{1.13}$$

with the understanding that two polynomials $f = \Sigma a_i x^i$ as in (1.13) and $g = \Sigma b_i x^i$ are equal if and only if $a_i = b_i$ $(i = 1, \ldots, n)$. These polynomials form a vector space with the usual addition and multiplication by scalars. Thus $\lambda f = \Sigma \lambda a_i x^i$ and if $g = b_0 + b_1 x + \cdots + b_m x^m$, then in forming $f + g$ we shall assume that $m = n$, by adding terms with zero coefficient to the polynomial with smaller degree. Then we have $f + g = \Sigma(a_i + b_i)x^i$. The rules V.1–9 are easily verified, so we again have a vector space. However, in contrast to our earlier examples the space of polynomials is infinite-dimensional. To obtain a finite-dimensional space we fix a positive integer n and consider all polynomials of degree at most n; they form a vector space with basis $1, x, x^2, \ldots, x^n$. By an easy extension of these ideas we may describe the space of all polynomials as a vector space with the infinite basis $1, x, x^2, \ldots$. This just means that every polynomial can be written as a unique linear combination of finitely many powers of x.

As another example of a vector space we have the set of all continuous real functions defined on the interval $[0, 1]$, with the usual addition and multiplication by scalars. Again it is straightforward to verify V.1–9. This space is again infinite-dimensional, but this time it is not so easy to find a basis. Later on we shall have to deal with some of its subspaces, but they are usually finite-dimensional and then finding a basis presents no great problem.

With the above examples in mind let us define an abstract **vector space** as a set whose members are called **vectors**, admitting an operation of addition of pairs of vectors, $\mathbf{u} + \mathbf{v}$, and an operation of multiplication by (real) scalars, satisfying V.1–9. To ensure that our spaces are finite-dimensional we shall need one further rule, which holds in most of our examples:

V.10 *There is a finite set of vectors spanning the whole vector space.*

For example, the differentiable functions of a real variable form a vector space which fails to satisfy V.10, but the functions satisfying a given linear differential equation form a subspace for which V.10 holds.

In the above definition of vector space we have tacitly assumed that the scalars are the real numbers **R**, though we could equally well have a complex vector space, with **C** as field of scalars, or indeed a vector space over any field, but we shall only be concerned with **R** and **C** as coefficient fields.

Let V be any vector space satisfying V.10. Our aim will be to show that the vectors of V can be represented by n-tuples as in section 1.4, for some integer n. To do so we take a finite spanning set $\mathbf{u}_1, \ldots, \mathbf{u}_m$ of V, which exists by V.10. By the process described in section 1.9 we can select from it a basis $\mathbf{v}_1, \ldots, \mathbf{v}_n$ for our space. Now every vector $\mathbf{x}$ in V can be written as a linear combination of the $\mathbf{v}_i$ with uniquely determined coefficients, by the properties of a basis established in section 1.8:

$$\mathbf{x} = \xi_1\mathbf{v}_1 + \cdots + \xi_n\mathbf{v}_n$$

We shall represent the vector $\mathbf{x}$ by the n-tuple $\boldsymbol{\xi} = (\xi_1, \ldots, \xi_n)$. If $\mathbf{y}$ is another vector in V, let $\mathbf{y} = \Sigma\, \eta_i\mathbf{v}_i$, so that $\mathbf{y}$ is represented by $\boldsymbol{\eta} = (\eta_1, \ldots, \eta_n)$. Then we have $\mathbf{x} + \mathbf{y} = \Sigma(\xi_i + \eta_i)\mathbf{v}_i$, hence $\mathbf{x} + \mathbf{y}$ is represented by $\boldsymbol{\xi} + \boldsymbol{\eta} = (\xi_1 + \eta_1, \ldots, \xi_n + \eta_n)$; further, $\lambda\mathbf{x}$ is represented by $\lambda\boldsymbol{\xi} = (\lambda\xi_1, \ldots, \lambda\xi_n)$, so for any calculation with the vectors in V we may take their representatives in $\mathbf{R}^n$. More formally, we have a correspondence between V and $\mathbf{R}^n$ which is one-one in the sense that to each vector in V there corresponds one n-tuple and to each n-tuple there corresponds one vector in V. Moreover, what is most important, this one-one correspondence preserves addition and multiplication by scalars, as we have just seen. Thus if $\mathbf{x}$ corresponds to $\boldsymbol{\xi} = (\xi_1, \ldots, \xi_n)$ and $\mathbf{y}$ corresponds to $\boldsymbol{\eta} = (\eta_1, \ldots, \eta_n)$, then $\mathbf{x} + \mathbf{y}$ corresponds to $\boldsymbol{\xi} + \boldsymbol{\eta}$ and $\lambda\mathbf{x}$ corresponds to $\lambda\boldsymbol{\xi}$. Such a one-one correspondence preserving all the vector space operations is called an **isomorphism** and the existence of an isomorphism between V and $\mathbf{R}^n$ is expressed by saying that V is **isomorphic** to $\mathbf{R}^n$, in symbols $V \cong \mathbf{R}^n$.

Our conclusion can be stated as follows:

Theorem 1.1

Every vector space V over **R** *with a finite spanning set of vectors has a finite basis. If $\mathbf{v}_1, \ldots, \mathbf{v}_n$ is a basis, then V is isomorphic to $\mathbf{R}^n$, via the isomorphism $\Sigma\, \xi_i\mathbf{v}_i \leftrightarrow (\xi_1, \ldots, \xi_n)$.*

We have seen that every (real) vector space V satisfying V.10 is isomorphic to $\mathbf{R}^n$, for some n. It still remains to show that the number n depends only on V and not on the choice of basis in V; this amounts to showing that for $m \neq n$, $\mathbf{R}^m$ and $\mathbf{R}^n$ are not isomorphic, a task left for Chapter 2.

The fact just proved, that every vector space satisfying V.1–10 is isomorphic to a vector space of the form $\mathbf{R}^n$, means that any results proved about $\mathbf{R}^n$ can be applied to all such vector spaces. But most results are just as easily proved directly. For example, the process of choosing a basis from a given spanning set (section 1.9) can be carried out in any vector space. As a kind of dual result it can be shown that any linearly independent set of vectors can be completed to a basis. Thus let $\mathbf{u}_1, \ldots, \mathbf{u}_r$ be a linearly independent set of vectors in a vector space V, which has a spanning set $\mathbf{v}_1, \ldots, \mathbf{v}_m$. Then the vectors $\mathbf{u}_1, \ldots, \mathbf{u}_r, \mathbf{v}_1, \ldots, \mathbf{v}_m$ form a spanning set and our aim will be to discard **v**s until we are left with a linearly independent spanning set, i.e. a basis. If $\mathbf{v}_1$ is a linear combination of $\mathbf{u}_1, \ldots, \mathbf{u}_r$, we omit $\mathbf{v}_1$; otherwise we claim that the set $\mathbf{u}_1, \ldots, \mathbf{u}_r, \mathbf{v}_1$ is linearly independent. For suppose there is a relation

$$\alpha_1 \mathbf{u}_1 + \cdots + \alpha_r \mathbf{u}_r + \beta \mathbf{v}_1 = \mathbf{0} \tag{1.14}$$

If $\beta \neq 0$, we can divide this relation by β and so express $\mathbf{v}_1$ as a linear combination of $\mathbf{u}_1, \ldots, \mathbf{u}_r$. But this case is excluded, so $\beta = 0$ and (1.14) represents a linear dependence relation between $\mathbf{u}_1, \ldots, \mathbf{u}_r$, which must be the trivial one, because of their linear independence. Thus the relation (1.14) is trivial and so $\mathbf{u}_1, \ldots, \mathbf{u}_r, \mathbf{v}_1$ are linearly independent. We now continue in this way. At each stage $\mathbf{v}_i$ either depends linearly on the preceding vectors and so can be omitted, or adjoining it to the set gives again a linearly independent set of vectors. Eventually we are left with a linearly independent set consisting of $\mathbf{u}_1, \ldots, \mathbf{u}_r$ and some of the **v**s, whose linear span includes all the **v**s and hence is the whole space V. We record the result as a formal statement:

Theorem 1.2
Let $\mathbf{u}_1, \ldots, \mathbf{u}_r$ be a linearly independent set of vectors in a space V and let $\mathbf{v}_1, \ldots, \mathbf{v}_m$ be any spanning set. Then V has a basis consisting of $\mathbf{u}_1, \ldots, \mathbf{u}_r$ and some of $\mathbf{v}_1, \ldots, \mathbf{v}_m$.

To give an example, suppose we have the linearly independent vectors $\mathbf{u}_1 = (2, 3, -1, 5)$, $\mathbf{u}_2 = (0, 2, 0, 3)$ in $\mathbf{R}^4$. As our spanning set we take $\mathbf{v}_1 = (1, 1, 1, 1)$, $\mathbf{v}_2 = (0, 1, 1, 1)$, $\mathbf{v}_3 = (0, 0, 1, 1)$, $\mathbf{v}_4 = (0, 0, 0, 1)$. A few trials show that $\mathbf{v}_1$ is not a linear combination of $\mathbf{u}_1$ and $\mathbf{u}_2$, but we have

$$\mathbf{u}_1 - 2\mathbf{u}_2 - 2\mathbf{v}_1 + 3\mathbf{v}_2 = \mathbf{0}$$

hence $\mathbf{v}_2$ depends linearly on $\mathbf{u}_1, \mathbf{u}_2, \mathbf{v}_1$ and so can be omitted. Next $\mathbf{u}_1, \mathbf{u}_2, \mathbf{v}_1, \mathbf{v}_3$ are linearly independent, while $\mathbf{v}_4$ depends on them. Thus $\mathbf{u}_1$, $\mathbf{u}_2, \mathbf{v}_1, \mathbf{v}_3$ is a basis of $\mathbf{R}^4$. This example could still be solved by trial and error, but it is clearly desirable to have a more systematic way of proceeding. This will form the subject of Chapter 2.

The idea of representing points in the plane by pairs of numbers, which lies at the basis of coordinate geometry, goes back to René Descartes (1596–1650) and Pierre de Fermat (1601–65). But the notion of n-dimensional space, for $n > 3$, was only developed in the nineteenth century. Sir William Rowan Hamilton (1805–65), Astronomer Royal of Ireland, followed his discovery of quaternions in 1843 by an elaborate geometrical treatment of vectors. At about the same time Hermann Günther Grassmann (1809–77), an autodidact who worked most of his life as a schoolmaster, wrote *Lineale Ausdehnungslehre* in 1844, dealing with 'linear extension-calculus', which proposed the idea of vectors in n-dimensional space, but which was so obscurely written as to be ignored for many years. The general notion of a vector space gradually crystallized from these beginnings. One of the first books presenting the axiomatic foundations of the subject is the *Elementi di calcolo vettoriale con numerose applicazioni alla geometria, alla meccánica e alla fisica matematica*, by C. Burali Forti (1861–1931) and R. Marcolongo, (1862–1943) which appeared in 1909.

EXERCISES

1.1. Write the following expressions out in full:

$$\sum_{i=1}^{3} a_i b_i, \quad \sum_{j=1}^{2} a_{ij} x_j, \quad \sum_{j=1}^{2} a_{ij} b_{jk}, \quad \sum_{i=1}^{2}\sum_{j=1}^{2} u_i b_{ij} x_j, \quad \sum_{j=1}^{2}\sum_{i=1}^{2} u_i b_{ij} x_j$$

Write the following expressions in abbreviated form:

$$u_1 x_1 + u_2 x_2, \quad a_{i1} x_1 + a_{i2} x_2 + a_{i3} x_3, \quad a_{i1} y_1 + a_{i2} y_1 + a_{i3} y_1$$

1.2. Which of the following sets of vectors are linearly dependent? In each case find a linearly independent subset and express the other vectors in terms of this subset: (i) (0, 1, 1), (1, 0, 1), (1, 1, 0); (ii) (1, 2, 3), (4, 5, 6), (7, 8, 9); (iii) (13, 7, 9, 2), (0, 0, 0, 0), (3, − 2, 5, 8); (iv) (4, 0, 3, 2), (2, 1, − 1, 3), (− 2, − 3, 6, − 7).

1.3. Show that the following set is linearly dependent, although the first vector does not depend on the remaining two:

$$(16, 12, 15, -10), \quad (9, -15, -6, 24), \quad (-15, 25, 10, -40).$$

1.4. Show that **C** may be regarded as a vector space over **R** and find a basis. Find a basis for $\mathbf{C}^3$ as **R**-space.

1.5. Interpret the statement that two 3-vectors are linearly independent geometrically. Similarly, interpret the statement that three 3-vectors are linearly dependent.

1.6. Show that the vectors (2, − 1, 0, 0), (1, 1, 1, 1), (− 1, 1, 1, − 1), (0, 0, 1, 3) span the space of all 4-vectors.

1.7. Find a basis for $\mathbf{R}^4$ consisting of (3, − 5, − 4, 1), (8, 0, 1, 6) and some of the vectors of the spanning set in Exercise 1.6.

1.8. Show that the vectors (x_1, x_2, x_3, x_4) satisfying the conditions $x_1 = 0$, $x_3 = x_4$ form a subspace of $\mathbf{R}^4$ and find a basis for this subspace. Show also that the vectors satisfying the condition $x_3 = 1$ do not form a subspace.

1.9. Show that in the second form of the definition of subspace in section 1.7, S.1 can be replaced by the condition: U is not empty (that is, it contains at least one vector).

1.10. Show that the polynomials in x^2 form a vector space. Do the polynomials in x that are of even degree form a vector space?

1.11. If S is a linearly independent set of vectors in a vector space V, spanning a subspace U, and $\mathbf{x}$ is a vector not lying in U, show that the set $S \cup \{\mathbf{x}\}$ consisting of the vectors in S as well as the vector $\mathbf{x}$ is linearly independent.

1.12. Let V be a vector space and U', U'' subspaces of V. Show that the set $U' \cap U''$ of all vectors in both U' and U'' is again a subspace (the **intersection** of U' and U''), but the **union** $U' \cup U''$, consisting of all vectors either in U' or in U'' is not a subspace, unless one of U', U'' is contained in the other.

1.13. Verify that the linear span of a subset S of a vector space V is the least subspace of V containing S.

1.14. Let V be a vector space and U', U'' subspaces of V. Show that the set of all vectors of the form $\mathbf{u}' + \mathbf{u}''$, where $\mathbf{u}' \in U'$, $\mathbf{u}'' \in U''$, is a subspace (it is denoted by $U' + U''$ and is called the **sum** of U' and U'').

1.15. Let V be a vector space and U a subspace. By taking a basis of U and enlarging it to a basis of V, find a subspace U' such that $U \cap U' = 0$, $U + U' = V$.

1.16. Show that V.8 is a consequence of V.1–4 and V.6, but that V.9 is not a consequence of the other axioms. (*Hint*: Find a meaning for $\lambda\mathbf{x}$ which satisfies V.5–8 but not V.9.)

1.17. Show that the infinite sequences of real numbers $(a_0, a_1, a_2, \ldots)$ form a vector space (under componentwise addition and multiplication by scalars), V_1 say. Show further that the sequences for which $\lim a_n$ exists form a subspace V_2, the sequences in V_2 such that $\lim a_n = 0$ form a subspace V_3 and the sequences with only finitely many non-zero terms form a subspace V_4, and that $V_1 \supset V_2 \supset V_3 \supset V_4$.

1.18. Verify that the vector space V_1 in exercise 1.17 has an infinite ascending sequence of subspaces: $U_1 \subset U_2 \subset \cdots$ and show that such a vector space cannot satisfy V.10. Show that V_4 has an infinite descending sequence of subspaces $V_4 \supset W_1 \supset W_2 \supset \cdots$.

2

The solution of a system of equations: the regular case

2.1 THE USE OF VECTOR NOTATION

With the help of vector notation any $m \times n$ system of linear equations may be written as a single equation. Thus the 2×2 system

$$3x + 2y = 7$$
$$2x + 5y = 12$$

encountered in the Introduction may be written as

$$\begin{pmatrix} 3 \\ 2 \end{pmatrix} x + \begin{pmatrix} 2 \\ 5 \end{pmatrix} y = \begin{pmatrix} 7 \\ 12 \end{pmatrix}$$

where the vectors have been written as columns, instead of rows as in Chapter 1. This way of writing the equations puts the problem in a different light: to solve this vector equation is to express the vector on the right as a linear combination of the two vectors on the left.

We can treat the general $m \times n$ system

$$\begin{aligned} a_{11}x_1 + \cdots + a_{1n}x_n &= k_1 \\ \cdots \quad \cdots \quad &\cdots \\ a_{m1}x_1 + \cdots + a_{mn}x_n &= k_m \end{aligned} \tag{2.1}$$

similarly. Let us write

$$\mathbf{a}_j = \begin{pmatrix} a_{1j} \\ \vdots \\ a_{mj} \end{pmatrix} \quad (j = 1, \ldots, n)$$

and call $\mathbf{a}_1, \ldots, \mathbf{a}_n$ the **column vectors** or simply the **columns** of the system (2.1). In order to save space we often write the components of a column vector in a horizontal row, with a superscript T (to indicate **transpose**) to show that it is really a column. The space of column vectors with n real components will be written $^n\mathbf{R}$, just as the space of row vectors is written $\mathbf{R}^n$. The reason for this notation will become clear later, when it will also become clear why these vectors should be considered as columns rather than rows. The definition of $\mathbf{a}_j$ can now be written

$$\mathbf{a}_j = (a_{1j}, \ldots, a_{mj})^{\mathrm{T}} \quad (j = 1, \ldots, n)$$

If we put $\mathbf{k} = (k_1, \ldots, k_m)^{\mathrm{T}}$, then (2.1) may be written as a single vector equation

$$\mathbf{a}_1 x_1 + \cdots + \mathbf{a}_n x_n = \mathbf{k} \tag{2.2}$$

The problem is to determine the possible values for the x_i, that is, all the possible ways of expressing $\mathbf{k}$ as a linear combination of $\mathbf{a}_1, \ldots, \mathbf{a}_n$.

2.2 DEFINITION OF A REGULAR SYSTEM AND STATEMENT OF THE RESULTS

In this chapter we shall consider an especially important case of the system (2.1), namely the case where $m = n$ and the n columns on the left of (2.1) are linearly independent. Such a system will be called a **regular system** of **order** n. Our aim will be to prove

Theorem 2.1
A regular system has exactly one solution.

Before proving Theorem 2.1 we remark, first, that whether a given $n \times n$ system is regular or not depends only on the coefficients a_{ij} on the left-hand side, not on the coefficients k_i on the right, because the ks are not used in the definition of regularity. So if a given system is regular, then Theorem 2.1 shows that it has a unique solution for every value of the right-hand side. In particular, when $k_1 = \cdots = k_m = 0$, then the unique solution is $\mathbf{x} = (0, \ldots, 0)$; this is called the **trivial solution**.

Second, to say that the system (2.2) is regular is to say that the vectors $\mathbf{a}_1, \ldots, \mathbf{a}_n$ are n-dimensional and are linearly independent. This means that there is no non-trivial linear relation between them; we can therefore restate Theorem 2.1 and at the same time strengthen it as follows:

Theorem 2.2
If $\mathbf{a}_1, \ldots, \mathbf{a}_n$ are any n vectors (of dimension n) such that

$$\mathbf{a}_1 x_1 + \cdots + \mathbf{a}_n x_n = \mathbf{0} \tag{2.3}$$

has only the trivial solution $(x_1, \ldots, x_n) = (0, \ldots, 0)$, then the system

$$\mathbf{a}_1 x_1 + \cdots + \mathbf{a}_n x_n = \mathbf{k} \tag{2.4}$$

is regular and has, for any vector $\mathbf{k}$, *just one solution* $(x_1, \ldots, x_n)$. *Conversely, if the system (2.4) has exactly one solution for any vector* $\mathbf{k}$, *then it must be regular.*

Here the last part follows because when there is only one solution for (2.3), it must be the zero solution, hence the vectors $\mathbf{a}_1, \ldots, \mathbf{a}_n$ are then linearly independent.

2.3 ELEMENTARY OPERATIONS

The proof of Theorem 2.1, which will at the same time give a practical means of solving the equations, amounts to showing that the traditional process of eliminating the unknowns one by one can be carried out. This process of elimination consists in adding multiples of one equation to another. In other words, we modify the system of equations in a way which does not affect the solutions. All such ways of operating on such a system can be obtained as a succession of the following 'elementary' operations:

α. *interchanging two of the equations;*
β. *multiplying an equation by a non-zero scalar;*
γ. *adding a multiple of one equation to another equation of the system.*

Let us call two systems of equations in the same unknowns $x_1, \ldots, x_n$ **equivalent** if they each contain the same number of equations and have the same solutions. We shall show that the operations α, β, γ, applied to any system, produce an equivalent system.

This is clear for the operation α, for it simply amounts to renumbering the equations. To show that β and γ yield equivalent systems, let us abbreviate the ith equation by writing

$$f_i(\mathbf{x}) = a_{i1}x_1 + \cdots + a_{in}x_n - k_i \quad (i = 1, \ldots, n)$$

To say that an n-vector $\mathbf{u}$ is a solution of the original equations just means that

$$f_1(\mathbf{u}) = f_2(\mathbf{u}) = \cdots = f_n(\mathbf{u}) = 0 \tag{2.5}$$

If we apply an operation β, say by multiplying the first equation by λ, where $\lambda \neq 0$, we obtain the system

$$\lambda f_1(\mathbf{u}) = f_2(\mathbf{u}) = \cdots = f_n(\mathbf{u}) = 0 \tag{2.6}$$

and this shows that $\mathbf{u}$ satisfies the new equations. Conversely, if (2.6) holds, then on dividing the first equation by λ – which we may do because $\lambda \neq 0$ – we obtain (2.5), so that every solution of the new system also satisfies the old system. Similarly, if (2.5) holds and we apply an operation γ, say by replacing the left-hand side of the first equation by $f_1(\mathbf{u}) + \mu f_2(\mathbf{u})$, then we have, by (2.5),

$$f_1(\mathbf{u}) + \mu f_2(\mathbf{u}) = f_2(\mathbf{u}) = f_3(\mathbf{u}) = \cdots = f_n(\mathbf{u}) = 0 \tag{2.7}$$

Conversely, when (2.7) holds, we can get back to (2.5) by subtracting μf_2 from $f_1 + \mu f_2$. The same reasoning applies when the operations involve equations other than the first or second, and it is clear that the number of equations remains unchanged. Thus we obtain an equivalent system by the application of any of the operations α, β, γ.

We remark that this argument does not make use of the regularity of the system, so that it applies to any $m \times n$ system.

As we are only concerned with finding the solution of our system, we may apply the operations α, β, γ, which we saw do not affect the solutions. But we have to show also that they do not affect the property of being regular. To say that the original system is regular is to say that when the vector $\mathbf{k}$ on the right is zero, the system has only the trivial solution (see section 2.2). Applying one of α, β, γ to such a system, we again get zero on the right, and since the new system is equivalent to the old, it has again only the trivial solution. Moreover, the number of equations or unknowns has not been changed by these operations, therefore the new system is again regular.

2.4 PROOF OF THE MAIN THEOREM (THEOREM 2.1)

Let us write down our system

$$\begin{aligned} a_{11}x_1 + \cdots + a_{1n}x_n &= k_1 \\ \cdots \quad \cdots \quad & \cdots \\ a_{n1}x_1 + \cdots + a_{nn}x_n &= k_n \end{aligned} \tag{2.8}$$

and suppose the equations are numbered so that $a_{11} \neq 0$; if the coefficient of x_1 in the first equation is zero, we interchange it (by an operation α) against an equation with a non-zero coefficient of x_1. This is possible provided that the first column contains a non-zero element. But that is always the case, for if not, then the first column would be the zero column and the columns of our system would not be linearly independent (which we know they are, by regularity). We can reduce the coefficient of x_1 in the first equation to 1 by multiplying that equation by $1/a_{11}$ (operation β). If we now subtract a_{i1} times the first equation from the ith equation, for $i = 2, \ldots, n$ (operation γ), we obtain a system in which x_1 occurs with the coefficient 1 in the first equation and with coefficient 0 in the remaining ones. Thus the new system has the form

$$\begin{aligned} x_1 + b_{12}x_2 + \cdots + b_{1n}x_n &= l_1 \\ b_{22}x_2 + \cdots + b_{2n}x_n &= l_2 \\ \cdots \quad \cdots \quad & \cdots \\ b_{n2}x_2 + \cdots + b_{nn}x_n &= l_n \end{aligned} \tag{2.9}$$

where the coefficients are related to those of (2.8) by the equations

$$b_{ij} = a_{ij} - \frac{a_{i1}}{a_{11}} a_{1j}, \quad l_i = k_i - \frac{a_{i1}}{a_{11}} k_1 \quad (i, j = 2, \ldots, n)$$

The effect of these operations has been to eliminate x_1 from all equations except the first. The term $a_{11}x_1$ which has been used to accomplish this elimination is called the **pivot**. Since we have used only the operations α, β, γ, this system is equivalent to the original system, and it is again regular, by the result of section 2.3.

We continue the elimination by applying the same process to the last $n-1$ equations of system (2.9), which constitute an $(n-1)\times(n-1)$ system, but before we can do so we must show that this system of order $n-1$ is regular.

We have to show that when we replace l_i by 0 in (2.9), the last $n-1$ equations have only the trivial solution. Suppose this is not so and let $(u_2, \ldots, u_n)$ be a non-trivial solution of these $n-1$ equations. Then we obtain a solution $(u_1, u_2, \ldots, u_n)$ of the whole system of n equations by taking

$$u_1 = -b_{12}u_2 - \cdots - b_{1n}u_n$$

For this satisfies the first equation, and the others are satisfied by hypothesis. Further, the solution is non-trivial, because at least one of $u_2, \ldots, u_n$ is non-zero, by hypothesis. This would mean that the n columns of (2.9) are linearly dependent, in contradiction to the fact that this system is regular. So the last $n-1$ equations in (2.9) do constitute a regular system.

We have now reduced the proof of Theorem 2.1, for a system of order n, to the proof of Theorem 2.1 for a system of order $n-1$, and we can complete the proof by using **mathematical induction**. This may be used whenever we want to prove a proposition P_n for all positive integers n. We first prove P_1 and then, for any $n \geqslant 2$, show that P_n follows from P_{n-1}, the **induction hypothesis**. The principle of induction then states that P_n is true for all n.

To apply this remark to our case we first have to prove Theorem 2.1 for systems of order 1, that is, single equations in one unknown. The equation $ax = k$ is regular precisely when $a \neq 0$, and in that case the equation has the unique solution $x = k/a$, so our theorem holds for systems of order 1. Now we use the induction hypothesis on (2.9): the last $n-1$ equations have the unique solution

$$\begin{aligned} x_2 &= r_2 \\ &\vdots \\ x_n &= r_n \end{aligned}$$

so that we obtain the following system equivalent to (2.9):

$$\begin{array}{lllll} x_1 + b_{12}x_2 + & \cdots & + b_{1n}x_n & = l_1 \\ \phantom{x_1 + b_{12}}x_2 & & & = r_2 \\ & x_3 & & = r_3 \\ & & \cdots & \\ & & x_n & = r_n \end{array}$$

Thus $x_2, \ldots, x_n$ are uniquely determined, and the first equation gives the value

$r_1 = l_1 - \sum_2^n b_{1j}r_j$ for x_1. Hence the original system (2.8) has the unique solution $\mathbf{x} = (r_1, \ldots, r_n)^T$ and this completes the proof of the theorem. Now Theorem 2.2 is an immediate consequence.

2.5 GAUSSIAN ELIMINATION AND THE GAUSS–JORDAN REDUCTION

To illustrate Theorem 2.1 let us take the system

$$\begin{aligned} x_1 + 5x_2 + 2x_3 &= 9 \\ x_1 + x_2 + 7x_3 &= 6 \\ -3x_2 + 4x_3 &= -2 \end{aligned} \tag{2.10}$$

and apply the method described in section 2.4. For brevity we shall not write down the equations in full each time, but merely give the scheme of coefficients. The given system may be written

$$\begin{array}{rrr|r} 1 & 5 & 2 & 9 \\ 1 & 1 & 7 & 6 \\ 0 & -3 & 4 & -2 \end{array}$$

where the position of the equals sign is indicated by a vertical line. The coefficient of x_1 in the first equation is already 1; taking it as a pivot, we replace row 2 of the scheme (which we will henceforth abbreviate as R2) by R2 − R1 and obtain

$$\begin{array}{rrr|r} 1 & 5 & 2 & 9 \\ 0 & -4 & 5 & -3 \\ 0 & -3 & 4 & -2 \end{array}$$

Next we divide R2 by -4 and then replace R3 by R3 + 3.R2:

$$\begin{array}{rrr|r} 1 & 5 & 2 & 9 \\ 0 & 1 & -5/4 & 3/4 \\ 0 & 0 & 1/4 & 1/4 \end{array} \tag{2.11}$$

From this **triangular form** (so called because all coefficients outside a certain triangle are zero), it is easy to obtain the explicit solution, starting from the last equation, which is $x_3 = 1$, if we omit the factor 1/4. Substituting in the next equation, we find $x_2 = 3/4 + (5/4)x_3 = 2$, and finally $x_1 = 9 - 5x_2 - 2x_3 = -3$. Thus our system (2.10) has the unique solution $\mathbf{x} = (-3, 2, 1)^T$. As a check on our calculation we should of course verify that it is a solution; this is easily done, by substituting the values found for x_1, x_2, x_3 in (2.10).

This process of solving the system (2.10) is called **Gaussian elimination**, after Carl Friedrich Gauss (1777–1855), who first used it in calculations on the asteroid Pallas at the beginning of the nineteenth century (Gauss, one of the greatest mathematicians of all time, has made fundamental contributions to almost all parts of the subject).

Sometimes it is convenient to continue the elimination beyond (2.11) by subtracting 5.R2 from R1:

$$\begin{array}{rrr|r} 1 & 0 & 33/4 & 21/4 \\ 0 & 1 & -5/4 & 3/4 \\ 0 & 0 & 1/4 & 1/4 \end{array}$$

Finally if we multiply R3 by 4 and subtract appropriate multiples from R1 and R2, we obtain a system in diagonal form:

$$\begin{array}{rrr|r} 1 & 0 & 0 & -3 \\ 0 & 1 & 0 & 2 \\ 0 & 0 & 1 & 1 \end{array}$$

from which the solution $\mathbf{x} = (-3, 2, 1)^T$ can be directly read off. This reduction process is known as **Gauss–Jordan reduction**, after Wilhelm Jordan (1842–99), a geodesist who described the method in his *Handbuch der Geodäsie* in 1873.

It is not necessary to follow the procedure exactly, as long as we use only the operations α, β, γ. By putting in extra steps we can sometimes make the calculation easier. Thus in the above example we could avoid the occurrence of fractions by proceeding as follows (after the reduction of the first column):

$$\begin{array}{rrr|r} 1 & 5 & 2 & 9 \\ 0 & -4 & 5 & -3 \\ 0 & -3 & 4 & -2 \end{array} \quad R2 \to R2 - R3 \quad \begin{array}{rrr|r} 1 & 5 & 2 & 9 \\ 0 & -1 & 1 & -1 \\ 0 & -3 & 4 & -2 \end{array} \quad R2 \to -R2$$

$$\begin{array}{rrr|r} 1 & 5 & 2 & 9 \\ 0 & 1 & -1 & 1 \\ 0 & -3 & 4 & -2 \end{array} \quad \begin{array}{l} R1 \to R1 - 5.R2 \\ R3 \to R3 + 3.R2 \end{array} \quad \begin{array}{rrr|r} 1 & 0 & 7 & 4 \\ 0 & 1 & -1 & 1 \\ 0 & 0 & 1 & 1 \end{array} \quad \begin{array}{l} R1 \to R1 - 7.R3 \\ R2 \to R2 + R3 \end{array}$$

$$\begin{array}{rrr|r} 1 & 0 & 0 & -3 \\ 0 & 1 & 0 & 2 \\ 0 & 0 & 1 & 1 \end{array}$$

The operations performed are indicated before each scheme, thus $R2 \to R2 - R3$ before the second scheme means that this scheme is obtained from the first by replacing R2 by R2 − R3. Similarly, $R1 \leftrightarrow R2$ before a scheme would mean that it has been obtained by from the previous scheme by interchanging R1 and R2.

2.6 ILLUSTRATIONS

In solving the system considered in section 2.5 we have not at any stage verified its regularity, although once solved, this followed from the uniqueness of the solution (Theorem 2.2). In solving a system by the method of Theorem 2.1 it is not necessary to check for regularity; if the system is regular, the procedure leads to a solution, while a non-regular system fails to give a unique solution.

As an example consider the system with the first two equations as in (2.10) and the third equation $4x_2 - 5x_3 = 2$. The reduction runs as follows:

$$\begin{array}{rrr|r} 1 & 5 & 2 & 9 \\ 1 & 1 & 7 & 6 \\ 0 & 4 & -5 & 2 \end{array} \quad R2 \to R2 - R1 \quad \begin{array}{rrr|r} 1 & 5 & 2 & 9 \\ 0 & -4 & 5 & -3 \\ 0 & 4 & -5 & 2 \end{array}$$

$$R3 \to R3 + R2 \quad \begin{array}{rrr|r} 1 & 5 & 2 & 9 \\ 0 & -4 & 5 & -3 \\ 0 & 0 & 0 & -1 \end{array}$$

The third equation of the last system reads $0 = -1$, and this makes it clear that the system has no solution. If the third equation of our system had been $4x_2 - 5x_3 = 3$, then the same reduction would lead to

$$\begin{array}{rrr|r} 1 & 5 & 2 & 9 \\ 0 & -4 & 5 & -3 \\ 0 & 0 & 0 & 0 \end{array}$$

The row of zeros corresponds to the equation $0 = 0$, which may be omitted without affecting the solution, so we are left with 2 equations in 3 unknowns. This system cannot have a unique solution, because we can assign x_3 arbitrarily and then solve for x_1, x_2 as before; therefore the system cannot be regular.

As a further illustration we consider the 4×4 system

$$\begin{aligned} x_1 - 2x_2 - 7x_3 + 7x_4 &= 5 \\ -x_1 + 2x_2 + 8x_3 - 5x_4 &= -7 \\ 3x_1 - 4x_2 - 17x_3 + 13x_4 &= 14 \\ 2x_1 - 2x_2 - 11x_3 + 8x_4 &= 7 \end{aligned}$$

We shall write down the scheme with the successive reductions:

$$\begin{array}{rrrr|rl} 1 & -2 & -7 & 7 & 5 & \\ -1 & 2 & 8 & -5 & -7 & R2 \to R2 + R1 \\ 3 & -4 & -17 & 13 & 14 & R3 \to R3 - 3.R1 \\ 2 & -2 & -11 & 8 & 7 & R4 \to R4 - 2.R1 \end{array}$$

$$\begin{array}{rrrr|r} 1 & -2 & -7 & 7 & 5 \\ 0 & 0 & 1 & 2 & -2 \\ 0 & 2 & 4 & -8 & -1 \\ 0 & 2 & 3 & -6 & -3 \end{array} \quad R2 \leftrightarrow R3$$

$$\begin{array}{rrrr|r} 1 & -2 & -7 & 7 & 5 \\ 0 & 2 & 4 & -8 & -1 \\ 0 & 0 & 1 & 2 & -2 \\ 0 & 2 & 3 & -6 & -3 \end{array} \qquad \mathrm{R2} \to \tfrac{1}{2}.\mathrm{R2}$$

$$\begin{array}{rrrr|r} 1 & -2 & -7 & 7 & 5 \\ 0 & 1 & 2 & -4 & -\frac{1}{2} \\ 0 & 0 & 1 & 2 & -2 \\ 0 & 2 & 3 & -6 & -3 \end{array} \qquad \mathrm{R4} \to \mathrm{R4} - 2.\mathrm{R2}$$

$$\begin{array}{rrrr|r} 1 & -2 & -7 & 7 & 5 \\ 0 & 1 & 2 & -4 & -\frac{1}{2} \\ 0 & 0 & 1 & 2 & -2 \\ 0 & 0 & -1 & 2 & -2 \end{array} \qquad \mathrm{R4} \to \mathrm{R4} + \mathrm{R3}$$

$$\begin{array}{rrrr|r} 1 & -2 & -7 & 7 & 5 \\ 0 & 1 & 2 & -4 & -\frac{1}{2} \\ 0 & 0 & 1 & 2 & -2 \\ 0 & 0 & 0 & 4 & -4 \end{array}$$

From the triangular form found we obtain $4x_4 = -4$, hence $x_4 = -1$, $x_3 = -2 - 2x_4 = 0$, $x_2 = -\frac{1}{2} - 2x_3 + 4x_4 = -4\frac{1}{2}$, $x_1 = 5 + 2x_2 + 7x_3 - 4x_4 = 3$, so the unique solution is $\mathbf{x} = (3, -4\frac{1}{2}, 0, -1)^{\mathrm{T}}$. Here we had to renumber the equations before we could reduce the second column. Evidently we cannot avoid fractions in the process this time, since they appear in the solution.

To obtain the Gauss–Jordan reduction, we continue from the last scheme:

$$\begin{array}{l rrrr|r l} \mathrm{R1} \to \mathrm{R1} + 2.\mathrm{R2} & 1 & 0 & -3 & -1 & 4 & \mathrm{R1} \to \mathrm{R1} + 3.\mathrm{R3} \\ \mathrm{R2} \to \mathrm{R2} - 2.\mathrm{R3} & 0 & 1 & 0 & -8 & 3\frac{1}{2} & \mathrm{R2} \to \mathrm{R2} + 8.\mathrm{R4} \\ & 0 & 0 & 1 & 2 & -2 & \mathrm{R3} \to \mathrm{R3} - 2.\mathrm{R4} \\ \mathrm{R4} \to \frac{1}{4}.\mathrm{R4} & 0 & 0 & 0 & 1 & -1 & \end{array}$$

$$\begin{array}{rrrr|r} 1 & 0 & 0 & 5 & -2 \\ 0 & 1 & 0 & 0 & -4\frac{1}{2} \\ 0 & 0 & 1 & 0 & 0 \\ 0 & 0 & 0 & 1 & -1 \end{array} \qquad \mathrm{R1} \to \mathrm{R1} - 5.\mathrm{R4}$$

$$\begin{array}{rrrr|r} 1 & 0 & 0 & 0 & 3 \\ 0 & 1 & 0 & 0 & -4\frac{1}{2} \\ 0 & 0 & 1 & 0 & 0 \\ 0 & 0 & 0 & 1 & -1 \end{array}$$

Now the solution appears in the last column.

2.7 THE CHOICE OF PIVOT

We have seen that a zero coefficient cannot be used as a pivot, and this suggests that in numerical work it is inexpedient to work with a small pivot. We illustrate this by a simple example. Consider the system

$$\begin{aligned} 0.1x_1 + 100x_2 &= 50 \\ 50x_1 - \ \ 20x_2 &= 40 \end{aligned} \qquad (2.12)$$

The reduction to triangular form, using the term $0.1x_1$ as a pivot, looks as follows:

$$\text{R1} \to 10.\text{R1} \quad \left.\begin{array}{rr} 1 & 1000 \\ 50 & -20 \end{array}\right| \begin{array}{r} 500 \\ 40 \end{array} \quad \text{R2} \to \text{R2} - 50.\text{R1} \quad \left.\begin{array}{rr} 1 & 1000 \\ 0 & -50\,020 \end{array}\right| \begin{array}{r} 500 \\ -24\,960 \end{array}$$

Suppose that our calculator rounds off to three significant figures. Then the second equation becomes $-50\,000x_2 = -25\,000$, i.e. $x_2 = 0.5$. Now the first equation shows that $x_1 = 500 - 500 = 0$. If we check this solution $(0, 0.5)$ by substitution, we find that

$$\begin{aligned} 0.1x_1 + 100x_2 &= 50 \\ 50x_1 - \ \ 20x_2 &= -10 \end{aligned}$$

which is wide of the mark. Let us now instead use the term $50x_1$ as a pivot; we obtain

$$\text{R1} \leftrightarrow \text{R2} \quad \left.\begin{array}{rr} 50 & -20 \\ 0.1 & 100 \end{array}\right| \begin{array}{r} 40 \\ 50 \end{array} \quad \begin{array}{l} \text{R1} \to 0.02.\text{R1} \\ \text{R2} \to 10.\text{R2} \end{array} \quad \left.\begin{array}{rr} 1 & -0.4 \\ 1 & 1000 \end{array}\right| \begin{array}{r} 0.8 \\ 500 \end{array}$$

$$\text{R2} \to \text{R2} - \text{R1} \quad \left.\begin{array}{rr} 1 & -0.4 \\ 0 & 1000 \end{array}\right| \begin{array}{r} 0.8 \\ 501 \end{array}$$

and this leads to the solution $x_2 = 0.5$, $x_1 = 0.8 + 0.2 = 1$. This time a check gives

$$\begin{aligned} 0.1x_1 + 100x_2 &= 50.1 \\ 50x_1 - \ \ 20x_2 &= 40 \end{aligned}$$

which is in much closer agreement. A more accurate calculation shows the solution to be $\boldsymbol{x} = (1.000, 0.499)$ to four significant figures.

This method, of renumbering the equations so as to use the term with the largest coefficient in the first column as pivot, is known as **partial pivoting**. By contrast, **full pivoting** consists in choosing the term with the largest coefficient in any row or column as pivot; this may require a renumbering of the unknowns as well as the equations. Some judgement is needed using partial pivoting, for if in our system (2.12) the first equation is multiplied by 500, we have an equivalent system, but now both coefficients in the first column are 50 and there is nothing to choose between them, although as we know, the second equation rather than the first should be used. We shall not discuss this point further, beyond remarking that the aim should be to choose as pivot a term whose coefficient occurs in a large term of the determinant of the system (cf. Chapter 5).

2.8 THE LINEAR DEPENDENCE OF $n+1$ VECTORS IN n DIMENSIONS

Let us now return to the question left open from Chapter 1 about the invariance of the number of elements in a basis of a vector space. We begin with a general remark.

An $n \times n$ system in which all the coefficients $a_{n1}, a_{n2}, \ldots, a_{nn}$ of the last equation are zero, cannot be regular. For if all the coefficients on the left of the last equation vanish, then for the system to have a solution, the term on the right of the equation must vanish too, i.e. $k_n = 0$. So the system has no solution when $k_n \neq 0$, whereas a regular system has a solution whatever the vector k on the right (by Theorem 2.2), hence the system cannot be regular. The result may also be stated as follows, if we remember that a regular system is just one whose n columns are linearly independent:

If n vectors of dimension n have their last component equal to zero, then they are linearly dependent.

From this proposition we can easily deduce the important

Theorem 2.3
Any $n+1$ vectors in n dimensions are linearly dependent.

For if $\mathbf{u}_i = (u_{i1}, \ldots, u_{in})$ $(i = 1, \ldots, n+1)$ are the given n-vectors, we can consider the $n+1$ vectors $\mathbf{u}_i^* = (u_{i1}, \ldots, u_{in}, 0)$ of dimension $n+1$. These vectors have 0 as their last component and therefore, by the result just proved, they are linearly dependent, say

$$\mathbf{u}_{n+1}^* = \lambda_1 \mathbf{u}_1^* + \cdots + \lambda_n \mathbf{u}_n^*$$

Equating components in this relation, we find

$$u_{n+1,j} = \lambda_1 u_{1j} + \cdots + \lambda_n u_{nj} \quad (j = 1, \ldots, n)$$

whence

$$\mathbf{u}_{n+1} = \lambda_1 \mathbf{u}_1 + \cdots + \lambda_n \mathbf{u}_n$$

and this is the dependence relation we wished to find.

2.9 THE UNIQUENESS OF THE DIMENSION

As a first consequence of Theorem 2.3 we note that in n dimensions any n vectors which are linearly independent form a basis for the space of all n-vectors. For let $\mathbf{v}_1, \ldots, \mathbf{v}_n$ be any linearly independent set of n vectors in the space of all n-vectors and consider any vector $\mathbf{x}$ in this space. The $n+1$ vectors $\mathbf{x}, \mathbf{v}_1, \ldots, \mathbf{v}_n$ are linearly dependent, by Theorem 2.3; thus we have a relation

$$\alpha\mathbf{x} + \alpha_1\mathbf{v}_1 + \cdots + \alpha_n\mathbf{v}_n = \mathbf{0} \tag{2.13}$$

where not all of $\alpha, \alpha_1, \ldots, \alpha_n$ are zero. If α were zero, then (2.13) would be a non-trivial relation between $\mathbf{v}_1, \ldots, \mathbf{v}_n$, which contradicts the fact that these vectors are linearly independent. So $\alpha \neq 0$, and writing $\xi_i = -\alpha_i/\alpha_1$, we have the relation

$$\mathbf{x} = \xi_1\mathbf{v}_1 + \cdots + \xi_n\mathbf{v}_n$$

which expresses $\mathbf{x}$ as a linear combination of $\mathbf{v}_1, \ldots, \mathbf{v}_n$. Hence the $\mathbf{v}_i$ span the space of all n-vectors, and so form a basis for it.

We can now achieve our aim of proving

Theorem 2.4
Any two bases of a given vector space have the same number of elements.

In proving this result we shall limit ourselves to vector spaces spanned by a finite set (i.e. satisfying V.10), although as a matter of fact the result holds quite generally.

Let V be our vector space; in the case considered there is a finite spanning set, from which we can select a basis $\mathbf{v}_1, \ldots, \mathbf{v}_n$ as in section 1.9. Now every vector $\mathbf{x}$ of V can be uniquely expressed as a linear combination of the basis vectors $\mathbf{v}_1, \ldots, \mathbf{v}_n$:

$$\mathbf{x} = \xi_1\mathbf{v}_1 + \cdots + \xi_n\mathbf{v}_n$$

and we may take $\xi_1, \ldots, \xi_n$ to be the components of $\mathbf{x}$ as in section 1.10. In terms of these components the vectors of V are n-dimensional, and therefore any $n+1$ vectors of V are linearly dependent. Thus no linearly independent set of vectors in V can contain more than n members, and in particular, no basis of V can have more than n elements. If there is a basis with m elements, where $m \leqslant n$, then the same argument applied with the new basis shows that $n \leqslant m$, so $m = n$ and all bases of V have the same number of elements. This number will be denoted by $\dim V$.

This result shows that the dimension of any vector space, such as $\mathbf{R}^n$, is uniquely determined and may be found by counting the number of elements in any basis of the space.

EXERCISES

2.1 The following schemes represent regular systems of equations, in the notation of section 2.5. Find the solution in each case:

$$\text{(i)}\ \begin{array}{rrr|r} 2 & 4 & 1 & 5 \\ 1 & 1 & 1 & 6 \\ 2 & 3 & 1 & 6 \end{array} \qquad \text{(ii)}\ \begin{array}{rrr|r} 5 & 15 & -10 & 0 \\ -2 & -2 & -4 & 4 \\ 3 & 4 & 1 & -5 \end{array} \qquad \text{(iii)}\ \begin{array}{rrr|r} 1 & 4 & 11 & 7 \\ 2 & 8 & 16 & 8 \\ 1 & 6 & 17 & 9 \end{array}$$

(iv)

$$\begin{array}{rrr|r} 3 & -3 & 5 & 6 \\ 1 & 7 & 5 & 4 \\ 5 & 10 & 15 & 9 \end{array}$$

(v)

$$\begin{array}{rrr|r} 3 & 2 & -1 & 0 \\ 1 & 7 & 5 & 0 \\ -1 & 0 & 1 & 0 \end{array}$$

(vi)

$$\begin{array}{rrrr|r} 2 & 1 & 2 & -1 & 0 \\ 6 & 8 & 12 & -13 & -21 \\ 10 & 2 & 2 & 3 & 21 \\ -4 & 0 & 1 & -3 & -13 \end{array}$$

(vii)

$$\begin{array}{rrrr|r} 2 & -1 & 4 & 7 & 10 \\ 0 & 1 & 0 & -5 & -6 \\ 0 & 0 & -3 & 2 & 8 \\ 0 & 0 & 0 & 1 & 1 \end{array}$$

(viii)

$$\begin{array}{rrrr|r} 4 & -8 & 4 & -20 & 0 \\ 3 & -6 & 4 & -14 & 4 \\ -2 & 4 & -4 & 9 & 5 \\ 4 & -7 & 5 & -8 & 6 \end{array}$$

2.2. Show that the triangular system

$$\begin{array}{ccccc|c} a_{11} & a_{12} & a_{13} & \cdots & a_{1n} & k_1 \\ 0 & a_{22} & a_{23} & \cdots & a_{2n} & k_2 \\ & \cdots & \cdots & \cdots & & \\ 0 & 0 & 0 & \cdots & a_{nn} & k_n \end{array}$$

is regular if and only if the diagonal coefficients $a_{11}, \ldots, a_{nn}$ are all different from zero.

2.3. Show that a system in which two rows are the same, or two columns are the same, cannot be regular.

2.4. Show that a system (2.8) is regular if and only if the rows (on the left of the equations) are linearly independent.

2.5. Show that any rearrangement of n equations can be accomplished by successively interchanging a pair of equations. Show that it can be done even by interchanging each time a pair of neighbouring equations. (*Hint*: See section 5.2 below.)

2.6. Show that an elementary operation α can be accomplished by performing suitable operations β and γ. Are operations of type γ enough?

2.7. Find the flaw in the following argument: In the system

$$\begin{array}{cc} 3 & 5 \\ 2 & 1 \end{array}$$

subtract the first row from the second and the second row from the first, giving

$$\begin{array}{rr} 1 & 4 \\ -1 & -4 \end{array}$$

Now add the first row to the second to reduce the latter to zero. Hence the given system is equivalent to a system with a zero row and so is not regular.

2.8. Let V be the vector space consisting of all polynomials of degree at most $n-1$ in x. Given any polynomial f of degree $n-1$, show that f together with its derivatives $f', f'', \ldots, f^{(n-1)}$ is a basis of V.

2.9. Show that any basis of a vector space V is a minimal spanning set, i.e. it spans V but fails to do so if any of its members is removed. Show that conversely, any minimal spanning set of V is a basis.

2.10. Show that any basis of a vector space V is a maximal linearly independent set, i.e. it is linearly independent but becomes linearly dependent once it is enlarged. Show conversely that any maximal linearly independent set in V is a basis.

3

Matrices

3.1 DEFINITION OF A MATRIX

There is yet another way of regarding an $m \times n$ system of equations, which will lead to an even shorter way of writing it than the vector notation used in Chapter 2. Let us first write the system as

$$\begin{aligned} a_{11}x_1 + \cdots + a_{1n}x_n &= y_1 \\ \cdots \quad \cdots \quad &\cdots \\ a_{m1}x_1 + \cdots + a_{mn}x_n &= y_m \end{aligned} \tag{3.1}$$

where we have replaced the constants k_i on the right by variables y_i. If we take $\mathbf{x} = (x_1, \ldots, x_n)^{\mathrm{T}}$ and $\mathbf{y} = (y_1, \ldots, y_m)^{\mathrm{T}}$ to be column vectors, we may regard the set of coefficients (a_{ij}) in (3.1) as an operator which acts on the n-vector $\mathbf{x}$ to produce the m-vector $\mathbf{y}$. In this sense we write (3.1) as

$$\begin{pmatrix} a_{11} & a_{12} & \cdots & a_{1n} \\ a_{21} & a_{22} & \cdots & a_{2n} \\ & \cdots & \cdots & \\ a_{m1} & a_{m2} & \cdots & a_{mn} \end{pmatrix} \begin{pmatrix} x_1 \\ x_2 \\ \vdots \\ x_n \end{pmatrix} = \begin{pmatrix} y_1 \\ y_2 \\ \vdots \\ y_m \end{pmatrix} \tag{3.2}$$

The rectangular array of coefficients

$$\begin{pmatrix} a_{11} & a_{12} & \cdots & a_{1n} \\ a_{21} & a_{22} & \cdots & a_{2n} \\ & \cdots & \cdots & \\ a_{m1} & a_{m2} & \cdots & a_{mn} \end{pmatrix} \tag{3.3}$$

is called a **matrix** with the **entries** a_{ij} and the left-hand side of (3.2) may be regarded as the 'product' of this matrix by $\mathbf{x}$. A comparison with (3.1) shows that this product is evaluated by multiplying the elements of the jth column of (3.3) by x_j and adding the column vectors so obtained. The sum of these vectors has its ith component equal to $a_{i1}x_1 + \cdots + a_{in}x_n$ $(i = 1, \ldots, m)$, and the vector equation in (3.2) states that this vector is equal to the column vector $\mathbf{y} = (y_1, \ldots, y_m)^{\mathrm{T}}$, which is in fact a restatement of the equations (3.1).

A matrix with m rows and n columns as in (3.3) is also called an $m \times n$ matrix, and the set of all $m \times n$ matrices with entries in $\mathbf{R}$ will be denoted by ${}^m\mathbf{R}^n$. When $m = 1$, this reduces to the vector space of rows ${}^1\mathbf{R}^n$, also written $\mathbf{R}^n$, while $n = 1$ gives the space ${}^m\mathbf{R}^1$ of columns, also written ${}^m\mathbf{R}$, in agreement with our earlier notation $\mathbf{R}^n$ for the space of rows and ${}^m\mathbf{R}$ for the space of columns of the indicated dimensions.

If we denote the matrix (3.3) by $\mathbf{A}$, we may write (3.2) more briefly as

$$\mathbf{Ax} = \mathbf{y}$$

In (3.2) we applied the matrix $\mathbf{A}$ to $\mathbf{x}$ and obtained the result $\mathbf{y}$. From the meaning given to (3.2) it follows that an $m \times n$ matrix can only be applied to an n-vector, and the result will be an m-vector. In the special case $m = n$ we call $\mathbf{A}$ a **square matrix** of **order** n.

The introduction of matrices enables us to write many formulae more compactly and in a more suggestive way, just as vectors already helped us in Chapter 2 to obtain a clearer picture of the solutions of linear equations. We shall therefore in this chapter introduce the basic rules of calculating with matrices.

Since the coefficients a_{ij} in (3.1) were arbitrary real numbers, it follows that any mn real numbers, arranged in a rectangle as in (3.3), form a matrix. Matrices will usually be denoted by capitals in bold type: $\mathbf{A}, \mathbf{B}, \ldots$; more explicitly we may write the matrix (3.3) as (a_{ij}), where the first suffix (in this case i) indicates the row and the second suffix (here j) indicates the column. Of course, if the entries of the matrix have definite numerical values, the matrix must be written out in full. Thus the matrix of the first example considered in the Introduction is

$$\begin{pmatrix} 3 & 2 \\ 2 & 5 \end{pmatrix}$$

3.2 THE MATRIX AS A LINEAR OPERATOR

A most important property of matrices is expressed in the equations

$$\mathbf{A}(\mathbf{u} + \mathbf{v}) = \mathbf{Au} + \mathbf{Av} \tag{3.4a}$$

$$\mathbf{A}(\lambda\mathbf{u}) = \lambda\mathbf{Au} \tag{3.4b}$$

where $\mathbf{A}$ is an $m \times n$ matrix, $\mathbf{u}$ and $\mathbf{v}$ are n-vectors and λ is a scalar. Their vertification is easy and is left to the reader. The properties expressed in (3.4) are described by saying that the expression $\mathbf{Au}$ is **linear** in $\mathbf{u}$, or also that $\mathbf{A}$ is a **linear operator**.

For example, when $m = n = 2$, then (3.4b) states that

$$\begin{pmatrix} a_{11}\lambda u_1 + a_{12}\lambda u_2 \\ a_{21}\lambda u_1 + a_{22}\lambda u_2 \end{pmatrix} = \begin{pmatrix} \lambda(a_{11}u_1 + a_{12}u_2) \\ \lambda(a_{21}u_1 + a_{22}u_2) \end{pmatrix}$$

Equations (3.4) may be used to express $\mathbf{Ax}$ as a linear combination of the vectors $\mathbf{Ae}_i$, where the $\mathbf{e}_i$ are the n standard basis vectors introduced in section 1.8, but now thought of as columns. For if $\mathbf{x} = (x_1, \ldots, x_n)^{\mathrm{T}}$, then $\mathbf{x} = x_1\mathbf{e}_1 + \cdots + x_n\mathbf{e}_n$ and hence

$$\begin{aligned}\mathbf{Ax} = \mathbf{A}(x_1\mathbf{e}_1 + \cdots + x_n\mathbf{e}_n) &= \mathbf{A}(x_1\mathbf{e}_1) + \cdots + \mathbf{A}(x_n\mathbf{e}_n)\\ &= x_1\mathbf{Ae}_1 + \cdots + x_n\mathbf{Ae}_n\end{aligned}$$

by repeated use of (3.4). Thus we have expressed $\mathbf{Ax}$ as a linear combination of the n vectors $\mathbf{Ae}_1, \ldots, \mathbf{Ae}_n$, which of course do not depend on the choice of $\mathbf{x}$. These vectors $\mathbf{Ae}_j$ are nothing but the columns of $\mathbf{A}$: $\mathbf{Ae}_j = (a_{1j}, a_{2j}, \ldots, a_{mj})^{\mathrm{T}}$.

3.3 EQUALITY OF MATRICES

Two matrices $\mathbf{A}$ and $\mathbf{B}$ are said to be **equal**, $\mathbf{A} = \mathbf{B}$, if they have the same number of rows and columns – say, they are both $m \times n$ matrices – and if corresponding entries are equal:

$$a_{ij} = b_{ij} \quad (i = 1, \ldots, m, j = 1, \ldots, n)$$

Thus a matrix equation between $m \times n$ matrices is equivalent to mn scalar equations. In fact we may look on $m \times n$ matrices as vectors in mn dimensions, whose components are arranged not as a row or column but as a rectangle. This interpretation is further justified by the operations of addition and multiplication by scalars to be defined for matrices in section 3.4 below.

It follows from the definition that if $\mathbf{A} = \mathbf{B}$, then $\mathbf{Au} = \mathbf{Bu}$ for any vector $\mathbf{u}$ of dimension equal to the number of columns of $\mathbf{A}$ or $\mathbf{B}$. Conversely, if $\mathbf{A}$ and $\mathbf{B}$ are $m \times n$ matrices such that $\mathbf{Au} = \mathbf{Bu}$ for every n-vector $\mathbf{u}$, then $\mathbf{A} = \mathbf{B}$. For if we take the standard basis vector $\mathbf{e}_j$ for $\mathbf{u}$ we see that the jth column of $\mathbf{A}$ must equal the jth column of $\mathbf{B}$, i.e. $a_{ij} = b_{ij}$ $(i = 1, \ldots, m)$. Since this holds for $j = 1, \ldots, n$, it follows that corresponding entries of $\mathbf{A}$ and $\mathbf{B}$ are equal and so $\mathbf{A} = \mathbf{B}$. Our findings can be expressed as follows:

Two $m \times n$ matrices $\mathbf{A}$ and $\mathbf{B}$ are equal if and only if

$$\mathbf{Ae}_j = \mathbf{Be}_j \quad \textit{for} \quad j = 1, \ldots, n$$

where $\mathbf{e}_1, \ldots, \mathbf{e}_n$ are the standard basis vectors.

More generally, let $\mathbf{u}_1, \ldots, \mathbf{u}_n$ be any basis for the space of n-vectors. If $\mathbf{A}, \mathbf{B}$ are $m \times n$ matrices such that $\mathbf{Au}_i = \mathbf{Bu}_i$ $(i = 1, \ldots, n)$, then $\mathbf{A} = \mathbf{B}$. For any n-vector can be written as a linear combination of $\mathbf{u}_1, \ldots, \mathbf{u}_n$, say $\mathbf{v} = \Sigma \lambda_i \mathbf{u}_i$, hence $\mathbf{Av} = \mathbf{A}(\Sigma \lambda_i \mathbf{u}_i) = \Sigma \lambda_i(\mathbf{Au}_i)$ by linearity; similarly $\mathbf{Bv} = \Sigma \lambda_i(\mathbf{Bu}_i)$ and since $\mathbf{Au}_i = \mathbf{Bu}_i$, it follows that $\mathbf{Av} = \mathbf{Bv}$. This holds for all vectors $\mathbf{v}$, so we conclude that $\mathbf{A} = \mathbf{B}$.

3.4 THE VECTOR SPACE OF MATRICES

If we think of matrices as forming a vector space, it is clear how their addition and multiplication by a scalar are to be defined. Given two $m \times n$ matrices $\mathbf{A} = (a_{ij})$, $\mathbf{B} = (b_{ij})$ and a scalar λ, we put

$$\mathbf{A} + \mathbf{B} = (a_{ij} + b_{ij})$$
$$\lambda\mathbf{A} = (\lambda a_{ij})$$

Thus we add (and multiply by scalars) 'componentwise', just as for vectors, and the same rules hold for the addition of matrices as for vectors (cf. sections 1.3–1.4). For example:

$$2\begin{pmatrix} 3 & -2 & 8 \\ 1 & 0 & 5 \end{pmatrix} = \begin{pmatrix} 6 & -4 & 16 \\ 2 & 0 & 10 \end{pmatrix}$$

$$\begin{pmatrix} 4 & 7 & -1 \\ 3 & -2 & 5 \end{pmatrix} + \begin{pmatrix} 3 & -5 & 6 \\ -8 & 1 & 9 \end{pmatrix} = \begin{pmatrix} 7 & 2 & 5 \\ -5 & -1 & 14 \end{pmatrix}$$

The $m \times n$ matrix whose entries are all zero is called the **zero matrix** (of m rows and n columns) and is denoted by $\mathbf{0}$, the number of rows and columns usually being understood from the context.

The operations just defined satisfy the important rules

$$(\mathbf{A} + \mathbf{B})\mathbf{u} = \mathbf{A}\mathbf{u} + \mathbf{B}\mathbf{u}$$
$$(\lambda\mathbf{A})\mathbf{u} = \lambda(\mathbf{A}\mathbf{u}) \tag{3.5}$$

The proof is quite straightforward: it is only necessary to show that the two sides agree when $\mathbf{u} = \mathbf{e}_j$ $(j = 1, \ldots, n)$, and this may be left to the reader. We content ourselves with an illustration: If

$$\mathbf{A} = \begin{pmatrix} 3 & 1 \\ -2 & 5 \end{pmatrix} \quad \text{and} \quad \mathbf{B} = \begin{pmatrix} 2 & 4 \\ 6 & -3 \end{pmatrix},$$

then

$$2\mathbf{A} - 3\mathbf{B} = \begin{pmatrix} 0 & -10 \\ -22 & 19 \end{pmatrix}$$

Further, if $\mathbf{u} = (1, 1)^{\mathrm{T}}$, then

$$2(\mathbf{A}\mathbf{u}) - 3(\mathbf{B}\mathbf{u}) = 2\begin{pmatrix} 3 & 1 \\ -2 & 5 \end{pmatrix}\begin{pmatrix} 1 \\ 1 \end{pmatrix} - 3\begin{pmatrix} 2 & 4 \\ 6 & -3 \end{pmatrix}\begin{pmatrix} 1 \\ 1 \end{pmatrix} = 2\begin{pmatrix} 4 \\ 3 \end{pmatrix} - 3\begin{pmatrix} 6 \\ 3 \end{pmatrix} = \begin{pmatrix} -10 \\ -3 \end{pmatrix},$$

$$(2\mathbf{A} - 3\mathbf{B})\mathbf{u} = \begin{pmatrix} 0 & -10 \\ -22 & 19 \end{pmatrix}\begin{pmatrix} 1 \\ 1 \end{pmatrix} = \begin{pmatrix} -10 \\ -3 \end{pmatrix}$$

3.5 MULTIPLICATION OF SQUARE MATRICES

For matrices there is a further important operation: the multiplication of matrices. To describe it we shall first take the special case of square matrices.

Let $\mathbf{A} = (a_{ij})$ and $\mathbf{B} = (b_{ij})$ be square matrices of order n; then the **product** of **A** and **B**, denoted by **AB**, is defined to be the $n \times n$ matrix $\mathbf{C} = (c_{ij})$ with entries

$$c_{ij} = \Sigma_\nu a_{i\nu} b_{\nu j} \tag{3.6}$$

As an illustration we write out the case $n = 2$:

$$\mathbf{AB} = \begin{pmatrix} a_{11}b_{11} + a_{12}b_{21} & a_{11}b_{12} + a_{12}b_{22} \\ a_{21}b_{11} + a_{22}b_{21} & a_{21}b_{12} + a_{22}b_{22} \end{pmatrix}$$

In the general case the definition (3.6) of **C** ($=$ **AB**) may be described as follows: c_{11}, the (1, 1)th entry of **AB**, is obtained by multiplying the elements of the *first row* of **A** by the corresponding elements of the *first column* of **B** and adding the results. The (1, 2)th entry c_{12} is obtained by multiplying the elements of the *first row* of **A** by the corresponding elements from the *second column* of **B** and so on; generally, to obtain the (i, j)th entry c_{ij} of **AB** we use the ith row of **A** and the jth column of **B**. To give a numerical example, let

$$\mathbf{A} = \begin{pmatrix} 3 & 1 \\ -2 & 5 \end{pmatrix}, \quad \mathbf{B} = \begin{pmatrix} 2 & 4 \\ 6 & -3 \end{pmatrix}; \quad \text{then } \mathbf{AB} = \begin{pmatrix} 12 & 9 \\ 26 & -23 \end{pmatrix}.$$

In order to understand the idea behind this definition, let us take an n-vector **u** and form **Bu**. This is again an n-vector, and so we can apply **A**; the result is the vector $\mathbf{v} = \mathbf{A}(\mathbf{Bu})$, with the components

$$v_i = \Sigma_\nu a_{i\nu}(\Sigma_j b_{\nu j} u_j) \tag{3.7}$$

for $\Sigma b_{\nu j} u_j$ is just the νth component of **Bu**. On the right-hand side of (3.7) we have a sum of n expressions, each of which is a sum of n terms, thus altogether we have n^2 terms to add. Now it does not matter in which order we add them, so we may first add all the terms with $j = 1$, then all terms with $j = 2$, and so on. In this way we obtain, for the sum on the right-hand side of (3.7),

$$\Sigma_\nu a_{i\nu} b_{\nu 1} u_1 + \Sigma_\nu a_{i\nu} b_{\nu 2} u_2 + \cdots + \Sigma_\nu a_{i\nu} b_{\nu n} u_n$$

or more briefly, $\Sigma_k(\Sigma_\nu a_{i\nu} b_{\nu k}) u_k$. Thus we can write (3.7) as

$$v_i = \Sigma_k c_{ik} u_k \tag{3.8}$$

where the c_{ik} are the entries of the product **AB**. In vector notation (3.8) reads

$$\mathbf{A}(\mathbf{Bu}) = (\mathbf{AB})\mathbf{u} \tag{3.9}$$

and it was with this equation in mind that the definition of the matrix product was chosen.

Again there are a number of laws satisfied by matrix multiplication.

M.1 $(\mathbf{AB})\mathbf{C} = \mathbf{A}(\mathbf{BC})$ *(associative law)*

M.2 $\mathbf{A}(\mathbf{B} + \mathbf{D}) = \mathbf{AB} + \mathbf{AD}$
M.3 $(\mathbf{B} + \mathbf{D})\mathbf{C} = \mathbf{BC} + \mathbf{DC}$ *(distributive laws)*

These rules are proved by showing that the two sides of each equation, when applied to an arbitrary vector **u**, give the same result (cf. section 3.3). For example, M.1 follows because

$$[(\mathbf{AB})\mathbf{C}]\mathbf{u} = (\mathbf{AB})(\mathbf{Cu}) = \mathbf{A}[\mathbf{B}(\mathbf{Cu})] = \mathbf{A}[(\mathbf{BC})\mathbf{u}] = [\mathbf{A}(\mathbf{BC})]\mathbf{u}$$

where we have used (3.9) at each step. M.2 and M.3 may be proved in the same way, using (3.4) and (3.5), respectively. Of course, all these rules may also be verified by direct calculation, using the definition (3.6) of the product **AB**.

It is important to note that the order of the factors in a matrix product is material, for we have

$$\mathbf{AB} \neq \mathbf{BA}$$

except in special cases. Thus taking the matrices used in the last numerical example, if we compute **BA**, we find

$$\begin{pmatrix} 2 & 4 \\ 6 & -3 \end{pmatrix}\begin{pmatrix} 3 & 1 \\ -2 & 5 \end{pmatrix} = \begin{pmatrix} -2 & 22 \\ 24 & -9 \end{pmatrix}$$

which is different from **AB**. An even simpler example is the following:

$$\mathbf{A} = \begin{pmatrix} 1 & 0 \\ 0 & 0 \end{pmatrix} \quad \mathbf{B} = \begin{pmatrix} 0 & 1 \\ 0 & 0 \end{pmatrix}$$

$$\mathbf{AB} = \begin{pmatrix} 0 & 1 \\ 0 & 0 \end{pmatrix} \quad \mathbf{BA} = \begin{pmatrix} 0 & 0 \\ 0 & 0 \end{pmatrix}$$

Matrices were introduced by Arthur Cayley (1821–95) in the 1850s. The name is due to James Joseph Sylvester (1814–97). (Both were prolific writers in algebra; they practiced law as well as mathematics and their meeting in 1850 resulted in a lifelong and very fruitful collaboration.)

3.6 THE ZERO MATRIX AND THE UNIT MATRIX

The behaviour of the zero matrix resembles that of the scalar zero under multiplication as well as addition. Thus we have

$$\mathbf{0A} = \mathbf{A0} = \mathbf{0}$$

for any $n \times n$ matrix **A**, where **0** is the $n \times n$ zero matrix. However, the product of two matrices may be **0** even when neither factor is **0**, as the example at the end of section 3.5 shows.

There is also an analogue of the number 1. This is the square matrix

$$\begin{pmatrix} 1 & 0 & 0 & \cdots & 0 \\ 0 & 1 & 0 & \cdots & 0 \\ 0 & 0 & 1 & \cdots & 0 \\ & \cdots & & \cdots & \\ 0 & 0 & 0 & \cdots & 1 \end{pmatrix}$$

whose ith column has 1 in the ith place and 0 elsewhere. This matrix is denoted by $\mathbf{I}$ and is called the **unit matrix**. The diagonal containing the 1s is called the **main diagonal**. The unit matrix may also be defined as

$$\mathbf{I} = (\delta_{ij}), \quad \text{where} \quad \delta_{ij} = \begin{cases} 1 & \text{if} \quad i = j, \\ 0 & \text{if} \quad i \neq j. \end{cases} \tag{3.10}$$

The symbol δ defined in (3.10) occurs frequently and is called the **Kronecker delta symbol** (after Leopold Kronecker, 1823–91).

The columns of the unit matrix are just the standard basis vectors $\mathbf{e}_j$, therefore we have $\mathbf{I}\mathbf{e}_j = \mathbf{e}_j$ $(j = 1, \ldots, n)$. More generally, we have

$$\mathbf{I}\mathbf{u} = \mathbf{u}, \quad \text{for every vector } \mathbf{u} \tag{3.11}$$

as follows by writing down the system of equations corresponding to the left-hand side of (3.11). We also note that for any $n \times n$ matrix $\mathbf{A}$,

$$\mathbf{AI} = \mathbf{IA} = \mathbf{A}.$$

The verification of these equations is a simple application of the multiplication rule (3.6) and may be left to the reader.

More generally, one can define for each scalar λ (and each positive integer n) a scalar $n \times n$ matrix $\lambda\mathbf{I} = (\lambda\delta_{ij})$, and the multiplication by a scalar may now be regarded as a special case of matrix multiplication: $\lambda\mathbf{A} = (\lambda\mathbf{I})\mathbf{A}$. Still more generally, a **diagonal matrix** is a matrix of the form $(a_i\delta_{ij})$.

3.7 MULTIPLICATION OF GENERAL MATRICES

In the definition of the product $\mathbf{AB}$ given above we have taken $\mathbf{A}$ and $\mathbf{B}$ to be square matrices of the same order n. But this is not necessary; the same definition can be used for any matrices $\mathbf{A}$, $\mathbf{B}$, provided only that the number of columns of $\mathbf{A}$ equals the number of rows of $\mathbf{B}$. If $\mathbf{A}$ is an $m \times n$ matrix and $\mathbf{B}$ an $n \times p$ matrix, then for any p-vector $\mathbf{u}$, $\mathbf{Bu}$ is defined and is an n-vector, hence $\mathbf{A}(\mathbf{Bu})$ is defined and is an m-vector. This shows that the product $\mathbf{AB}$ must be an $m \times p$ matrix, for it operates on a p-vector and the result is an m-vector. This relation is easily remembered if it is written thus:

$$\underset{m \times n}{\mathbf{A}} \cdot \underset{n \times p}{\mathbf{B}} = \underset{m \times p}{\mathbf{C}} \tag{3.12}$$

To give an example, a 1×2 matrix multiplied by a 2×1 matrix is a 1×1 matrix, i.e. a scalar, whereas the product of a 2×1 matrix by a 1×2 matrix is a 2×2 matrix:

$$(5 \quad 1)\begin{pmatrix} 3 \\ -2 \end{pmatrix} = 13$$

$$\begin{pmatrix} 3 \\ -2 \end{pmatrix}(5 \quad 1) = \begin{pmatrix} 15 & 3 \\ -10 & -2 \end{pmatrix}$$

The laws written down previously for the addition and multiplication of matrices all apply to rectangular matrices whose dimensions are such that either side of the equation is defined. Here the fact that, in general, $\mathbf{AB} \neq \mathbf{BA}$ becomes even more obvious: $\mathbf{AB}$ may be defined without $\mathbf{BA}$ being defined and even when both $\mathbf{AB}$ and $\mathbf{BA}$ are defined, they need not have the same number of rows or of columns, for if $\mathbf{A}$ is $m \times n$ and $\mathbf{B}$ is $n \times m$, as it must be for both $\mathbf{AB}$ and $\mathbf{BA}$ to be defined, then $\mathbf{AB}$ is $m \times m$ and $\mathbf{BA}$ is $n \times n$. In particular, $\mathbf{A}$ and $\mathbf{B}$ need not be of the same size for their product to be defined; by contrast, the sum $\mathbf{A} + \mathbf{B}$ is defined precisely when $\mathbf{A}$ and $\mathbf{B}$ have the same number of rows and the same number of columns.

Whereas the zero matrix is defined for any number of rows and columns, the unit matrix $\mathbf{I}$ is necessarily square, and if in the equations

$$\mathbf{IA} = \mathbf{AI} = \mathbf{A}, \tag{3.13}$$

$\mathbf{A}$ is an $m \times n$ matrix, then the first $\mathbf{I}$ must be of order m and the second $\mathbf{I}$ of order n, if the equations are to have a meaning. Whenever the dimensions conform in this way, the equations (3.13) are true. In cases of doubt the order of any $\mathbf{I}$ occurring could be indicated by a suffix; the above equations, for an $m \times n$ matrix $\mathbf{A}$, would then read

$$\mathbf{I}_m\mathbf{A} = \mathbf{AI}_n = \mathbf{A}$$

3.8 BLOCK MULTIPLICATION OF MATRICES

Consider the product of the following 4×4 matrices:

$$\mathbf{A} = \begin{pmatrix} 0 & 0 & 2 & -1 \\ 0 & 0 & 1 & 3 \\ 4 & 5 & 0 & 0 \\ -3 & 2 & 0 & 0 \end{pmatrix} \quad \mathbf{B} = \begin{pmatrix} 2 & 3 & 0 & 0 \\ 5 & 1 & 0 & 0 \\ 0 & 0 & -2 & 6 \\ 0 & 0 & 5 & 1 \end{pmatrix}$$

In carrying out the multiplication we realize that we are multiplying certain 2×2 matrix blocks. The process can be described by partitioning $\mathbf{A}$ and $\mathbf{B}$ into blocks of second-order matrices:

$$\mathbf{A} = \begin{pmatrix} \mathbf{A}_{11} & \mathbf{A}_{12} \\ \mathbf{A}_{21} & \mathbf{A}_{22} \end{pmatrix}$$

$$\mathbf{B} = \begin{pmatrix} \mathbf{B}_{11} & \mathbf{B}_{12} \\ \mathbf{B}_{21} & \mathbf{B}_{22} \end{pmatrix} \tag{3.14}$$

In our particular case $\mathbf{A}_{11} = \mathbf{A}_{22} = \mathbf{0} = \mathbf{B}_{12} = \mathbf{B}_{21}$, so we have

$$\mathbf{AB} = \begin{pmatrix} \mathbf{0} & \mathbf{A}_{12}\mathbf{B}_{22} \\ \mathbf{A}_{21}\mathbf{B}_{11} & \mathbf{0} \end{pmatrix} = \begin{pmatrix} 0 & 0 & -9 & 11 \\ 0 & 0 & 13 & 9 \\ 33 & 17 & 0 & 0 \\ 4 & -7 & 0 & 0 \end{pmatrix}$$

For a general pair of 4×4 matrices $\mathbf{A}, \mathbf{B}$ their product $\mathbf{AB}$ can be expressed in terms of the partitioning (3.14) as

$$\mathbf{AB} = \begin{pmatrix} \mathbf{A}_{11}\mathbf{B}_{11} + \mathbf{A}_{12}\mathbf{B}_{21} & \mathbf{A}_{11}\mathbf{B}_{12} + \mathbf{A}_{12}\mathbf{B}_{22} \\ \mathbf{A}_{21}\mathbf{B}_{11} + \mathbf{A}_{22}\mathbf{B}_{21} & \mathbf{A}_{21}\mathbf{B}_{12} + \mathbf{A}_{22}\mathbf{B}_{22} \end{pmatrix} \tag{3.15}$$

We note that the blocks in (3.14) are multiplied as if they were scalars. In other words, a 4×4 matrix can be regarded as a 2×2 matrix whose entries are 2×2 matrices of numbers.

More generally, let $\mathbf{A}$, $\mathbf{B}$ be any matrices for which the product $\mathbf{AB}$ can be formed, say $\mathbf{A}$ is $m \times n$ and $\mathbf{B}$ is $n \times p$. If we split m, n, p into sums of positive integers in any way: $m = \Sigma\, m_\mu$, $n = \Sigma\, n_\nu$, $p = \Sigma\, p_\pi$, we can then partition $\mathbf{A}$ into $m_\mu \times n_\nu$ matrix blocks and $\mathbf{B}$ into $n_\nu \times p_\pi$ matrix blocks and carry out the multiplication on these blocks as if they were numbers. For example, if $m = 4 = 3 + 1$, $n = 5 = 2 + 3$ and $p = 6 = 4 + 2$, then $\mathbf{A}$ is a 4×5 matrix, $\mathbf{B}$ is a 5×6 matrix and they are partitioned as in (3.14), where $\mathbf{A}_{11}$ is 3×2, $\mathbf{A}_{12}$ is 3×3, $\mathbf{A}_{21}$ is 1×2 and $\mathbf{A}_{22}$ is 1×3, while $\mathbf{B}_{11}$ is 2×4, $\mathbf{B}_{12}$ is 2×2, $\mathbf{B}_{21}$ is 3×4 and $\mathbf{B}_{22}$ is 3×2. This process of splitting a matrix into blocks is called **block decomposition.**

A block decomposition is often helpful in carrying out the multiplication of matrices with many entries. For example, a 60×60 matrix has 3600 entries and each entry in the product of two such matrices is a sum of 60 terms. This may exceed the storage capacity of our computer, but if we partition our matrices as 3×3 blocks of square matrices of order 20, we reduce the problem to a series of multiplications of pairs of matrices with 400 entries each, where each entry in the product is a sum of 20 terms.

Often we need to consider matrices in which after partitioning all the non-zero blocks are along the main diagonal. If $\mathbf{A}_1, \ldots, \mathbf{A}_r$ are any matrices, not necessarily of the same size, then the matrix

$$\mathbf{A} = \begin{pmatrix} \mathbf{A}_1 & \mathbf{0} & \mathbf{0} & \cdots & \mathbf{0} \\ \mathbf{0} & \mathbf{A}_2 & \mathbf{0} & \cdots & \mathbf{0} \\ & \cdots & & \cdots & \\ \mathbf{0} & \mathbf{0} & \cdots & \mathbf{0} & \mathbf{A}_r \end{pmatrix}$$

is called the **diagonal sum** (or also **direct sum**) of the $\mathbf{A}_i$ and is denoted by $\mathbf{A}_1 \oplus \mathbf{A}_2 \oplus \cdots \oplus \mathbf{A}_r$. Occasionally we also write $\mathrm{diag}(\mathbf{A}_1, \ldots, \mathbf{A}_r)$; this notation is used mainly when the $\mathbf{A}_i$ are all 1×1 matrices, so that $\mathbf{A}$ is in fact a diagonal matrix.

3.9 THE INVERSE OF A MATRIX

For real numbers there is a fourth operation besides addition, subtraction and multiplication, namely division. For matrices this is more problematic, because of the non-commutativity of multiplication: one would have to distinguish between $\mathbf{A}/\mathbf{B}$ and $\mathbf{B}\backslash\mathbf{A}$, say, but one can at least consider the inverse of a matrix. A square matrix $\mathbf{A}$ is said to be **invertible** if there exists a matrix $\mathbf{B}$ such that

$$\mathbf{AB} = \mathbf{BA} = \mathbf{I} \tag{3.16}$$

Such a matrix $\mathbf{B}$ is called the **inverse** of $\mathbf{A}$ and is denoted by $\mathbf{A}^{-1}$. It is unique, if it exists at all, for if we also had $\mathbf{AC} = \mathbf{CA} = \mathbf{I}$, then

$$\mathbf{C} = \mathbf{CI} = \mathbf{C}(\mathbf{AB}) = (\mathbf{CA})\mathbf{B} = \mathbf{IB} = \mathbf{B}$$

where we have used the associative law. This means that once we have found (by whatever means) a matrix $\mathbf{B}$ satisfying (3.16), this $\mathbf{B}$ must be the inverse of $\mathbf{A}$. We shall use this remark to verify the following rules:

I.1 *If* $\mathbf{A}^{-1}$ *exists, it has the inverse* $\mathbf{A}$: $(\mathbf{A}^{-1})^{-1} = \mathbf{A}$.
I.2 *If* $\mathbf{A}, \mathbf{B}$ *are invertible, then so is* $\mathbf{AB}$ *and* $(\mathbf{AB})^{-1} = \mathbf{B}^{-1}\mathbf{A}^{-1}$.

We note the reversal of the factors in I.2, an instance of the 'sock-and-shoe rule' (cf. Exercise 3.2). Rule I.1 follows immediately from the uniqueness of inverses, writing (3.16) as

$$\mathbf{AA}^{-1} = \mathbf{A}^{-1}\mathbf{A} = \mathbf{I} \tag{3.17}$$

To prove I.2, we have $\mathbf{AB}\cdot\mathbf{B}^{-1}\mathbf{A}^{-1} = \mathbf{AA}^{-1} = \mathbf{I}$, $\mathbf{B}^{-1}\mathbf{A}^{-1}\mathbf{AB} = \mathbf{B}^{-1}\mathbf{B} = \mathbf{I}$.

Of course, not every matrix has an inverse. To find the precise conditions for an inverse to exist, let us call a square matrix $\mathbf{A}$ **regular** if its columns are linearly independent, and **singular** when they are linearly dependent. This means that $\mathbf{A}$ is regular precisely when the system of equations $\mathbf{Ax} = \mathbf{k}$ is regular, as we saw in section 2.1. We can now give a criterion for a matrix to have an inverse:

Theorem 3.1
A matrix is invertible if and only if it is square and regular. The inverse $\mathbf{A}^{-1}$ *of* $\mathbf{A}$ *satisfies*

$$\mathbf{AA}^{-1} = \mathbf{I} = \mathbf{A}^{-1}\mathbf{A} \tag{3.18}$$

and is uniquely determined by these equations.

Proof
Let us first assume that $\mathbf{A}$ is square and regular. Then the system

$$\mathbf{Ax} = \mathbf{k} \tag{3.19}$$

has a unique solution for every value of $\mathbf{k}$. Let $\mathbf{b}_j$ be the unique vector such that $\mathbf{Ab}_j = \mathbf{e}_j$, where $\mathbf{e}_1, \ldots, \mathbf{e}_n$ is the standard basis, and write $\mathbf{B}$ for the matrix whose columns are $\mathbf{b}_1, \ldots, \mathbf{b}_n$. Then the columns of $\mathbf{AB}$ are $\mathbf{Ab}_1, \ldots, \mathbf{Ab}_n$, while $\mathbf{e}_1, \ldots, \mathbf{e}_n$ are the columns of the unit matrix. Hence we conclude that

$$\mathbf{AB} = \mathbf{I} \tag{3.20}$$

We claim that $\mathbf{A}^{-1} = \mathbf{B}$; to prove it we still have to show that $\mathbf{BA} = \mathbf{I}$. We have $\mathbf{A}\cdot\mathbf{BA} = \mathbf{AB}\cdot\mathbf{A} = \mathbf{IA} = \mathbf{A}$, and so

$$\mathbf{A}\cdot\mathbf{BA} = \mathbf{A} \tag{3.21}$$

Now the matrix equation

$$\mathbf{AX} = \mathbf{A} \tag{3.22}$$

has as solution a unique $n \times n$ matrix $\mathbf{X}$, by the regularity of $\mathbf{A}$, applied to the columns in (3.22). By (3.21) one solution is $\mathbf{X} = \mathbf{BA}$ and clearly $\mathbf{X} = \mathbf{I}$ is another, so by uniqueness, $\mathbf{BA} = \mathbf{I}$ as claimed; hence we have found an inverse for $\mathbf{A}$. That (3.18) determines it uniquely, we have already noted.

Conversely, assume that $\mathbf{A}^{-1}$ exists, satisfying (3.18). We shall verify the regularity of $\mathbf{A}$ by showing that the system (3.19) has a unique solution for any value of $\mathbf{k}$. If there is a solution $\mathbf{u}$ of (3.19), then on multiplying the equation $\mathbf{Au} = \mathbf{k}$ on the left by $\mathbf{A}^{-1}$, we obtain

$$\mathbf{u} = \mathbf{Iu} = \mathbf{A}^{-1}\mathbf{Au} = \mathbf{A}^{-1}\mathbf{k}$$

where we have used the right-hand equality in (3.18). This shows that the only possible solution of (3.19) is $\mathbf{x} = \mathbf{A}^{-1}\mathbf{k}$, and it is indeed a solution, since (by the left-hand equality in (3.18)),

$$\mathbf{AA}^{-1}\mathbf{k} = \mathbf{Ik} = \mathbf{k}$$

Thus the system (3.19) is regular, hence so is $\mathbf{A}$.

Finally we note that an invertible matrix must be square, for if $\mathbf{A}$ is $m \times n$, then $m \geqslant n$ because the n columns of the regular system (3.19) are linearly independent, so there cannot be more than m, by Theorem 2.3. By the symmetry of the situation it follows that $m \leqslant n$, so $m = n$ and $\mathbf{A}$ is square. This completes the proof.

We remark that the appeal to symmetry could be put more formally by using the transpose matrix, to be defined in section 3.11 below. Alternatively we can argue that if an $m \times n$ system (3.19) has a unique solution for any $\mathbf{k}$, then the columns of $\mathbf{A}$ span the m-dimensional space of columns and hence $n \geqslant m$, because any spanning set of an m-space has at least m vectors.

For a square matrix $\mathbf{A}$ to have an inverse it is enough to know that $\mathbf{B}$ exists to satisfy $\mathbf{BA} = \mathbf{I}$. For then $\mathbf{A}$ must be regular, because $\mathbf{Au} = \mathbf{0}$ implies that $\mathbf{0} = \mathbf{B}\cdot\mathbf{Au} = \mathbf{Iu} = \mathbf{u}$, so we can apply Theorem 3.1 to conclude that $\mathbf{A}^{-1}$ exists; in fact $\mathbf{A}^{-1} = \mathbf{IA}^{-1} = (\mathbf{BA})\mathbf{A}^{-1} = \mathbf{B}(\mathbf{AA}^{-1}) = \mathbf{BI} = \mathbf{B}$. By symmetry, the same conclusion holds if there exists $\mathbf{B}$ such that $\mathbf{AB} = \mathbf{I}$.

3.10 CALCULATION OF THE INVERSE

The proof of Theorem 3.1 also provides us with a practical means of finding the inverse matrix, when it exists. We know that its columns are given by the solutions of the equations $\mathbf{Ax} = \mathbf{e}_j$, so we apply the Gauss–Jordan reduction to the corresponding schemes $\mathbf{A}|\mathbf{e}_1$, $\mathbf{A}|\mathbf{e}_2,\ldots,$, $\mathbf{A}|\mathbf{e}_n$, or writing all columns on the right of the bar in a single scheme, we have

$$\mathbf{A}|\mathbf{I} \tag{3.23}$$

If we reduce this system by elementary operations to the form $\mathbf{I}|\mathbf{B}$, then $\mathbf{B}=\mathbf{A}^{-1}$. To see this in detail, we remark that any elementary operation α,β or γ corresponds to left multiplication by a certain matrix, namely the matrix obtained by applying the given elementary operation to the unit matrix. Taking 3×3 matrices for illustration, for an operation of type α, say interchanging the first and second row, we have to multiply on the left by

$$\begin{pmatrix} 0 & 1 & 0 \\ 1 & 0 & 0 \\ 0 & 0 & 1 \end{pmatrix} \tag{3.24}$$

For an operation β let us take the multiplication of row1 (R1) by λ and for type γ, adding μ.R2 to R1. These operations are accomplished by left multiplication by $\begin{pmatrix} \lambda & 0 & 0 \\ 0 & 1 & 0 \\ 0 & 0 & 1 \end{pmatrix}$ and $\begin{pmatrix} 1 & \mu & 0 \\ 0 & 1 & 0 \\ 0 & 0 & 1 \end{pmatrix}$, respectively.

We shall call these matrices corresponding to elementary operations **elementary**. These matrices are invertible, with inverses again elementary matrices:

$$\begin{pmatrix} 0 & 1 & 0 \\ 1 & 0 & 0 \\ 0 & 0 & 1 \end{pmatrix}^{-1} = \begin{pmatrix} 0 & 1 & 0 \\ 1 & 0 & 0 \\ 0 & 0 & 1 \end{pmatrix} \qquad \begin{pmatrix} \lambda & 0 & 0 \\ 0 & 1 & 0 \\ 0 & 0 & 1 \end{pmatrix}^{-1} = \begin{pmatrix} \lambda^{-1} & 0 & 0 \\ 0 & 1 & 0 \\ 0 & 0 & 1 \end{pmatrix}$$

$$\begin{pmatrix} 1 & \mu & 0 \\ 0 & 1 & 0 \\ 0 & 0 & 1 \end{pmatrix}^{-1} = \begin{pmatrix} 1 & -\mu & 0 \\ 0 & 1 & 0 \\ 0 & 0 & 1 \end{pmatrix}$$

Now the process of solving a regular system of equations $\mathbf{Ax}=\mathbf{k}$ by Gauss–Jordan reduction consisted in applying elementary operations on the rows so as to reduce $\mathbf{A}$ to $\mathbf{I}$. If the matrices corresponding to these elementary operations are $\mathbf{P}_1,\mathbf{P}_2,\ldots,\mathbf{P}_r$, then by applying the operations we obtain in turn $\mathbf{P}_1\mathbf{Ax}=\mathbf{P}_1\mathbf{k}$, $\mathbf{P}_2\mathbf{P}_1\mathbf{Ax}=\mathbf{P}_2\mathbf{P}_1\mathbf{k},\ldots,\mathbf{BAx}=\mathbf{Bk}$, where $\mathbf{B}=\mathbf{P}_r\mathbf{P}_{r-1}\cdots\mathbf{P}_1$. Since $\mathbf{A}$ has been reduced to the unit matrix, we have $\mathbf{BA}=\mathbf{I}$, and the uniqueness of the solution shows that $\mathbf{AB}=\mathbf{I}$. In this way we can calculate $\mathbf{A}^{-1}$ and we see, moreover, that it is a product of elementary matrices. By applying this result to $\mathbf{A}^{-1}$ in place of $\mathbf{A}$, we obtain the following

Corollary

Every invertible matrix is a product of elementary matrices.

We illustrate the actual calculation by using the matrix

$$\mathbf{A} = \begin{pmatrix} 1 & 5 & 2 \\ 1 & 1 & 7 \\ 0 & -3 & 4 \end{pmatrix}$$

which has already occurred in section 2.5. We start from the scheme

$$\begin{array}{rrr|rrr} 1 & 5 & 2 & 1 & 0 & 0 \\ 1 & 1 & 7 & 0 & 1 & 0 \\ 0 & -3 & 4 & 0 & 0 & 1 \end{array}$$

If we operate on the rows of this scheme as in section 2.5, we obtain

$$\begin{array}{rrr|rrr} 1 & 5 & 2 & 1 & 0 & 0 \\ 1 & 1 & 7 & 0 & 1 & 0 \\ 0 & -3 & 4 & 0 & 0 & 1 \end{array} \quad \begin{array}{l} \\ R2 \to R2 - R1 \\ \\ \end{array}$$

$$\begin{array}{rrr|rrr} 1 & 5 & 2 & 1 & 0 & 0 \\ 0 & -4 & 5 & -1 & 1 & 0 \\ 0 & -3 & 4 & 0 & 0 & 1 \end{array} \quad \begin{array}{l} \\ R2 \to R2 - R3 \\ \\ \end{array}$$

$$\begin{array}{rrr|rrr} 1 & 5 & 2 & 1 & 0 & 0 \\ 0 & -1 & 1 & -1 & 1 & -1 \\ 0 & -3 & 4 & 0 & 0 & 1 \end{array} \quad \begin{array}{l} \\ R2 \to -R2 \\ \\ \end{array}$$

$$\begin{array}{rrr|rrr} 1 & 5 & 2 & 1 & 0 & 0 \\ 0 & 1 & -1 & 1 & -1 & 1 \\ 0 & -3 & 4 & 0 & 0 & 1 \end{array} \quad \begin{array}{l} R1 \to R1 - 5.R2 \\ R3 \to R3 + 3.R2 \end{array}$$

$$\begin{array}{rrr|rrr} 1 & 0 & 7 & -4 & 5 & -5 \\ 0 & 1 & -1 & 1 & -1 & 1 \\ 0 & 0 & 1 & 3 & -3 & 4 \end{array} \quad \begin{array}{l} R1 \to R1 - 7.R3 \\ R2 \to R2 + R3 \end{array} \quad \begin{array}{rrr|rrr} 1 & 0 & 0 & -25 & 26 & -33 \\ 0 & 1 & 0 & 4 & -4 & 5 \\ 0 & 0 & 1 & 3 & -3 & 4 \end{array}$$

Hence we have

$$\mathbf{A}^{-1} = \begin{pmatrix} -25 & 26 & -33 \\ 4 & -4 & 5 \\ 3 & -3 & 4 \end{pmatrix}$$

As a check on our calculations we can now work out $\mathbf{AA}^{-1}$ or $\mathbf{A}^{-1}\mathbf{A}$ to see that we get the value expected.

The process may be described as follows. If by a series of elementary operations α, β, γ on the rows of the scheme $\mathbf{A}|\mathbf{I}$ we can reduce it to the scheme $\mathbf{I}|\mathbf{B}$, then $\mathbf{A}$ is regular and $\mathbf{A}^{-1} = \mathbf{B}$. The inverse steps are again of the form α, β or γ and performed in the opposite order, they lead from $\mathbf{I}|\mathbf{B}$ to $\mathbf{A}|\mathbf{I}$ or, what is the same, from $\mathbf{B}|\mathbf{I}$ to $\mathbf{I}|\mathbf{A}$. Hence $\mathbf{B}$ $(= \mathbf{A}^{-1})$ is also regular and $\mathbf{B}^{-1} = \mathbf{A}$. This shows again that if $\mathbf{A}$ is a regular matrix, then $\mathbf{A}^{-1}$ is again regular and $(\mathbf{A}^{-1})^{-1} = \mathbf{A}$ (cf. I.1).

3.11 THE TRANSPOSE OF A MATRIX

An extreme case of a rectangular matrix is a vector: a column vector of dimension n may be considered as an $n \times 1$ matrix, and from this point of view the way in which a matrix operates on a column vector (as in (3.2)) is just an instance of matrix multiplication. The equation

$$\underset{m\times n}{\mathbf{A}}\ \underset{n\times 1}{\mathbf{x}} = \underset{m\times 1}{\mathbf{y}}$$

illustrates the rule (3.12). It also shows the advantage of writing vectors as columns (on the right of the matrix). For this leads us to adopt the 'row-by-column' rule of multiplication, which in turn is responsible for the very simple form of the associative law of multiplication (cf. M.1).

Nevertheless row vectors – that is, vectors written as horizontal rows – can also be fitted into this scheme. A row vector of dimension n is a $1 \times n$ matrix and if we want to multiply it by an $n \times n$ matrix we have to put the vector on the left. For example, for $n = 2$ we could write

$$(x_1 \quad x_2)\begin{pmatrix} a & b \\ c & d \end{pmatrix} = (y_1 \quad y_2) \tag{3.25}$$

This gives $y_1 = x_1 a + x_2 c$, $y_2 = x_1 b + x_2 d$ or, in matrix form,

$$\begin{pmatrix} a & c \\ b & d \end{pmatrix}\begin{pmatrix} x_1 \\ x_2 \end{pmatrix} = \begin{pmatrix} y_1 \\ y_2 \end{pmatrix} \tag{3.26}$$

Thus equation (3.25) between row vectors is equivalent to equation (3.26) between column vectors. The square matrix in (3.25) appears again in (3.26), but with its columns written as rows and its rows written as columns. The operation of deriving a new matrix $\mathbf{A}^{\mathrm{T}}$ from $\mathbf{A}$ by taking for the columns of $\mathbf{A}^{\mathrm{T}}$ the rows of $\mathbf{A}$ is called **transposition** and $\mathbf{A}^{\mathrm{T}}$ is called the **transpose** of $\mathbf{A}$. If $\mathbf{A}$ is an $m \times n$ matrix, its transpose $\mathbf{A}^{\mathrm{T}}$ is $n \times m$, for example

$$\mathbf{A} = \begin{pmatrix} 2 & 5 & -1 \\ 3 & 0 & 4 \end{pmatrix} \quad \mathbf{A}^{\mathrm{T}} = \begin{pmatrix} 2 & 3 \\ 5 & 0 \\ -1 & 4 \end{pmatrix}$$

In particular, the transpose of a column vector is the corresponding row vector and vice versa. We shall always denote the transpose of a matrix $\mathbf{A}$ by $\mathbf{A}^{\mathrm{T}}$; this agrees with the notation introduced in section 2.1, where we wrote column vectors in the form $(x_1, \ldots, x_n)^{\mathrm{T}}$.

We record the following rules obeyed by the transpose.

T.1 $(\mathbf{A}^{\mathrm{T}})^{\mathrm{T}} = \mathbf{A}$
T.2 $(\mathbf{A} + \mathbf{B})^{\mathrm{T}} = \mathbf{A}^{\mathrm{T}} + \mathbf{B}^{\mathrm{T}}$
T.3 $(\lambda\mathbf{A})^{\mathrm{T}} = \lambda\mathbf{A}^{\mathrm{T}}$
T.4 $(\mathbf{AB})^{\mathrm{T}} = \mathbf{B}^{\mathrm{T}}\mathbf{A}^{\mathrm{T}}$

Of these rules T.1 follows because transposing $\mathbf{A}^T$ means taking the rows of $\mathbf{A}$ as rows and the columns as columns. T.2 and T.3 are immediate, and T.4 follows by considering the (i, k)th entry on both sides: on the left it is $\Sigma_j a_{kj} b_{ji}$, on the right it is $\Sigma_j b_{ji} a_{kj}$ and these two expressions are clearly equal. An illustration of T.4 is provided by a comparison of (3.25) and (3.26). If we write (3.25) as $\mathbf{xA} = \mathbf{y}$, then (3.26) reads $\mathbf{A}^T\mathbf{x}^T = \mathbf{y}^T$.

A matrix $\mathbf{A}$ is said to be **symmetric** if $\mathbf{A}^T = \mathbf{A}$; this just means that $\mathbf{A} = (a_{ij})$, where $a_{ij} = a_{ji}$. For example, if $\mathbf{B}$ is any matrix (not necessarily square), then $\mathbf{B}^T\mathbf{B}$ and $\mathbf{B}\mathbf{B}^T$ are symmetric.

3.12 AN APPLICATION TO GRAPH THEORY

Matrices arise quite naturally in other contexts, for example in the study of graphs. A (finite) **graph** is a collection of points or **vertices** $P_1, \ldots, P_n$ together with lines or **edges** joining certain pairs of vertices. We shall assume that the endpoints of each edge are distinct, thus there are no loops at any vertex. Examples are given in Figs. 3.1–3.3.

A graph with n vertices can be specified by an $n \times n$ matrix, its **adjacency matrix** $\mathbf{A} = (a_{ij})$, where $a_{ij} = 1$ if there is an edge from P_i to P_j, and $a_{ij} = 0$

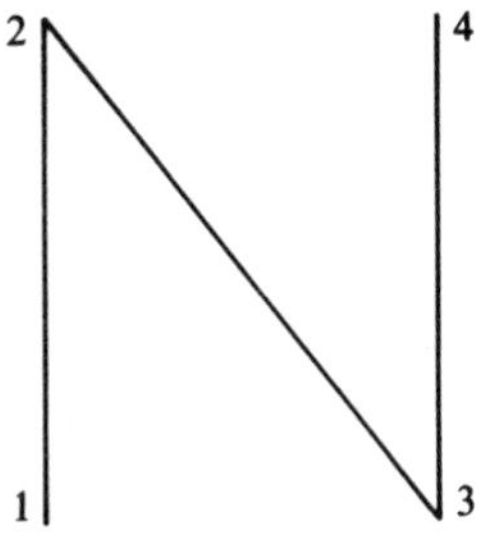

Fig. 3.1

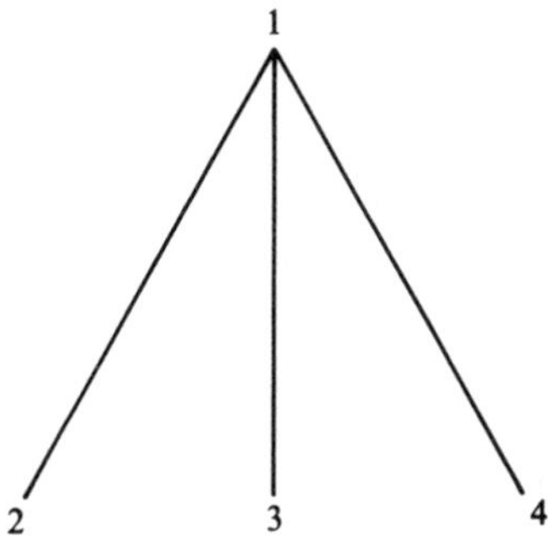

Fig. 3.2

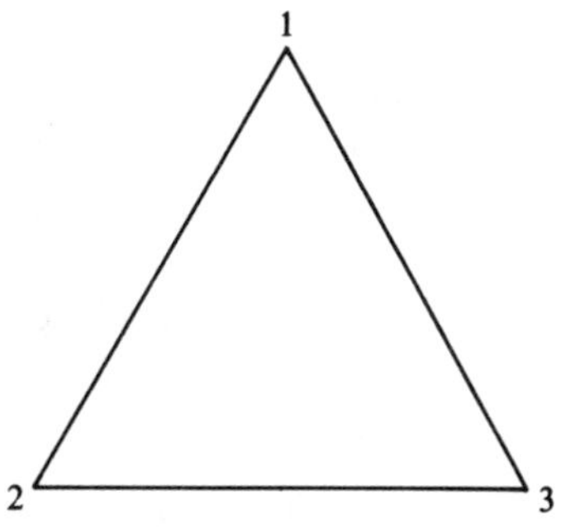

Fig. 3.3

otherwise. For example, the graph in Fig. 3.1 has matrix

$$\begin{pmatrix} 0 & 1 & 0 & 0 \\ 1 & 0 & 1 & 0 \\ 0 & 1 & 0 & 1 \\ 0 & 0 & 1 & 0 \end{pmatrix}$$

that in Fig. 3.2 is represented by

$$\begin{pmatrix} 0 & 1 & 1 & 1 \\ 1 & 0 & 0 & 0 \\ 1 & 0 & 0 & 0 \\ 1 & 0 & 0 & 0 \end{pmatrix}$$

while that in Fig. 3.3 corresponds to

$$\begin{pmatrix} 0 & 1 & 1 \\ 1 & 0 & 1 \\ 1 & 1 & 0 \end{pmatrix}$$

If a graph Γ has the adjacency matrix $\mathbf{A}$, then the number of paths of length r from P_i to P_j is given by the (i,j)th entry of $\mathbf{A}^r$. Put differently, the number of paths of length r from P_i to P_j is equal to the ith component of $\mathbf{A}^r\mathbf{e}_j$. These facts are not hard to check and will be left to the reader to verify.

With each graph Γ we associate a second matrix, its **incidence matrix B**; this is an $m \times n$ matrix $(b_{\lambda j})$, where n is the number of vertices and m the number of edges of Γ, and $b_{\lambda j} = 1$ if the λth edge has P_j as a vertex, and $b_{\lambda j} = 0$ otherwise. Further, for each vertex P_i we define its **order** d_i as the number of edges ending at P_i. Let us put $\mathbf{D} = \mathrm{diag}(d_1, \dots, d_n)$. Then the matrices $\mathbf{A}, \mathbf{B}, \mathbf{D}$ are related by the equation

$$\mathbf{B}^{\mathrm{T}}\mathbf{B} = \mathbf{A} + \mathbf{D} \tag{3.27}$$

For the (i,j)th entry of $\mathbf{B}^{\mathrm{T}}\mathbf{B}$ is the sum of the products of the components of the ith row of $\mathbf{B}^{\mathrm{T}}$, i.e. the ith column of $\mathbf{B}$, by the corresponding components of the jth column of $\mathbf{B}$: $\Sigma_\lambda b_{\lambda i} b_{\lambda j}$. Suppose first that $i \neq j$ and consider the λth entry in this sum, corresponding to the λth edge. It is 1 if the edge joins P_i and P_j, and 0 otherwise, so in all the total is just 1 if there is an edge joining P_i and P_j, and 0 otherwise, and this is just a_{ij}. When $i = j$, we have the sum $\Sigma_\lambda b_{\lambda i}^2$, which is the number of edges ending at P_i and this is just the order d_i at P_i, while $a_{ii} = 0$, by hypothesis. This completes the proof of (3.27).

To illustrate the result, we have for the incidence matrices of the graphs in Figs. 3.1–3.3:

$$\begin{pmatrix} 1 & 1 & 0 & 0 \\ 0 & 1 & 1 & 0 \\ 0 & 0 & 1 & 1 \end{pmatrix}, \quad \begin{pmatrix} 1 & 1 & 0 & 0 \\ 1 & 0 & 1 & 0 \\ 1 & 0 & 0 & 1 \end{pmatrix}, \quad \begin{pmatrix} 1 & 1 & 0 \\ 0 & 1 & 1 \\ 1 & 0 & 1 \end{pmatrix},$$

and for the products $\mathbf{B}^T\mathbf{B}$ we obtain

$$\begin{pmatrix} 1 & 1 & 0 & 0 \\ 1 & 2 & 1 & 0 \\ 0 & 1 & 2 & 1 \\ 0 & 0 & 1 & 1 \end{pmatrix}, \quad \begin{pmatrix} 3 & 1 & 1 & 1 \\ 1 & 1 & 0 & 0 \\ 1 & 0 & 1 & 0 \\ 1 & 0 & 0 & 1 \end{pmatrix}, \quad \begin{pmatrix} 2 & 1 & 1 \\ 1 & 2 & 1 \\ 1 & 1 & 2 \end{pmatrix}$$

The diagonal terms give the order matrices, while the non-diagonal terms give the adjacency matrices.

EXERCISES

3.1. Verify M.1–3 for the following values of $\mathbf{A}, \mathbf{B}, \mathbf{C}, \mathbf{D}$:

(i) $\mathbf{A} = \begin{pmatrix} 0 & 3 \\ 4 & 5 \end{pmatrix}$, $\mathbf{B} = \begin{pmatrix} 2 & -1 \\ 3 & 2 \end{pmatrix}$, $\mathbf{C} = \begin{pmatrix} 1 & 4 \\ 0 & -2 \end{pmatrix}$, $\mathbf{D} = \begin{pmatrix} -1 & 1 \\ 2 & -2 \end{pmatrix}$

(ii) $\mathbf{A} = \begin{pmatrix} 1 & -2 & 5 \\ 3 & 0 & 4 \end{pmatrix}$, $\mathbf{B} = \begin{pmatrix} 2 & 1 \\ 3 & 6 \\ 1 & 5 \end{pmatrix}$, $\mathbf{C} = \begin{pmatrix} 3 \\ -3 \end{pmatrix}$, $\mathbf{D} = \begin{pmatrix} 1 & 3 \\ 2 & 4 \\ -1 & 0 \end{pmatrix}$

3.2. In which of the following cases are $\mathbf{AB}$ and $\mathbf{BA}$ (a) both defined, (b) of the same number of rows and of columns, (c) equal? (i) $\mathbf{A}$, $\mathbf{B}$ as in Exercise 3.1(i); (ii) $\mathbf{A}, \mathbf{B}$ as in Exercise 3.1(ii);

(iii) $\mathbf{A} = \begin{pmatrix} 2 & -1 \\ 3 & 2 \end{pmatrix}$, $\mathbf{B} = \begin{pmatrix} 1 & -4 \\ 12 & 1 \end{pmatrix}$, (iv) $\mathbf{A} = \begin{pmatrix} 3 & 1 & -4 \\ -2 & 0 & 5 \\ 1 & -2 & 3 \end{pmatrix}$,

$$\mathbf{B} = \begin{pmatrix} 2 & 0 & 0 \\ 0 & 5 & 0 \\ 0 & 0 & -1 \end{pmatrix}$$

An example from everyday life of operations which can be performed in succession is the following: let $\mathbf{A}$ be the operation of putting a sock on the right foot and $\mathbf{B}$ the operation of putting a shoe on the right foot. Let $\mathbf{AB}$ mean 'first $\mathbf{A}$, then $\mathbf{B}$'; similarly, $\mathbf{BA}$ means 'first $\mathbf{B}$, then $\mathbf{A}$'. Then we have $\mathbf{AB} \neq \mathbf{BA}$ (as may be verified experimentally). Further, both $\mathbf{A}$ and $\mathbf{B}$ are sometimes invertible, and when they are, then $(\mathbf{AB})^{-1} = \mathbf{B}^{-1}\mathbf{A}^{-1}$ (this can again be checked experimentally). On the other hand, if $\mathbf{C}$ is the operation of putting a sock on the left foot, then $\mathbf{BC} = \mathbf{CB}$, $\mathbf{AC} = \mathbf{CA}$.

3.3. Expand $(\mathbf{A} - \mathbf{B})^2$ and check your answer by taking $\mathbf{A}, \mathbf{B}$ as in Exercise 3.1(i).

3.4. Show that the matrix

$$\mathbf{A} = \begin{pmatrix} 6 & -4 \\ 9 & -6 \end{pmatrix}$$

satisfies the equation $\mathbf{A}^2 = \mathbf{0}$, and find all 2×2 matrices satisfing this equation. Show that any 2×2 matrix $\mathbf{A}$ satisfying the equation $\mathbf{A}^3 = \mathbf{0}$ also satisfies $\mathbf{A}^2 = \mathbf{0}$. (*Hint*: See section 8.3.)

3.5. Evaluate $\mathbf{A}^2$, where

$$\mathbf{A} = \begin{pmatrix} 2 & -5 \\ 3 & 1 \end{pmatrix},$$

and find scalars α, β, γ, not all zero, such that $\alpha\mathbf{I} + \beta\mathbf{A} + \gamma\mathbf{A}^2 = \mathbf{0}$.

3.6. Verify T.4 for the matrices $\mathbf{A}, \mathbf{B}$ as in Exercises 3.1(i) and 3.1(ii).

3.7 Show that for any matrix $\mathbf{A}$, the products $\mathbf{A}^T\mathbf{A}$ and $\mathbf{A}\mathbf{A}^T$ are defined, and evaluate $\mathbf{u}\mathbf{u}^T$ and $\mathbf{u}^T\mathbf{u}$, where $\mathbf{u} = (3, -2, 1, 4)^T$.

3.8. The **trace** of a square matrix $\mathbf{A}$, $\mathrm{tr}(\mathbf{A})$, is defined as the sum of its diagonal entries; thus the trace of $\mathbf{A} = (a_{ij})$ is $\Sigma_i a_{ii}$. Show that for any 3×3 matrices $\mathbf{A}$ and $\mathbf{B}$, $\mathbf{AB}$ and $\mathbf{BA}$ have the same trace.

3.9. If $\mathbf{A}$ and $\mathbf{B}$ are $n \times n$ matrices and the columns of $\mathbf{B}$ are denoted by $\mathbf{b}_1, \ldots, \mathbf{b}_n$, show that the columns of $\mathbf{AB}$ are $\mathbf{A}\mathbf{b}_1, \ldots, \mathbf{A}\mathbf{b}_n$, in symbols:

$$\mathbf{A}(\mathbf{b}_1, \ldots, \mathbf{b}_n) = (\mathbf{A}\mathbf{b}_1, \ldots, \mathbf{A}\mathbf{b}_n).$$

3.10. Show that the rows of an $m \times n$ matrix $\mathbf{A}$ are $\mathbf{e}_i^T\mathbf{A}$, where $\mathbf{e}_1, \ldots, \mathbf{e}_m$ is the standard basis of column m-vectors.

3.11. Find the inverse of each of the matrices belonging to the systems in Exercise 2.1.

3.12. Prove that any regular matrix $\mathbf{A}$ satisfies $(\mathbf{A}^{-1})^T = (\mathbf{A}^T)^{-1}$. Verify this rule for the matrices of Exercise 3.11.

3.13. Adapt the method of sections 3.9–3.10 to calculate $\mathbf{A}^{-1}\mathbf{B}$, where $\mathbf{A}$ and $\mathbf{B}$ are square matrices of the same order and $\mathbf{A}$ is regular. Find a similar method for $\mathbf{B}\mathbf{A}^{-1}$. (*Hint*: To obtain $\mathbf{A}^{-1}\mathbf{B}$, reduce the scheme $\mathbf{A}|\mathbf{B}$ by elementary row operations until the matrix on the left of the bar is the unit matrix. For $\mathbf{B}\mathbf{A}^{-1}$ start from $\mathbf{A}^T|\mathbf{B}^T$ and transpose the result.)

3.14. If $\mathbf{A}$ is an $m \times n$ matrix and $\mathbf{B}$ is an $n \times m$ matrix such that $\mathbf{AB} = \mathbf{I}_m$, show that $m \leqslant n$ and give an example where $m < n$.

3.15. Let $\mathbf{E}_{ij}$ be the $n \times n$ matrix whose (i, j)th entry is 1 while all other entries are zero. Verify that $\mathbf{E}_{ij}\mathbf{E}_{kl} = \delta_{jk}\mathbf{E}_{il}$, $\Sigma\mathbf{E}_{ii} = \mathbf{I}$. ($\mathbf{E}_{ij}$ is called a **matrix unit**.)

3.16. Show that any $n \times n$ matrix $\mathbf{A} = (a_{ij})$ can be expressed as $\mathbf{A} = \Sigma a_{ij}\mathbf{E}_{ij}$, where the $\mathbf{E}_{ij}$ are as in Exercise 3.15. What can be said about $\mathbf{A}$ if $\mathbf{A}\mathbf{E}_{ij} = \mathbf{E}_{ij}\mathbf{A}$ for $i, j = 1, \ldots, n$? What if $\mathbf{A}\mathbf{E}_{ij} = \mathbf{E}_{ji}\mathbf{A}$ for $i, j = 1, \ldots, n$, where $n > 1$?

3.17. Write $\mathbf{B}_{ij}(a)$ for the $n \times n$ matrix which differs from the unit matrix only in the (i, j)th entry which is a (where $i \neq j$). Show that $\mathbf{B}_{ij}(a)$ represents an elementary operation γ, and that $\mathbf{B}_{12}(1)\mathbf{B}_{21}(-1)\mathbf{B}_{12}(1)$ represents an elementary operation α except for a change in sign. Show that $\mathrm{diag}(\lambda, \lambda^{-1})$ can be written as a product of matrices $\mathbf{B}_{ij}(a)$.

3.18. Find the product $\mathbf{E}^{-1}\mathbf{A}\mathbf{E}$, where $\mathbf{A}$ is a given matrix and $\mathbf{E}$ is an elementary matrix.

3.19. Show that the adjacency matrix of a graph is symmetric, with zero diagonal entries and other entries 0 or 1. Conversely, given such a matrix **A**, let $\mathbf{D} = \text{diag}(d_1, \ldots, d_n)$ be the diagonal matrix with d_i equal to the sum of the entries in the ith column of **A**. Show that **A** is the adjacency matrix of a graph and find a matrix **B** such that $\mathbf{A} + \mathbf{D} = \mathbf{B}^{\mathrm{T}}\mathbf{B}$.

3.20. Establish the interpretation of $\mathbf{A}^r$ (the power of the adjacency matrix) given in the text.

4

The solution of a system of equations: the general case

4.1 THE GENERAL LINEAR SYSTEM AND THE ASSOCIATED HOMOGENEOUS SYSTEM

In the case of the special systems of equations treated in Chapter 2, namely the regular systems, we saw that there was always just one solution, irrespective of the value of the right-hand side. In the general case this need not be so: whether there is a solution at all depends generally on the value of the right-hand side, and when there is a solution it need not be the only one.

Let us consider a general $m \times n$ system, written as

$$\mathbf{Ax} = \mathbf{k} \tag{4.1}$$

in the notation of Chapter 3. Here $\mathbf{A}$ is the $m \times n$ matrix of coefficients, $\mathbf{k}$ is a given m-vector and we are looking for all n-vectors $\mathbf{x}$ which satisfy the equation.

The special case of (4.1) in which $\mathbf{k} = \mathbf{0}$,

$$\mathbf{Ax} = \mathbf{0} \tag{4.2}$$

is called a **homogeneous system**. Every homogeneous system has at least one solution, namely the trivial solution $\mathbf{x} = \mathbf{0}$; it turns out to be a little easier to deal with than the general system (4.1) which may have no solution at all.

With any system of the form (4.1) we associate the homogeneous system (4.2) obtained by replacing the right-hand side of (4.1) by $\mathbf{0}$. Then we have the following rules relating the solutions of (4.1) to the solutions of the associated homogeneous system (4.2):

H.1 *If* $\mathbf{x} = \mathbf{v}$, $\mathbf{x} = \mathbf{v}'$ *are any two solutions of (4.1), then* $\mathbf{x} = \mathbf{v} - \mathbf{v}'$ *is a solution of (4.2).*

H.2 *If* $\mathbf{x} = \mathbf{u}$ *is a solution of (4.2) and* $\mathbf{x} = \mathbf{v}$ *is a solution of (4.1), then* $\mathbf{x} = \mathbf{u} + \mathbf{v}$ *is a solution of (4.1).*

These rules follow by linearity (equations (3.4) in section 3.2). If $\mathbf{v}$ and $\mathbf{v}'$ are solutions of (4.1), then $\mathbf{Av} = \mathbf{Av}' = \mathbf{k}$, hence $\mathbf{A}(\mathbf{v} - \mathbf{v}') = \mathbf{Av} - \mathbf{Av}' = \mathbf{k} - \mathbf{k} = \mathbf{0}$ and so $\mathbf{v} - \mathbf{v}'$ is a solution of (4.2). Similarly, if $\mathbf{v}$ is a solution of (4.1) and $\mathbf{u}$ a solution of (4.2), so that $\mathbf{Au} = \mathbf{0}$, $\mathbf{Av} = \mathbf{k}$, then $\mathbf{A}(\mathbf{u} + \mathbf{v}) = \mathbf{Au} + \mathbf{Av} = \mathbf{0} + \mathbf{k} = \mathbf{k}$, so $\mathbf{u} + \mathbf{v}$ is a solution of (4.1).

Expressed differently, H.1 states that any two solutions of (4.1) differ only by a solution of (4.2), therefore if we know a single solution of (4.1), we can find all solutions of (4.1) by adding to it the different solutions of (4.2). H.2 states that when we add a solution of (4.2) to a solution of (4.1), the result is again a solution of (4.1). So in order to find the complete solution of (4.1) we need only know a single solution of (4.1) (if one exists), together with the most general solution of (4.2).

4.2 THE RANK OF A SYSTEM

Before embarking on the solution of (4.1) and (4.2), we recall that for a regular system, $\mathbf{Ax} = \mathbf{k}$, the unique solution can be expressed in the form $\mathbf{x} = \mathbf{A}^{-1}\mathbf{k}$, and a system $\mathbf{Ax} = \mathbf{k}$ is regular precisely when $\mathbf{A}$ is square and all its columns are linearly independent (cf. section 3.9). For general systems we shall now introduce a numerical invariant, the rank, which will help us to describe the solution. We begin by defining this term for matrices.

Let $\mathbf{A}$ be an $m \times n$ matrix. Then the columns of $\mathbf{A}$ span a subspace of m-space, called the **column space** of $\mathbf{A}$; its dimension is called the **rank** of $\mathbf{A}$, written $\text{rk}(\mathbf{A})$. More precisely, this number is the **column rank**. The **row rank** may be defined similarly as the dimension of the **row space**, i.e. the space spanned by the rows of $\mathbf{A}$, but we shall soon see (section 4.9) that the row rank and column rank of $\mathbf{A}$ are always equal. As Theorem 3.1 shows, a square matrix is invertible if and only if its rank is equal to its order.

By sections 1.9 and 2.9, if $\mathbf{a}_1, \ldots, \mathbf{a}_n$ are the columns of $\mathbf{A}$ and if $\mathbf{a}_1, \ldots, \mathbf{a}_r$ are linearly independent and every column of $\mathbf{A}$ is linearly dependent on the first r, then $\text{rk}(\mathbf{A}) = r$.

It is clear that the rank r of an $m \times n$ matrix $\mathbf{A}$ satisfies $r \leqslant n$; we also have $r \leqslant m$, for the columns of $\mathbf{A}$ are m-vectors and we know from Theorem 2.3 that there cannot be more than m such vectors in a linearly independent set. Thus, if we write $\min(m, n)$ for the smaller of m and n, we have the relation

$$\text{rk}(\mathbf{A}) \leqslant \min(m, n) \quad \text{for any } m \times n \text{ matrix } \mathbf{A} \tag{4.3}$$

Now the **rank** of a system of equations $\mathbf{Ax} = \mathbf{k}$ is defined as the rank of $\mathbf{A}$.

4.3 THE SOLUTION OF HOMOGENEOUS SYSTEMS

The basic fact about the solution of homogeneous systems is expressed by the statement that they form a vector space:

Theorem 4.1
Let

$$\mathbf{Ax} = \mathbf{0} \tag{4.2}$$

be a homogeneous $m \times n$ system of rank r. Then the solutions of this system form a vector space of dimension $n - r$. That is, there exist linearly independent vectors $\mathbf{c}_1, \ldots, \mathbf{c}_{n-r}$, solutions of (4.2), such that every linear combination of these **c***s is a solution of (4.2) and conversely, every solution of (4.2) can be expressed as a linear combination of the* **c***s. In particular, (4.2) has only the trivial solution precisely when $r = n$.*

To show that the solutions of (4.2) form a subspace of the space of n-vectors we use the linearity of the matrix operation: if $\mathbf{u}, \mathbf{u}'$ are any solutions of (4.2), then for any scalars λ, λ' we have $\mathbf{A}(\lambda\mathbf{u} + \lambda'\mathbf{u}') = \mathbf{A}(\lambda\mathbf{u}) + \mathbf{A}(\lambda'\mathbf{u}') = \lambda\mathbf{Au} + \lambda'\mathbf{Au}' = \mathbf{0}$, so $\lambda\mathbf{u} + \lambda'\mathbf{u}'$ is again a solution.

It remains to find the dimension of this solution space, and this is accomplished by finding a basis of $n - r$ vectors. Let us denote the columns of $\mathbf{A}$ again by $\mathbf{a}_1, \ldots, \mathbf{a}_n$, so that the system becomes

$$\mathbf{a}_1 x_1 + \mathbf{a}_2 x_2 + \cdots + \mathbf{a}_n x_n = \mathbf{0} \tag{4.4}$$

The system is unchanged except for notation if we renumber the columns in any way and renumber the x_i in the same way; we may therefore suppose the columns numbered so that $\mathbf{a}_1, \ldots, \mathbf{a}_r$ are linearly independent, while $\mathbf{a}_{r+1}, \ldots, \mathbf{a}_n$ depend linearly on $\mathbf{a}_1, \ldots, \mathbf{a}_r$, so that the equation

$$\mathbf{a}_1 x_1 + \cdots + \mathbf{a}_r x_r = -\mathbf{a}_{r+i}$$

has a solution $(x_1, \ldots, x_r) = (c_{1i}, \ldots, c_{ri})$, say, for $i = 1, \ldots, n - r$. This means that (4.4) has a solution $\mathbf{x} = \mathbf{c}_i$, where

$$\mathbf{c}_i = (c_{1i}, \ldots, c_{ri}, 0, 0, \ldots, 0, 1, 0, \ldots, 0)^{\mathrm{T}}$$

where the last $n - r$ components are zero except for the $(r + i)$th which is 1. We complete the proof by showing that $\mathbf{c}_1, \ldots, \mathbf{c}_{n-r}$ is a basis of the solution space.

In the first place each $\mathbf{c}_i$ is a solution of (4.4) because $\mathbf{Ac}_i = \mathbf{0}$, by construction. Conversely, if $\mathbf{x} = \mathbf{u} = (u_1, \ldots, u_n)^{\mathrm{T}}$ is any solution of (4.4), then $\mathbf{x} = u_{r+1}\mathbf{c}_1 + \cdots + u_n\mathbf{c}_{n-r}$ is a solution of (4.4) whose last $n - r$ components are $u_{r+1}, \ldots, u_n$, and it therefore agrees with $\mathbf{u}$ in these components. In other words, the last $n - r$ components of the difference

$$\mathbf{v} = \mathbf{u} - u_{r+1}\mathbf{c}_1 - \cdots - u_n\mathbf{c}_{n-r} \tag{4.5}$$

are zero. As a linear combination of the solutions, $\mathbf{v}$ is again a solution of (4.4), so if $\mathbf{v} = (v_1, \ldots, v_r, 0, \ldots, 0)^{\mathrm{T}}$, then

$$\mathbf{a}_1 v_1 + \cdots + \mathbf{a}_r v_r = \mathbf{0}$$

Since $\mathbf{a}_1, \ldots, \mathbf{a}_r$ are linearly independent, it follows that $v_1 = \cdots = v_r = 0$. So the first r components of $\mathbf{v}$ are zero as well, $\mathbf{v} = \mathbf{0}$ and we have

$$\mathbf{u} = u_{r+1}\mathbf{c}_1 + \cdots + u_n\mathbf{c}_{n-r}$$

The coefficients u_i in this equation are uniquely determined as the last $n-r$ components of $\mathbf{u}$. Thus $\mathbf{c}_1, \ldots, \mathbf{c}_{n-r}$ form a linearly independent set spanning the solution space, and so form a basis. It follows that the solution space has dimension $n-r$, and the proof is complete.

We note the special cases when $r = \min(m, n)$. First, if $\mathrm{rk}(\mathbf{A}) = n \leqslant m$, then all the columns of $\mathbf{A}$ are linearly independent and (4.2) has only the zero solution; now (4.1) can be reduced to a regular system by discarding $m-n$ equations linearly dependent on the rest, whenever (4.1) has a solution at all. Second, if $\mathrm{rk}(\mathbf{A}) = m \leqslant n$, then among the n column vectors there are just m linearly independent ones, which therefore form a basis of the space of all m-vectors. This means that every vector can be written as a linear combination of the columns of $\mathbf{A}$. In other words, the system $\mathbf{Ax} = \mathbf{k}$ has a solution whatever the value of $\mathbf{k}$, although the solution will not be unique unless $m = n$.

4.4 ILLUSTRATIONS (HOMOGENEOUS CASE)

To apply Theorem 4.1 we need to find a maximal set of linearly independent columns. The simplest way of achieving this is to carry out the reduction as in Chapter 2. It may not be possible now to eliminate the unknowns in the natural order (corresponding to the fact that we had to renumber the columns in the proof of Theorem 4.1), but this does not matter. In general we shall not have to solve for all the unknowns, since some of them can take arbitrary values. The process of elimination, applied to a homogeneous system of rank r, is complete when all but r equations have been reduced to 0 by subtracting multiples of other equations, and when r of the columns, taken in a suitable order, form the $r \times r$ unit matrix. The unknowns corresponding to these r columns are then determined by the equations in terms of the remaining $n-r$ unknowns whose values are arbitrary. In the proof of Theorem 4.1, these arbitrary or 'disposable' unknowns were $x_{r+1}, \ldots, x_n$. They are also called **free variables**, while $x_1, \ldots, x_r$ are called **basic variables**. Of course, in general they need not be the first r variables. We note that when a_{ij} has been used as a pivot, x_j has been eliminated from all equations except the ith and so it is then a basic variable.

We give two examples to illustrate the process of solving homogeneous systems:

$$2x_1 + x_2 + 5x_3 = 0 \tag{4.6a}$$

$$x_1 - 3x_2 + 6x_3 = 0 \tag{4.6b}$$

$$3x_1 + 5x_2 + 4x_3 = 0 \tag{4.6c}$$

$$7x_2 - 7x_3 = 0 \tag{4.6d}$$

Writing down the scheme of coefficients and the reductions as in Chapter 2, we have

$$\begin{array}{rrr} 2 & 1 & 5 \\ 1 & -3 & 6 \\ 3 & 5 & 4 \\ 0 & 7 & -7 \end{array} \quad \mathrm{R1} \leftrightarrow \mathrm{R2} \quad \begin{array}{rrr} 1 & -3 & 6 \\ 2 & 1 & 5 \\ 3 & 5 & 4 \\ 0 & 7 & -7 \end{array} \quad \begin{array}{l} \\ \mathrm{R2} \rightarrow \mathrm{R2} - 2.\mathrm{R1} \\ \mathrm{R3} \rightarrow \mathrm{R3} - 3.\mathrm{R1} \\ \\ \end{array}$$

$$\begin{array}{rrr} 1 & -3 & 6 \\ 0 & 7 & -7 \\ 0 & 14 & -14 \\ 0 & 7 & -7 \end{array} \quad \begin{array}{l} \\ \mathrm{R3} \rightarrow \mathrm{R3} - 2.\mathrm{R2} \\ \mathrm{R4} \rightarrow \mathrm{R4} - \mathrm{R2} \\ \\ \end{array} \quad \begin{array}{rrr} 1 & -3 & 6 \\ 0 & 7 & -7 \\ 0 & 0 & 0 \\ 0 & 0 & 0 \end{array}$$

At this stage we can discard the last two rows; they mean that the third and fourth equations have been reduced to zero by subtracting multiples of the first two equations. In other words, (4.6c) and (4.6d) are linear combinations of (4.6a) and (4.6b) and so do not contribute to the solution.

We continue with

$$\begin{array}{rrr} 1 & -3 & 6 \\ 0 & 7 & -7 \end{array} \quad \mathrm{R2} \rightarrow \tfrac{1}{7}.\mathrm{R2} \quad \begin{array}{rrr} 1 & -3 & 6 \\ 0 & 1 & -1 \end{array} \quad \begin{array}{l} \mathrm{R1} \rightarrow \mathrm{R1} + 3.\mathrm{R2} \\ \\ \end{array} \quad \begin{array}{rrr} 1 & 0 & 3 \\ 0 & 1 & -1 \end{array}.$$

We have now reached the stage described earlier: all but two of the rows have been reduced to zero, and two of the columns, taken in a suitable order, constitute the columns of the 2×2 unit matrix, with x_1, x_2 as basic variables. The corresponding equations are

$$\begin{aligned} x_1 + 3x_3 &= 0 \\ x_2 - \ \ x_3 &= 0 \end{aligned}$$

These equations show that x_3 is a free variable; its value can be arbitrarily assigned and it then determines the solution uniquely. If we denote the value of x_3 by λ, we obtain as the general solution

$$\begin{pmatrix} x_1 \\ x_2 \\ x_3 \end{pmatrix} = \begin{pmatrix} -3\lambda \\ \lambda \\ \lambda \end{pmatrix} = \lambda \begin{pmatrix} -3 \\ 1 \\ 1 \end{pmatrix}$$

The solution contains one parameter, hence the solution space is 1-dimensional and $n - r = 1$. Since $n = 3$, we conclude that the rank of the given system is 2. Of course, we could see this already from the fact that two equations were left at the end (which would determine the values of the basic variables).

As a second illustration we take a 3×4 system

$$4x_1 + 12x_2 - 7x_3 + \ 6x_4 = 0 \tag{4.7a}$$

$$x_1 + \ 3x_2 - 2x_3 + \ \ x_4 = 0 \tag{4.7b}$$

$$3x_1 + \ 9x_2 - 2x_3 + 11x_4 = 0 \tag{4.7c}$$

We have the reduction

$$\begin{matrix} 4 & 12 & -7 & 6 \\ 1 & 3 & -2 & 1 \\ 3 & 9 & -2 & 11 \end{matrix} \quad R1 \leftrightarrow R2 \quad \begin{matrix} 1 & 3 & -2 & 1 \\ 4 & 12 & -7 & 6 \\ 3 & 9 & -2 & 11 \end{matrix} \quad \begin{matrix} \\ R2 \to R2 - 4.R1 \\ R3 \to R3 - 3.R1 \end{matrix}$$

$$\begin{matrix} 1 & 3 & -2 & 1 \\ 0 & 0 & 1 & 2 \\ 0 & 0 & 4 & 8 \end{matrix} \quad \begin{matrix} \\ R1 \to R1 + 2.R2 \\ R3 \to R3 - 4.R2 \end{matrix} \quad \begin{matrix} 1 & 3 & 0 & 5 \\ 0 & 0 & 1 & 2 \\ 0 & 0 & 0 & 0 \end{matrix}$$

The first column shows that x_1 is determined by the first equation and the third column shows x_3 to be determined by the second equation. The third equation depends linearly on the first two and so can be omitted. Thus x_2 and x_4 are free variables; taking $x_2 = \lambda, x_4 = \mu$, we obtain as the complete solution

$$\begin{pmatrix} x_1 \\ x_2 \\ x_3 \\ x_4 \end{pmatrix} = \begin{pmatrix} -3\lambda - 5\mu \\ \lambda \\ -2\mu \\ \mu \end{pmatrix} = \lambda \begin{pmatrix} -3 \\ 1 \\ 0 \\ 0 \end{pmatrix} + \mu \begin{pmatrix} -5 \\ 0 \\ -2 \\ 1 \end{pmatrix}$$

4.5 THE SOLUTION OF GENERAL SYSTEMS

Consider now the general system

$$\mathbf{Ax} = \mathbf{k} \tag{4.1}$$

Taking $\mathbf{A}$ to be $m \times n$, we again denote the columns of $\mathbf{A}$ by $\mathbf{a}_1, \ldots, \mathbf{a}_n$ and write $(\mathbf{A}, \mathbf{k})$ for the $m \times (n+1)$ matrix with columns $\mathbf{a}_1, \ldots, \mathbf{a}_n, \mathbf{k}$; this is called the **augmented matrix** of the system (4.1).

To solve the system (4.1) is to express $\mathbf{k}$ as a linear combination of $\mathbf{a}_1, \ldots, \mathbf{a}_n$, which is possible if and only if $\mathbf{k}$ lies in the subspace spanned by $\mathbf{a}_1, \ldots, \mathbf{a}_n$, i.e. when the spaces spanned by $\mathbf{a}_1, \ldots, \mathbf{a}_n$ and $\mathbf{a}_1, \ldots, \mathbf{a}_n, \mathbf{k}$ respectively have the same dimension. In other words, (4.1) has a solution if and only if $\mathbf{A}$ and $(\mathbf{A}, \mathbf{k})$ have the same rank. If we recall the connexion between the system (4.1) and its associated homogeneous system explained in section 4.1, we can describe the nature of the solution as follows:

Theorem 4.2
The $m \times n$ system of equations

$$\mathbf{Ax} = \mathbf{k} \tag{4.1}$$

has a solution if and only if the rank of its matrix $\mathbf{A}$ *equals the rank of the augmented matrix* $(\mathbf{A}, \mathbf{k})$. *If this rank is r, then the complete solution of (4.1) contains $n - r$ independent parameters and is obtained by adding the general* *solution of the associated homogeneous system to a particular solution of (4.1).*

For we saw in section 4.1 that the general solution of (4.1) is obtained from any particular solution by adding the general solution of (4.2), and the latter is a linear combination of $n-r$ linearly independent solution vectors, by Theorem 4.1. Thus apart from solving (4.2) we still need to find a particular solution of (4.1).

4.6 ILLUSTRATIONS (GENERAL CASE)

To solve the general system (4.1) in practice, we apply again the method of elimination, writing down the column on the right as well. We carry out the reduction as in section 4.4. If at any stage the left-hand side of an equation has been reduced to zero and the right-hand side is non-zero, the system has no solution; if the right-hand side is zero too, the equation holds for any values and so can be discarded. For a consistent system of equations (i.e. a system having at least one solution) which is of rank r, the process comes to an end when all but r equations have been reduced to zero and so discarded, and among the columns of the remaining equations there are r which constitute the different columns of the $r \times r$ unit matrix (in a certain order). The corresponding unknowns are then determined by these r equations, while the remaining $n-r$ unknowns are free variables. To take an example, consider the system

$$x_1 - 2x_2 + 5x_3 = 1 \tag{4.8a}$$

$$2x_1 - 4x_2 + 8x_3 = 2 \tag{4.8b}$$

$$-3x_1 + 6x_2 + 7x_3 = 1 \tag{4.8c}$$

We have the reduction

$$\begin{array}{rrr|r} 1 & -2 & 5 & 1 \\ 2 & -4 & 8 & 2 \\ -3 & 6 & 7 & 1 \end{array} \quad \begin{array}{l} \\ R2 \to R2 - 2.R1 \\ R3 \to R3 + 3.R1 \end{array} \quad \begin{array}{rrr|r} 1 & -2 & 5 & 1 \\ 0 & 0 & -2 & 0 \\ 0 & 0 & 22 & 4 \end{array} \quad R2 \to -\tfrac{1}{2}.R2$$

$$\begin{array}{rrr|r} 1 & -2 & 5 & 1 \\ 0 & 0 & 1 & 0 \\ 0 & 0 & 22 & 4 \end{array} \quad R3 \to R3 - 22.R2 \quad \begin{array}{rrr|r} 1 & -2 & 5 & 1 \\ 0 & 0 & 1 & 0 \\ 0 & 0 & 0 & 4 \end{array}$$

This system has no solution, since the last equation now reads $0 = 4$. If in (4.8) the right-hand side had been $(1, 2, -3)^{\mathrm{T}}$, we would have found (with the same operations on the schemes as above)

$$\begin{array}{rrr|r} 1 & -2 & 5 & 1 \\ 2 & -4 & 8 & 2 \\ -3 & 6 & 7 & -3 \end{array} \qquad \begin{array}{rrr|r} 1 & -2 & 5 & 1 \\ 0 & 0 & -2 & 0 \\ 0 & 0 & 22 & 0 \end{array} \qquad \begin{array}{rrr|r} 1 & -2 & 5 & 1 \\ 0 & 0 & 1 & 0 \\ 0 & 0 & 22 & 0 \end{array} \qquad \begin{array}{rrr|r} 1 & -2 & 5 & 1 \\ 0 & 0 & 1 & 0 \\ 0 & 0 & 0 & 0 \end{array}$$

From the last scheme we obtain

$$\begin{aligned} x_1 - 2x_2 + 5x_3 &= 1 \\ x_3 &= 0 \end{aligned}$$

This is a consistent system of rank 2, with x_2 as a free variable. Thus the complete solution is

$$\begin{pmatrix} x_1 \\ x_2 \\ x_3 \end{pmatrix} = \begin{pmatrix} 1 \\ 0 \\ 0 \end{pmatrix} + \lambda \begin{pmatrix} 2 \\ 1 \\ 0 \end{pmatrix}$$

We see that $(1,0,0)^T$ is a particular solution, while $\lambda(2,1,0)^T$ is the general solution of the associated homogeneous system.

4.7 THE ROW-ECHELON FORM

The result of the reduction of a consistent system given in section 4.6 may be described as follows:

(i) *All zero rows occur at the bottom of the matrix.*

(ii) *In each row the first non-zero entry is 1 and it occurs to the right of all leading non-zero entries in earlier rows.*

A matrix in this form is said to be in **row-echelon form***. It replaces the triangular form found in the course of the Gauss elimination in section 2.5. Now we cannot be sure of finding a triangular form, because one of the unknowns may be zero in every solution; thus in the example in section 4.6 we saw that x_3 was zero and so could not be a free variable. However, by permuting the columns, with a corresponding renumbering of the unknowns, we can always bring our matrix to the form

$$\left.\begin{array}{ccccccc|c} 1 & a_{12} & a_{13} & \cdots & & \cdots & a_{1n} & k_1 \\ 0 & 1 & a_{23} & \cdots & & \cdots & a_{2n} & k_2 \\ & \cdots & \cdots & & & \cdots & & \\ 0 & 0 & \cdots & 0 & 1 & a_{r,r+1} & a_{rn} & k_r \\ 0 & 0 & \cdots & 0\cdots 0 & & \cdots & 0 & 0 \end{array}\right.$$

By subtracting suitable multiples of rows from earlier rows, and omitting the zero rows, we can now put our matrix in the form

$$\mathbf{I}_r\ \mathbf{C}|\mathbf{k}$$

corresponding to the Gauss–Jordan reduction. This form is sometimes called the **reduced** row-echelon form. For example, if we interchange the second

*Echelon: formation in parallel rows with the end of each row projecting further than the one in front.

and third columns in the system considered in section 4.6, we obtain

$$\begin{array}{rrr|r} 1 & 5 & -2 & 1 \\ 0 & 1 & 0 & 0 \\ 0 & 0 & 0 & 0 \end{array} \quad \mathrm{R1} \to \mathrm{R1} - 5.\mathrm{R2} \quad \begin{array}{rrr|r} 1 & 0 & -2 & 1 \\ 0 & 1 & 0 & 0 \end{array}$$

4.8 CONSTRUCTING A BASIS: A PRACTICAL METHOD

In section 1.9 we described a process for selecting from a given set of vectors a linearly independent set having the same linear span. The Gaussian elimination provides a practical means of construction. We write down our vectors as row vectors of a matrix and use the elementary row operations α, β, γ of section 2.3 to reduce the system to echelon form. At each stage the space spanned by all the vectors is the same, and the non-zero vectors of the system in echelon form are linearly independent. For by permuting the columns, if necessary (corresponding to a renumbering of the coordinates) we may take these vectors to have the form

$$\mathbf{u}_i = (0, 0, \ldots, 0, u_{ii}, u_{i,i+1}, \ldots, u_{in}) \quad (i = 1, \ldots, r)$$

where $u_{ii} \neq 0$. If we had a relation

$$\alpha_1 \mathbf{u}_1 + \cdots + \alpha_r \mathbf{u}_r = \mathbf{0} \tag{4.9}$$

where not all the α_i vanish, let α_h be the first non-zero coefficient. We equate h-components in (4.9); since $\alpha_1 = \cdots = \alpha_{h-1} = 0$, and $u_{h+1,h} = \cdots = u_{rh} = 0$, we find that $\alpha_h u_{hh} = 0$, which is a contradiction, because $\alpha_h \neq 0$ and $u_{hh} \neq 0$. Hence all the α_i vanish and $\mathbf{u}_1, \ldots, \mathbf{u}_r$ are linearly independent, as claimed.

As an example, let us take the 5-vectors $(2, 1, 3, 4, 0)$, $(1, 6, 3, 2, 1)$, $(-3, 4, 6, 2, 4)$, $(13, 12, 3, 10, -5)$. We write down the matrix with these rows and perform the reduction to echelon form:

$$\begin{array}{rrrrr} 2 & 1 & 3 & 4 & 0 \\ 1 & 6 & 3 & 2 & 1 \\ -3 & 4 & 6 & 2 & 4 \\ 13 & 12 & 3 & 10 & -5 \end{array} \to \begin{array}{rrrrr} 1 & 6 & 3 & 2 & 1 \\ 0 & -11 & -3 & 0 & -2 \\ 0 & 22 & 15 & 8 & 7 \\ 0 & -66 & -36 & -16 & -18 \end{array}$$

$$\to \begin{array}{rrrrr} 1 & 6 & 3 & 2 & 1 \\ 0 & 11 & 3 & 0 & 2 \\ 0 & 0 & 9 & 8 & 3 \\ 0 & 0 & -18 & -16 & -6 \end{array} \to \begin{array}{rrrrr} 1 & 6 & 3 & 2 & 1 \\ 0 & 11 & 3 & 0 & 2 \\ 0 & 0 & 9 & 8 & 3 \\ 0 & 0 & 0 & 0 & 0 \end{array}$$

This reduction shows that the subspace spanned by the given vectors has a basis $(1, 6, 3, 2, 1)$, $(0, 11, 3, 0, 2)$, $(0, 0, 9, 8, 3)$. If we examine the reduction process, we see that the last row has been reduced to zero by adding multiples of the first three rows. It follows that the vectors $(2, 1, 3, 4, 0)$, $(1, 6, 3, 2, 1)$, $(-3, 4, 6, 2, 4)$ also form a basis for our subspace.

As another example consider the vectors $(1,4,-2,3)$, $(3,12,-8,1)$, $(4,16,-10,4)$. Applying the same reduction process, we have

$$\begin{matrix} 1 & 4 & -2 & 3 \\ 3 & 12 & -8 & 1 \\ 4 & 16 & -10 & 4 \end{matrix} \quad \rightarrow \quad \begin{matrix} 1 & 4 & -2 & 3 \\ 0 & 0 & -2 & -8 \\ 0 & 0 & -2 & -8 \end{matrix} \quad \rightarrow \quad \begin{matrix} 1 & 4 & -2 & 3 \\ 0 & 0 & 1 & 4 \\ 0 & 0 & 0 & 0 \end{matrix}$$

We have echelon, but not triangular form; however, this does not affect the result, which provides the basis $(1,4,-2,3)$, $(0,0,1,4)$ for our space. Again the first two, or indeed any two, of the given vectors also form a basis for our subspace.

4.9 THE **PAQ**-REDUCTION OF MATRICES

The process of solving a general system of linear equations by row reduction leads to a reduction theorem for matrices which is often useful.

Theorem 4.3
Let $\mathbf{A}$ *be an* $m \times n$ *matrix of rank* r. *Then there exists an invertible* $m \times m$ *matrix* $\mathbf{P}$ *and an invertible* $n \times n$ *matrix* $\mathbf{Q}$ *such that*

$$\mathbf{PAQ} = \begin{pmatrix} \mathbf{I}_r & \mathbf{0} \\ \mathbf{0} & \mathbf{0} \end{pmatrix} \tag{4.10}$$

Proof
We remark that in the above reduction process we apply elementary row operations, corresponding to multiplication on the left by an invertible matrix $\mathbf{P}$, and permutations of columns, corresponding to multiplication on the right by an invertible matrix $\mathbf{Q}_1$. We thus find that

$$\mathbf{PAQ}_1 = \begin{pmatrix} \mathbf{I} & \mathbf{C} \\ \mathbf{0} & \mathbf{0} \end{pmatrix}$$

If we put $\mathbf{Q}_2 = \begin{pmatrix} \mathbf{I} & -\mathbf{C} \\ \mathbf{0} & \mathbf{I} \end{pmatrix}$, then $\mathbf{Q}_2$ is again invertible and we have

$$\mathbf{PAQ}_1\mathbf{Q}_2 = \begin{pmatrix} \mathbf{I} & \mathbf{0} \\ \mathbf{0} & \mathbf{0} \end{pmatrix}$$

Now (4.10) follows by taking $\mathbf{Q} = \mathbf{Q}_1\mathbf{Q}_2$.

This process of reduction to the form (4.10) is often called the **PAQ-reduction** or the reduction to **Smith normal form**, after Henry John Stephen Smith (1826–83), although, strictly speaking, this term refers to the **PAQ**-reduction over the integers as coefficients – with a more complicated right-hand side, of which (4.10) above may be regarded as a special case (cf. Cohn, 1982, section 10.5).

The expression (4.10) also shows that the notion of rank, originally defined in terms of the columns of **A**, could equally well have been defined in terms of rows, as the maximum number of linearly independent rows of **A**. This follows from the evident symmetry of (4.10).

This symmetry also follows from another way of defining the rank, which itself is more symmetric. Let **A** be any $m \times n$ matrix and consider the different ways of writing **A** as a product

$$\mathbf{A} = \mathbf{BC} \tag{4.11}$$

where **B** is $m \times t$ and **C** is $t \times n$. We can interpret (4.11) as saying that the columns of **A** are linear combinations of the columns of **B**, or also that the rows of **A** are linear combinations of the rows of **C**. Let us denote the rank of **A** by r; then we can express all the columns of **A** as a linear combination of r suitably chosen ones, hence we have an equation (4.11) with $t = r$. On the other hand, in any equation (4.11) all the columns of **A** have been expressed as a linear combination of the t columns of **B**, so we have $r \leqslant t$ in any factorization (4.11). It follows that the column rank of **A** is the least value of t occurring in factorizations (4.11) of **A**. Similarly the least value of t is also the row rank of **A** and this proves again that the row rank and column rank of **A** are equal. Suming up, we have:

Theorem 4.4
The rank of any matrix **A** *may be defined as the least value of* t *in any factorization* (4.11).

We have now three ways of defining the rank of a matrix, in terms of columns (as the maximum number of linearly independent columns), in terms of rows, by symmetry, or as the 'inner rank' as the least t occurring in factorizations (4.11). A fourth way, in terms of determinants, will be described in Chapter 5.

4.10 COMPUTATIONS OF RANK

In order to compute the rank of a matrix **A** we can use the method of section 2.10 for finding a basis for the row space of **A**, or by finding a basis for the column space, or by a judicious combination of the two. More formally, we can say that an elementary operation on the rows or columns does not change the rank of a matrix. Hence we can find the rank by reducing the matrix to triangular form, using any elementary row or column operations.

For example, let

$$\mathbf{A} = \begin{pmatrix} 5 & 3 & 6 & 8 & 7 \\ 2 & -1 & 4 & 0 & 3 \\ 4 & 9 & 0 & 16 & 5 \\ 9 & -2 & 3 & 1 & 4 \end{pmatrix}$$

Using the (2, 2)th entry as pivot, we reduce the other elements of the second row to 0:

$$\begin{array}{rrrrr} 5 & 3 & 6 & 8 & 7 \\ 2 & -1 & 4 & 0 & 3 \\ 4 & 9 & 0 & 16 & 5 \\ 9 & -2 & 3 & 1 & 4 \end{array} \rightarrow \begin{array}{rrrrr} 11 & 3 & 18 & 8 & 16 \\ 0 & -1 & 0 & 0 & 0 \\ 22 & 9 & 36 & 16 & 32 \\ 5 & -2 & -5 & 1 & -2 \end{array}$$

We can now omit the second row and second column and continue the reduction:

$$\begin{array}{rrrr} 11 & 18 & 8 & 16 \\ 22 & 36 & 16 & 32 \\ 5 & -5 & 1 & -2 \end{array} \rightarrow \begin{array}{rrrr} 11 & 18 & 8 & 16 \\ 0 & 0 & 0 & 0 \\ 5 & -5 & 1 & -2 \end{array} \rightarrow \begin{array}{rrrr} 1 & 28 & 6 & 20 \\ 5 & -5 & 1 & -2 \end{array}$$

$$\rightarrow \begin{array}{rrrr} 1 & 28 & 6 & 20 \\ 0 & -145 & -29 & -102 \end{array}$$

There are two non-zero rows left in the triangular form; together with the row put aside earlier this gives 3 as the rank.

4.11 SYLVESTER'S LAW OF NULLITY

The ranks of matrices in a product satisfy the following inequality:

$$\mathrm{rk}(\mathbf{AB}) \leqslant \min(\mathrm{rk}(\mathbf{A}), \mathrm{rk}(\mathbf{B})) \tag{4.12}$$

For the columns of $\mathbf{AB}$ are linear combinations of the columns of $\mathbf{A}$, hence $\mathrm{rk}(\mathbf{AB}) \leqslant \mathrm{rk}(\mathbf{A})$ and by symmetry, $\mathrm{rk}(\mathbf{AB}) \leqslant \mathrm{rk}(\mathbf{B})$. It is of interest to have an inequality going in the opposite direction. We first prove a special case:

Theorem 4.5
Let $\mathbf{A}$ *be an* $m \times n$ *matrix and* $\mathbf{B}$ *an* $n \times p$ *matrix. If* $\mathbf{AB} = \mathbf{0}$, *then*

$$\mathrm{rk}(\mathbf{A}) + \mathrm{rk}(\mathbf{B}) \leqslant n.$$

Proof
Let $\mathrm{rk}(\mathbf{A}) = r$. Then the homogeneous system $\mathbf{Ax} = \mathbf{0}$ has a solution space of dimension $n - r$, by Theorem 4.1. Thus all the columns of $\mathbf{B}$ lie in an $(n-r)$-dimensional space and so $\mathrm{rk}(\mathbf{B}) \leqslant n - r$, as we had to show.

As a consequence we have

Sylvester's law of nullity *If* $\mathbf{A}$ *is an* $m \times n$ *matrix and* $\mathbf{B}$ *is an* $n \times p$ *matrix, then*

$$\mathrm{rk}(\mathbf{A}) + \mathrm{rk}(\mathbf{B}) \leqslant n + \mathrm{rk}(\mathbf{AB}) \tag{4.13}$$

For if $\mathrm{rk}(\mathbf{AB}) = s$, then by the definition of inner rank we can write

$\mathbf{AB} = \mathbf{PQ}$, where $\mathbf{P}$ is $m \times s$ and $\mathbf{Q}$ is $s \times p$. This equation can be rewritten in block form:

$$(\mathbf{A} \quad \mathbf{P})\begin{pmatrix} \mathbf{B} \\ -\mathbf{Q} \end{pmatrix} = \mathbf{0}$$

Now $(\mathbf{A}\,\mathbf{P})$ is an $m \times (n + s)$ matrix and clearly $\mathrm{rk}(\mathbf{A}) \leqslant \mathrm{rk}(\mathbf{A}\,\mathbf{P})$; similarly

$$\mathrm{rk}(\mathbf{B}) \leqslant \mathrm{rk}\begin{pmatrix} \mathbf{B} \\ -\mathbf{Q} \end{pmatrix},$$

and so by Theorem 4.5,

$$\mathrm{rk}(\mathbf{A}) + \mathrm{rk}(\mathbf{B}) \leqslant \mathrm{rk}(\mathbf{A}\,\mathbf{P}) + \mathrm{rk}\begin{pmatrix} \mathbf{B} \\ -\mathbf{Q} \end{pmatrix} \leqslant n + s$$

This is the required inequality (4.13).

For a square matrix $\mathbf{A}$ of order n and rank r the **nullity** $n(\mathbf{A})$ is sometimes defined by the equation

$$n(\mathbf{A}) = n - r$$

In terms of this nullity the inequality (4.13) can be written for square matrices as:

$$n(\mathbf{AB}) \leqslant n(\mathbf{A}) + n(\mathbf{B})$$

and it was in this form that the law was originally stated by Sylvester in 1884.

EXERCISES

4.1. Prove the following rule (sometimes known as the **principle of superposition**): If $\mathbf{x} = \mathbf{v}_1$ is a solution of $\mathbf{Ax} = \mathbf{k}_1$ and $\mathbf{x} = \mathbf{v}_2$ is a solution of $\mathbf{Ax} = \mathbf{k}_2$, then for any scalars $\lambda_1, \lambda_2, \mathbf{x} = \lambda_1\mathbf{v}_1 + \lambda_2\mathbf{v}_2$ is a solution of $\mathbf{Ax} = \lambda_1\mathbf{k}_1 + \lambda_2\mathbf{k}_2$.

4.2. Show that an $m \times (m + 1)$ homogeneous system always has a non-trivial solution.

4.3. The following schemes represent systems of equations in the notation of section 2.5. Find the complete solution in the cases where there is one:

(i) $\begin{array}{rrr|r} 2 & 4 & 1 & 1 \\ 3 & 5 & 0 & 1 \\ 5 & 13 & 7 & 4 \end{array}$ (ii) $\begin{array}{rrr|r} 9 & -6 & 12 & 0 \\ -12 & 8 & -16 & 0 \\ -7 & 10 & -13 & 0 \end{array}$ (iii) $\begin{array}{rrr|r} 2 & 3 & 4 & 1 \\ 5 & 6 & 7 & 2 \\ 8 & 9 & 10 & 4 \end{array}$

(iv) $\begin{array}{rrrr|r} 2 & 1 & 3 & 5 & 6 \\ 3 & 2 & 4 & 6 & 8 \\ -1 & 3 & 2 & 7 & -3 \end{array}$ (v) $\begin{array}{rrrr|r} 3 & 1 & -2 & -1 & 2 \\ 15 & 5 & 4 & 3 & 4 \\ 6 & 2 & 3 & 1 & 0 \end{array}$

(vi) $\begin{array}{rrrr|r} 1 & -3 & 5 & -7 & -2 \\ -2 & 4 & -6 & 8 & 2 \\ 1 & 1 & 1 & 1 & 2 \\ 1 & 5 & 2 & 5 & 7 \end{array}$ (vii) $\begin{array}{rrrr|r} 3 & 1 & 2 & 4 & 3 \\ 5 & 2 & 3 & 6 & 5 \\ 4 & 1 & 3 & 6 & 4 \\ 5 & 1 & 4 & 8 & 5 \end{array}$

4.4. In the following schemes, determine all the values of t for which there is a solution, and give the complete solution in each case:

(i) $\begin{array}{rrrr|r} 1 & 4 & -2 & 3 & t \\ 3 & 5 & 0 & 2 & 5 \\ 0 & 7 & -6 & 7 & 13 \end{array}$ (ii) $\begin{array}{rrrr|r} 2 & 1 & 4 & 3 & 1 \\ 1 & 3 & 2 & -1 & 3t \\ 1 & 1 & 2 & 1 & t^2 \end{array}$

(iii) Determine t so that the system represented by the scheme

$$\begin{array}{rrr|r} 2 & 2 & 3 & 0 \\ 3 & t & 5 & 0 \\ 1 & 7 & 3 & 0 \end{array}$$

has a non-trivial solution, and then solve it completely.

4.5. Solve the systems (i) $5x_1 + 2x_2 = 10, x_1 + 6x_2 = 6$; (ii) $2x_1 + x_2 = 4$, $5x_1 + 3x_2 = 3$; (iii) $3x_1 - x_2 = 5$, $15x_1 - 5x_2 = 10$. In each case, plot the lines on a graph and obtain the solution geometrically.

4.6. What are the possible values for the rank of the matrix $\mathbf{uu}^{\mathrm{T}}$, where $\mathbf{u}$ is a column vector?

4.7. Let $\mathbf{A}$ be an $m \times n$ matrix of rank r. Show that if $r = m$, then there exists an $n \times m$ matrix $\mathbf{B}$ such that $\mathbf{AB} = \mathbf{I}$, while if $r = n$, there is an $n \times m$ matrix $\mathbf{C}$ such that $\mathbf{CA} = \mathbf{I}$.

4.8. Find the ranks of the following matrices:

$$\begin{pmatrix} 2 & 1 & 3 \\ 4 & 2 & 8 \\ 8 & 4 & 28 \end{pmatrix}, \begin{pmatrix} 5 & 1 & 4 & 9 \\ 2 & 3 & 1 & 5 \\ 6 & -1 & 4 & 7 \end{pmatrix}, \begin{pmatrix} 0 & 1 & 1 & 1 \\ 1 & 0 & 1 & 1 \\ 1 & 1 & 0 & 1 \end{pmatrix}$$

4.9. Find the ranks of the following matrices, for all values of t:

$$\begin{pmatrix} 3 & 1 & -2 & 4 \\ 6 & 2 & t & 8 \\ 0 & t & 0 & 0 \end{pmatrix}, \begin{pmatrix} t & 0 & 1 \\ 0 & t & 0 \\ 0 & 0 & t+3 \end{pmatrix}, \begin{pmatrix} 3 & 0 & 6 & 3t \\ t & 2 & 2(t+1) & 0 \\ -2 & 4 & 0 & -2t-2t^2 \end{pmatrix}$$

4.10. Let $\mathbf{Ax} = \mathbf{k}$ be an $m \times n$ system. Show that if this system has at least one solution for every vector $\mathbf{k}$, then $m \leqslant n$, and if the system has at most one solution for every $\mathbf{k}$, then $m \geqslant n$.

4.11. Let $\mathbf{A}, \mathbf{B}, \mathbf{C}$ be matrices such that $\mathbf{AB}$ and $\mathbf{BC}$ are defined, and let $\mathbf{v}_1, \ldots, \mathbf{v}_s$ be column vectors such that, together with a basis for the columns of $\mathbf{BC}$, they form a basis for the space spanned by the columns of $\mathbf{B}$. Show that $\mathbf{Av}_1, \ldots, \mathbf{Av}_s$, together with the columns of $\mathbf{ABC}$, span the space of columns of **AB**. Deduce that

$$\operatorname{rk}(\mathbf{AB}) + \operatorname{rk}(\mathbf{BC}) \leqslant \operatorname{rk}(\mathbf{B}) + \operatorname{rk}(\mathbf{ABC})$$

and obtain Sylvester's law of nullity as a corollary.

4.12. Given a matrix in block form

$$\begin{pmatrix} \mathbf{A} & \mathbf{B} \\ \mathbf{C} & \mathbf{D} \end{pmatrix}$$

show, by applying the **PAQ**-reduction to **A**, followed by elementary row and column operations, that

$$\operatorname{rk}\begin{pmatrix} \mathbf{A} & \mathbf{B} \\ \mathbf{C} & \mathbf{D} \end{pmatrix} - \operatorname{rk}\begin{pmatrix} \mathbf{A} \\ \mathbf{C} \end{pmatrix} \geqslant \operatorname{rk}(\mathbf{A}\ \mathbf{B}) - \operatorname{rk}(\mathbf{A})$$

4.13. Find bases for the subspaces spanned by (i) $(1, -1, 2, 4, 3)$, $(2, 5, -3, 5, 7)$, $(7, 7, 0, 22, 23)$, $(3, -10, 13, 15, 8)$; (ii) $(1, 1, 1, 1, 1)$, $(2, 2, 2, 2, 1)$, $(3, 3, 3, 1, 1)$, $(4, 4, 1, 1, 1)$.

4.14. Use the method of section 4.8 to solve Exercises 1.2 and 1.3 from Chapter 1.

5

Determinants

5.1 MOTIVATION

In Chapter 4 we described the form taken by the solution of an $m \times n$ system of linear equations. The form of this solution depended essentially on the rank of the system, which could be specified in three equivalent ways, as row rank, column rank or inner rank. But since the process of solving the equations also provided the value of the rank, we did not need any special methods for determining the rank. In fact, almost every method for determining the rank of a system is similar (and comparable in length) to the method of solving the equations given in Chapters 2 and 4. Nevertheless, it is often useful to have another characterization of the rank, and in particular to have a criterion for the linear dependence of n vectors in n dimensions. This is provided by the determinant; some preparation is necessary before we can give the general definition, but we shall begin by looking at the simplest cases, $n = 2$ and 3, where a direct definition is possible.

5.2 THE 2-DIMENSIONAL CASE

Consider the case of 2-vectors. If the vectors $\mathbf{u} = (u_1, u_2)^T$, $\mathbf{v} = (v_1, v_2)^T$ are linearly dependent then there exist two scalars λ, μ, not both zero, such that

$$\begin{aligned} \lambda u_1 + \mu v_1 &= 0 \\ \lambda u_2 + \mu v_2 &= 0 \end{aligned} \tag{5.1}$$

Eliminating μ and λ in turn, we find that

$$\begin{aligned} \lambda(u_1 v_2 - u_2 v_1) &= 0 \\ \mu(u_1 v_2 - u_2 v_1) &= 0 \end{aligned}$$

Since at least one of λ, μ is different from zero, we conclude that

$$u_1 v_2 - u_2 v_1 = 0 \tag{5.2}$$

Conversely, if (5.2) holds, then $\mathbf{u}$ and $\mathbf{v}$ are linearly dependent. For, either

$\mathbf{u} = \mathbf{0}$, so (5.1) holds with $\lambda = 1$, $\mu = 0$; or one of u_1, u_2 is non-zero, say $u_1 \neq 0$. Then (5.1) holds with $\lambda = v_1, \mu = -u_1$.

Thus the function $\Delta = u_1 v_2 - u_2 v_1$ of the components of $\mathbf{u}$ and $\mathbf{v}$ vanishes if and only if $\mathbf{u}$ and $\mathbf{v}$ are linearly dependent. It is called a **determinant** of **order** 2, and is denoted by

$$\begin{vmatrix} u_1 & v_1 \\ u_2 & v_2 \end{vmatrix} \tag{5.3}$$

The determinant (5.3) may be interpreted geometrically as twice the area of the triangle with the vertices $(0,0)$, (u_1, u_2), (v_1, v_2), with a sign depending on the order in which these points are taken. For if we complete the parallelogram by taking as fourth vertex the endpoint of the vector $\mathbf{u} + \mathbf{v}$, we have to show that the area of this parallelogram is $u_1 v_2 - u_2 v_1$ (Fig. 5.1). By drawing lines through the endpoint of $\mathbf{u} + \mathbf{v}$ parallel to the axes we enclose the parallelogram in a rectangle of area $(u_1 + v_1)(u_2 + v_2)$. To obtain our parallelogram, we have to subtract two rectangles, each of area $u_2 v_1$, and four triangles, two of area $(u_1 u_2)/2$ and two of area $(v_1 v_2)/2$. Thus the area of our parallelogram is

$$(u_1 + v_1)(u_2 + v_2) - 2u_2 v_1 - u_1 u_2 - v_1 v_2 = u_1 v_2 - u_2 v_1$$

as claimed.

We remark that if we had taken the vectors in the opposite order, $\mathbf{v}, \mathbf{u}$, we would get $v_1 u_2 - v_2 u_1$, which is the same expression with the opposite sign. This is related to the two possible orientations of the pair $(\mathbf{u}, \mathbf{v})$ and will be dealt with in section 7.3 below.

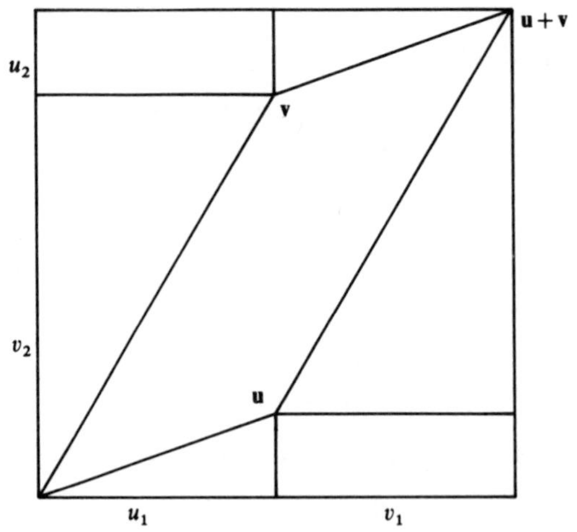

Fig. 5.1

5.3 THE 3-DIMENSIONAL CASE

Let us now take three vectors $\mathbf{u}, \mathbf{v}, \mathbf{w}$ in three dimensions. They are linearly dependent precisely when there exist λ, μ, ν, not all zero, such that

$$\lambda u_1 + \mu v_1 + \nu w_1 = 0$$
$$\lambda u_2 + \mu v_2 + \nu w_2 = 0$$
$$\lambda u_3 + \mu v_3 + \nu w_3 = 0$$

Eliminating μ and ν, we find

$$\lambda(u_1 v_2 w_3 - u_1 v_3 w_2 - u_2 v_1 w_3 + u_2 v_3 w_1 + u_3 v_1 w_2 - u_3 v_2 w_1) = 0$$

If instead we eliminate λ and ν, or λ and μ, we obtain the same equation, with λ replaced by μ or ν, respectively. Since at least one of λ, μ, ν is non-zero, we conclude that

$$u_1 v_2 w_3 + u_2 v_3 w_1 + u_3 v_1 w_2 - u_2 v_1 w_3 - u_1 v_3 w_2 - u_3 v_2 w_1 = 0 \qquad (5.4)$$

The expression on the left is written

$$\begin{vmatrix} u_1 & v_1 & w_1 \\ u_2 & v_2 & w_2 \\ u_3 & v_3 & w_3 \end{vmatrix} \qquad (5.5)$$

and is called a **determinant** of **order** 3. The first term, $u_1 v_2 w_3$, consists of entries from the main diagonal and is followed by terms in which the suffixes are permuted **cyclically**: $123 \rightarrow 231 \rightarrow 312$, followed by three terms with the suffixes in the remaining three orders, with a minus sign.

The expression (5.4) can easily be remembered by writing down the determinant (5.5), repeating the first two columns and taking three terms as indicated below, as products of the forward diagonals with a + sign and three terms as products of the backward diagonals with a $-$ sign:

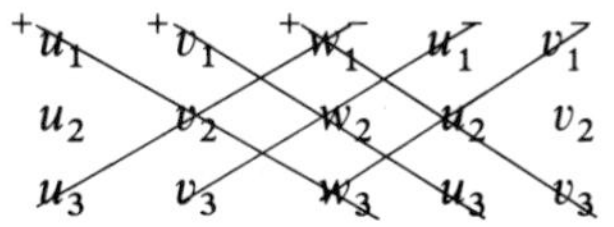

This is known as **Sarrus' rule** (after P.F. Sarrus, 1798–1861). It must be stressed that this rule applies only for $n = 3$; it has no analogue for $n > 3$.

We have seen that when $\mathbf{u}, \mathbf{v}$ and $\mathbf{w}$ are linearly dependent, then the determinant is zero, and it can be shown that the converse holds, as in the case of 2-vectors. Further, the value of the determinant is 6 times the volume of the tetrahedron with the vertices given by the vectors $\mathbf{0}, \mathbf{u}, \mathbf{v}$ and $\mathbf{w}$.

5.4 THE RULE OF SIGNS IN THE 3-DIMENSIONAL CASE

The preceding cases suggest that there is a function of the components of n vectors (in n dimensions) whose vanishing expresses the linear dependence of these vectors. In order to define such a function we again look at the left-hand side of (5.4), defining the third-order determinant. It is a sum of terms $\pm u_i v_j w_k$, where i, j, k represent the numbers 1, 2, 3 in some order, and the sign is + or −, depending on that order. More precisely, the sign is + when ijk are in the order 123, 231, 312, and − when the order is 213, 132, 321.

These six orders are all the possible ways of arranging 1, 2, 3 in some order, and each arrangement can be obtained from the natural order 123 by successively transposing certain pairs of numbers. For example, we get from 123 to 231 by first transposing 1 and 2: 213, and then 1 and 3: 231. This can be done in many different ways and the number of transpositions need not be the same each time; thus we can also get from 123 to 231 by using four transpositions instead of two: $123 \to 213 \to 312 \to 321 \to 231$. But whatever route we use to get from 123 to 231, the number of transpositions will always be even. And if we can get from 123 to ijk by an odd number of transpositions, then any method of getting from 123 to ijk by transpositions will involve an odd number of transpositions, as we shall see in section 5.5.

The rule of signs in (5.4) can now be expressed very simply as follows: $u_i v_j w_k$ appears in (5.4) with the sign + if ijk differs from 123 by an even number of transpositions and with the sign − if it differs by an odd number of transpositions.

5.5 PERMUTATIONS

It is worthwhile pausing briefly to examine the possible arrangements of n distinct numbers, and their division into two classes, as this will help us to understand the general definition of a determinant of order n.

Consider the number of ways in which the numbers $1, 2, \ldots, n$ can be ordered. If the number is $f(n)$, we have n choices for first place, and once this is fixed, we have $n - 1$ numbers left to distribute over $n - 1$ places, which can be done in $f(n-1)$ ways. Thus in all we have $f(n) = n.f(n-1)$ ways of ordering n numbers, hence by induction, $f(n) = n.(n-1)\ldots 2.1$. This number is usually denoted by $n!$ and called ***n* factorial**:

$$n! = n(n-1)\cdots 2.1$$

For example, $3! = 3.2.1 = 6$, as we saw in section 5.4.

The operation of passing from one arrangement $i_1 i_2 \cdots i_n$ to another, $j_1 j_2 \cdots j_n$, is called a **permutation**. Such a permutation can always be carried out by successive transpositions of pairs of numbers. For $n = 2$ this is clear, since there are only two numbers. In the general case we use induction on n (cf. section 2.4). We bring j_1 to the first place by transposing it with i_1; of course, if $i_1 = j_1$, this is not necessary. Having brought j_1 to its correct position, we need only apply a permutation to the remaining $n - 1$ numbers, and by our induction hypothesis this can be done by transpositions. Thus every permutation can be effected by a succession of transpositions.

For example, to get from 12345 to 31542 we have

$$12345 \to 32145 \to 31245 \to 31542$$

There are, of course, many other ways of expressing this permutation in terms of transpositions, but whether the number of transpositions needed is even or odd depends only on the permutation and not on the choice of transpositions.

To prove this assertion we take n variables $x_1, \ldots, x_n$ and form the product of all their differences $x_j - x_i$ for $i, j = 1, \ldots, n, j > i$:

$$\begin{aligned} \varphi = (x_2 - x_1)(x_3 - x_1)(x_4 - x_1)\cdots(x_n - x_1) \times \\ (x_3 - x_2)(x_4 - x_2)\cdots(x_n - x_2) \times \\ \cdots \quad \cdots \quad \cdots \quad \times \\ (x_n - x_{n-1}) \end{aligned}$$

When we interchange x_1 and x_2, only the first two rows in this product are affected and it is easily seen that φ changes sign; for the first factor is multiplied by -1 while the next $n-1$ factors are interchanged against the following $n-1$. Similarly, the interchange of any other pair of variables changes the sign of φ. Thus if we apply k transpositions, φ is multiplied by $(-1)^k$, which is 1 if k is even and -1 if k is odd. Since a permutation can always be effected by a succession of transpositions, any permutation, applied to $x_1, x_2, \ldots, x_n$, transforms φ into φ or $-\varphi$. Of these two possibilities only one is realized, because $\varphi \neq -\varphi$. In the first case the number of transpositions must be even, in the second case it is odd.

5.6 THE KRONECKER ε FUNCTION

We shall call a permutation **even** or **odd** according as the number of transpositions needed to effect the permutation is even or odd. Further, we shall write

$$\varepsilon(i_1 i_2 \cdots i_n) = \begin{cases} 1 & \text{if } i_1 i_2 \cdots i_n \text{ is obtained from } 12\cdots n \text{ by an even permutation} \\ -1 & \text{if } i_1 i_2 \cdots i_n \text{ is obtained from } 12\cdots n \text{ by an odd permutation} \end{cases}$$

The expression $\varepsilon(i_1 i_2 \cdots i_n)$ so defined is known as the **Kronecker ε symbol** or **function**. To calculate ε we simply reduce $i_1 i_2 \cdots i_n$ to $12 \cdots n$ by transpositions, changing the sign each time, and remembering that $\varepsilon(12\cdots n) = 1$. For example,

$$\varepsilon(31542) = -\varepsilon(31245) = \varepsilon(32145) = -\varepsilon(12345) = -1$$

We note the following properties of the ε function:

E.1 *If $j_1 j_2 \cdots j_n$ differs from $i_1 i_2 \cdots i_n$ by a single transposition, then $\varepsilon(j_1 j_2 \cdots j_n) = -\varepsilon(i_1 i_2 \cdots i_n)$.*

E.2 *If* $i_1i_2\cdots i_n$ *is obtained by a permutation of* $12\cdots n$ *and the permutation which takes* $i_1i_2\cdots i_n$ *to* $12\cdots n$ *takes* $12\cdots n$ *to* $j_1j_2\cdots j_n$, *then*

$$\varepsilon(j_1j_2\cdots j_n)=\varepsilon(i_1i_2\cdots i_n)$$

E.3 *If* $i_2i_3\cdots i_n$ *is obtained by a permutation of* $23\cdots n$, *then*

$$\varepsilon(1i_2i_3\cdots i_n)=\varepsilon(i_2i_3\cdots i_n)$$

Rule E.1 is obvious: if the two arrangements $i_1i_2\cdots i_n$ and $j_1j_2\cdots j_n$ differ by one transposition, then their ε functions must have opposite signs.

To prove E.2 we observe that the sequence of transpositions needed to get from $12\cdots n$ to $i_1i_2\cdots i_n$ when performed on $i_1i_2\cdots i_n$ in the opposite order, gets us back to $12\cdots n$. Therefore, we can get from $12\cdots n$ to $i_1i_2\cdots i_n$ and to $j_1j_2\cdots j_n$ by the same number of transpositions. As an illustration of E.2, we get from 31542 to 12345 by the transpositions (25), (12), (13). If we apply these transpositions to 12345, we get successively $12345\rightarrow 15342\rightarrow 25341\rightarrow 25143$. Hence by E.2, $\varepsilon(25143)=\varepsilon(31542)$.

Rule E.3 offers no difficulty: we have to apply a permutation to $1i_2i_3\cdots i_n$ and since 1 is in position, we need only use a permutation which sends $i_2i_3\cdots i_n$ to $23\cdots n$.

5.7 THE DETERMINANT OF AN $n\times n$ MATRIX

To illustrate the use of the ε function we can define the determinant of order 3 formed from the components of vectors $\mathbf{u},\mathbf{v}$ and $\mathbf{w}$ as

$$\sum\varepsilon(ijk)u_iv_jw_k \tag{5.6}$$

In this summation ijk runs over the 6 arrangements of 123. When we expand and substitute the values of ε, namely $\varepsilon(123)=\varepsilon(231)=\varepsilon(312)=1$, $\varepsilon(132)=\varepsilon(213)=\varepsilon(321)=-1$, we just obtain the left-hand side of (5.4). If we define $\varepsilon(i_1i_2\cdots i_n)$ to be zero when two of $i_1,\ldots,i_n$ are the same, so that ε is now defined for any numerical arguments, we can in (5.6) take the summation over i,j,k running independently from 1 to 3. Of the resulting 27 terms, 21 are zero and the remaining 6 give the determinant as before.

If for the three vectors we take the columns of a 3×3 matrix $\mathbf{A}=(a_{ij})$, the expression for the determinant of $\mathbf{A}$ reads

$$\sum\varepsilon(ijk)a_{i1}a_{j2}a_{k3}$$

where the summation is over i,j,k each running from 1 to 3. In analogy with this formula we define the determinant of an $n\times n$ matrix $\mathbf{A}=(a_{ij})$ as

$$\sum\varepsilon(i_1i_2\cdots i_n)a_{i_11}a_{i_22}\cdots a_{i_nn} \tag{5.7}$$

where the summation is over $i_1,\ldots,i_n$ each running from 1 to n. We obtain an expression with $n!$ terms, corresponding to the different arrangements of

$12\cdots n$. This expression (5.7) is called the **determinant** of the matrix $\mathbf{A}$; it is said to be of **order** n and is written $|\mathbf{A}|$ or $\det \mathbf{A}$, or $|\mathbf{a}_1,\ldots,\mathbf{a}_n|$, where the $\mathbf{a}_i$ are the columns of $\mathbf{A}$, or in full,

$$\begin{vmatrix} a_{11} & a_{12} & \cdots & a_{1n} \\ a_{21} & a_{22} & \cdots & a_{2n} \\ \cdots & & \cdots & \\ a_{n1} & a_{n2} & \cdots & a_{nn} \end{vmatrix}$$

5.8 COFACTORS AND EXPANSIONS

Let us consider the definition (5.7) for a moment. Clearly it generalizes the definition of second- and third-order determinants given earlier. As we have already noted above, the expression consists of $n!$ terms (ignoring terms with a vanishing ε factor). Each term includes just one element from each row and one element from each column.* For example, in any term containing a_{12} there are also $n-1$ other factors a_{ij}, which come from rows other than the first and from columns other than the second. If we collect the terms containing a_{12}, we can write their sum as $a_{12}A_{12}$, where A_{12}, the **cofactor** of a_{12} in (5.7), is a certain expression in the a_{ij} containing no element from the first row or the second column of $\mathbf{A}$. In an analogous way we can define the cofactor A_{ij} of any entry a_{ij} of $\mathbf{A}$.

Let us now fix a row, say the first row. Since each term in (5.7) contains just one element from the first row, every such term will occur exactly once in one of the expressions $a_{1j}A_{1j}$ ($j=1,\ldots,n$), and we can therefore express the determinant as

$$a_{11}A_{11}+a_{12}A_{12}+\cdots+a_{1n}A_{1n}$$

with the cofactors as coefficients. This is called the **expansion of $|\mathbf{A}|$ by its first row**. We can expand $|\mathbf{A}|$ by any of its rows in this way. The result is

$$a_{i1}A_{i1}+a_{i2}A_{i2}+\cdots+a_{in}A_{in}=|\mathbf{A}| \quad (i=1,\ldots,n) \tag{5.8}$$

As an example we give the expansion of a third-order determinant by its first row:

$$\begin{vmatrix} a_{11} & a_{12} & a_{13} \\ a_{21} & a_{22} & a_{23} \\ a_{31} & a_{32} & a_{33} \end{vmatrix} = a_{11}(a_{22}a_{33}-a_{32}a_{23})+a_{12}(-a_{21}a_{33}+a_{31}a_{23}) \\ +a_{13}(a_{21}a_{32}-a_{31}a_{22})$$

We notice that the cofactors in this expansion have the fofm of second-order

*Thus each term in the expression of an eighth-order determinant corresponds to a distribution of eight rooks on a chessboard in such a way that none can take any of the others.

determinants. In the general case we shall see that the cofactors can be expressed as determinants of order $n-1$ (cf. section 5.10).

We can also expand $|\mathbf{A}|$ by one of its columns, using the fact that each term of (5.7) contains just one entry from any given column. Thus the expansion by the jth column is

$$a_{1j}A_{1j} + a_{2j}A_{2j} + \cdots + a_{nj}A_{nj} = |\mathbf{A}| \quad (j = 1, \ldots, n) \tag{5.9}$$

where the coefficients are again the cofactors.

Determinants were first used by Gottfried Wilhelm von Leibniz (1646–1716) in letters in 1693 to Guillaume François Antoine de L'Hôpital (1661–1704), in which he wrote down third-order determinants and showed their use in solving systems of linear equations; these letters were only published in the nineteenth century and so had no influence on later developments. At about the same time determinants occur in the work of Seki Kōwa (also called Takakazu, 1642–1708). The general definition of determinants (with the rule for the signs of the terms) was given in 1750 by Gabriel Cramer (1704–52).

5.9 PROPERTIES OF DETERMINANTS

We now list the most important properties of determinants:

D.1 *The determinant of a square matrix* $\mathbf{A}$ *is a linear function of the entries of any column of* $\mathbf{A}$.

This means that for any n-vectors $\mathbf{a}_1, \mathbf{a}'_1, \mathbf{a}_2, \ldots, \mathbf{a}_n$ and any scalars λ, λ' we have

$$|\lambda\mathbf{a}_1 + \lambda'\mathbf{a}'_1, \mathbf{a}_2, \ldots, \mathbf{a}_n| = \lambda|\mathbf{a}_1, \mathbf{a}_2, \ldots, \mathbf{a}_n| + \lambda'|\mathbf{a}'_1, \mathbf{a}_2, \ldots, \mathbf{a}_n|$$

and similarly for the other columns.

D.2 *The determinant changes sign if two columns are interchanged; it is zero if two columns are the same.*

D.3 *The determinant is unchanged if a multiple of one column is added to another column.*

D.4 *The determinant is unchanged if the columns are written as rows, in symbols,*

$$|\mathbf{A}^{\mathrm{T}}| = |\mathbf{A}|$$

If we apply D.4 to D.1–3 we obtain

D.1^{T}–3^{T} *Rules D.1–3 apply when 'column' is replaced by 'row'.*

Rule D.1 is a consequence of equation (5.9), which expresses $|\mathbf{A}|$ as a linear function of the jth column ($j = 1, \ldots, n$). As an illustration,

$$\begin{vmatrix} \lambda a & \lambda b \\ \lambda c & \lambda d \end{vmatrix} = \lambda \begin{vmatrix} a & \lambda b \\ c & \lambda d \end{vmatrix} = \lambda^2 \begin{vmatrix} a & b \\ c & d \end{vmatrix}$$

In particular, a determinant vanishes if any column is the zero column. We note that D.1 may be used to express the determinant of a sum of two matrices $\mathbf{A}$ and $\mathbf{B}$ as a sum of 2^n determinants, the general determinant in this sum being formed by taking any number r of columns from $\mathbf{A}$ and the remaining $n - r$ columns from $\mathbf{B}$. Thus in general,

$$|\mathbf{A} + \mathbf{B}| \neq |\mathbf{A}| + |\mathbf{B}|$$

To prove D.2, let us take the second part first. If two columns of $\mathbf{A}$ are the same, say $\mathbf{a}_1 = \mathbf{a}_2$, then $a_{i1} = a_{i2}$ in (5.7). Consider a term with $i_1 < i_2$ in this expression; there is a corresponding term with i_1 and i_2 interchanged, and of opposite sign. Thus we can write (5.7) as

$$\sum_{i_1 < i_2} \varepsilon(i_1 i_2 i_3 \cdots i_n)(a_{i_1 1} a_{i_2 1} - a_{i_2 1} a_{i_1 1}) a_{i_3 3} \cdots a_{i_n n}$$

and this is clearly zero; similarly if two other columns coincide. Turning now to the first part, we can write $|\mathbf{A}|$ as a function of its first two columns, say $|\mathbf{A}| = f(\mathbf{a}_1, \mathbf{a}_2)$. By D.1, f is linear in each argument and $f(\mathbf{a}, \mathbf{a}) = 0$, by what has just been proved. Hence

$$\begin{aligned} 0 = f(\mathbf{a}_1 + \mathbf{a}_2, \mathbf{a}_1 + \mathbf{a}_2) &= f(\mathbf{a}_1, \mathbf{a}_1) + f(\mathbf{a}_1, \mathbf{a}_2) + f(\mathbf{a}_2, \mathbf{a}_1) + f(\mathbf{a}_2, \mathbf{a}_2) \\ &= f(\mathbf{a}_1, \mathbf{a}_2) + f(\mathbf{a}_2, \mathbf{a}_1) \end{aligned}$$

Now the desired relation

$$f(\mathbf{a}_1, \mathbf{a}_2) = -f(\mathbf{a}_2, \mathbf{a}_1)$$

follows. Similarly for any other pair of columns. We express D.2 by saying that $|\mathbf{A}|$ is an **alternating function** of its columns.

D.3 is a consequence of D.1–2. Consider, for example, the case where a multiple of the second column is added to the first. Using again our function f, we have as the value of the determinant,

$$\begin{aligned} f(\mathbf{a}_1 + \lambda \mathbf{a}_2, \mathbf{a}_2) &= f(\mathbf{a}_1, \mathbf{a}_2) + \lambda f(\mathbf{a}_2, \mathbf{a}_2) && \text{by D.1} \\ &= f(\mathbf{a}_1, \mathbf{a}_2) = |\mathbf{A}| && \text{by D.2} \end{aligned}$$

To prove D.4 we use E.2 from section 5.6. When we apply a permutation to change $i_1 i_2 \cdots i_n$ to $12 \cdots n$, then $12 \cdots n$ will change to $j_1 j_2 \cdots j_n$, say, and we can therefore write (5.7) as

$$\sum \varepsilon(i_1 i_2 \cdots i_n) a_{1 j_1} a_{2 j_2} \cdots a_{n j_n}$$

By E.2 we have $\varepsilon(i_1 i_2 \cdots i_n) = \varepsilon(j_1 j_2 \cdots j_n)$, hence

$$|\mathbf{A}| = \sum \varepsilon(j_1 j_2 \cdots j_n) a_{1 j_1} a_{2 j_2} \cdots a_{n j_n}$$

The right-hand side is just the expansion of $|\mathbf{A}^{\mathrm{T}}|$, hence $|\mathbf{A}| = |\mathbf{A}^{\mathrm{T}}|$.

5.10 AN EXPRESSION FOR THE COFACTORS

In order to be able to use (5.8) and (5.9) for the evaluation of determinants, we have to find an expression for the cofactors. Consider first A_{11}; this is the cofactor of a_{11} in (5.7), thus

$$A_{11} = \sum \varepsilon(1 i_2 \cdots i_n) a_{i_2 2} \cdots a_{i_n n}$$

where $i_2, \ldots, i_n$ run from 2 to n. By E.3, the ε function in this expression is equal to $\varepsilon(i_2 \cdots i_n)$, therefore

$$A_{11} = \sum \varepsilon(i_2 \cdots i_n) a_{i_2 2} \cdots a_{i_n n}$$

The expression on the right is a determinant of order $n-1$:

$$A_{11} = \begin{vmatrix} a_{22} & a_{23} & \cdots & a_{2n} \\ \cdots & & \cdots & \\ a_{n2} & a_{n3} & \cdots & a_{nn} \end{vmatrix} \tag{5.10}$$

Thus A_{11} is equal to the $(n-1)$th-order determinant obtained from $|\mathbf{A}|$ by omitting the first row and the first column. To evaluate A_{ij}, the cofactor of a_{ij}, we move the ith row up past row $i-1, \ldots, 1$ so that after $i-1$ interchanges of rows it becomes the new top row, and then move the jth column to the left past the first $j-1$ columns in turn, so that it becomes the initial column. We now have a matrix whose determinant is related to $|\mathbf{A}|$ by $i-1$ interchanges of rows and $j-1$ interchanges of columns; these $i-1+j-1$ interchanges produce $i+j-2$ changes of sign, or equivalently, $i+j$ changes of sign. Therefore the value of the new determinant is $(-1)^{i+j}|\mathbf{A}|$. The cofactor of a_{ij} in this determinant – in which a_{ij} is now the $(1,1)$th entry – is the determinant of order $n-1$ obtained from $|\mathbf{A}|$ by omitting the row and column intersecting in a_{ij}. If we denote this determinant by α_{ij}, we have

$$A_{ij} = (-1)^{i+j} \alpha_{ij} \tag{5.11}$$

The $(n-1)$th-order determinants α_{ij} are called the **minors** of the determinant $|\mathbf{A}|$, and (5.11) expresses the cofactors in terms of the minors, the distribution of signs being of the chess-board pattern

$$\begin{vmatrix} + & - & + & - & \cdots \\ - & + & - & + & \cdots \\ + & - & + & - & \cdots \\ \cdots & & \cdots & & \cdots \end{vmatrix}$$

5.11 EVALUATION OF DETERMINANTS

It is now an easy (though possibly lengthy) matter to evaluate determinants. We can expand the determinant by any row, according to (5.8), or by any column, according to (5.9), taking care to use the correct sign for the minors.

In our first example we use the expansion by the first row:

$$\begin{vmatrix} 7 & 2 & 4 \\ 3 & -4 & 5 \\ 1 & 3 & -2 \end{vmatrix} = 7(4.2 - 3.5) - 2(-3.2 - 5.1) + 4(3.3 + 4.1) = 25$$

For determinants of larger orders the task soon becomes unmanageable, unless one is able to use shortcuts offered by the special form of the matrix in each particular case. Thus if a row or column consists mainly of zeros, it is best to use it in the expansion, e.g.

$$\begin{vmatrix} 3 & 0 & 5 \\ 2 & 0 & 4 \\ 1 & 7 & 3 \end{vmatrix} = (-1)^{3+2} \cdot 7 \cdot \begin{vmatrix} 3 & 5 \\ 2 & 4 \end{vmatrix} = -14$$

As a special case, the determinant of a 'triangular' matrix, where all entries below the main diagonal are zero:

$$\begin{vmatrix} a_{11} & a_{12} & a_{13} & \cdots & a_{1n} \\ 0 & a_{22} & a_{23} & \cdots & a_{2n} \\ 0 & 0 & a_{33} & \cdots & a_{3n} \\ & \cdots & & \cdots & \\ 0 & 0 & \cdots & 0 & a_{nn} \end{vmatrix}$$

is $a_{11}a_{22} \cdots a_{nn}$. This follows on making an expansion by the first column, giving $a_{11}\alpha_{11}$, where the minor α_{11} is again triangular, and now using induction on n. In a similar way we can deal with a 'lower' triangular matrix, where the entries above the main diagonal are zero. In particular, this tells us that the determinant of a diagonal matrix

$$\mathbf{D} = \begin{pmatrix} d_1 & 0 & 0 & \cdots & 0 \\ 0 & d_2 & 0 & \cdots & 0 \\ \cdots & & \cdots & \cdots & \\ 0 & 0 & 0 & & d_n \end{pmatrix}$$

is just the product of the diagonal elements: $|\mathbf{D}| = d_1 d_2 \cdots d_n$. This also shows that the determinant of the unit matrix is 1; more generally, for a scalar matrix, i.e. a scalar multiple of the unit matrix, we have $|\lambda \mathbf{I}| = \lambda^n$, where n is the order of the matrix.

In the examples given in Chapter 2 we used the elementary operations α, β, γ to reduce the matrix of a regular system to the unit matrix. Whenever we carry out such a reduction we can find the value of the determinant at the same time by noting the effect of these operations on the determinant. Thus α – interchanging two rows – multiplies the determinant by -1; β – multiplying a row by a non-zero scalar – multiplies the determinant by that scalar; and γ – adding a multiple of one row to another – leaves the determinant unchanged. At the end of the reduction we have the unit matrix whose determinant is 1, therefore when we retrace our steps by taking the inverses of the operations α, β, γ used we obtain the determinant of the original matrix. It follows from this that the determinant of a regular matrix cannot be zero. For such a matrix can be reduced to the unit matrix by operations α, β, γ and when we evaluate the determinant by applying the inverse operations we obtain for each operation a non-zero factor.

As an illustration we use the matrix of the system discussed in section 2.6. The operations used were: γ, α (interchange of two rows), β (multiplication by 1/2), γ, γ, β (multiplication by 1/4). Hence the determinant is

$$\frac{1}{-1\cdot\frac{1}{2}\cdot\frac{1}{4}} = -8$$

The result of this reduction may also be described as follows. If a matrix $\mathbf{A}$ has been expressed as a product of elementary matrices, $\mathbf{A} = \mathbf{P}_1\mathbf{P}_2\cdots\mathbf{P}_r$, then

$$|\mathbf{A}| = |\mathbf{P}_1|\cdot|\mathbf{P}_2|\cdots|\mathbf{P}_r| \tag{5.12}$$

Now in section 3.10 we saw that every regular matrix can be written as a product of elementary matrices. We can use this remark to obtain a formula for the determinant of a product. Let $\mathbf{A}$, $\mathbf{B}$ be any regular matrices and express them as products of elementary matrices: $\mathbf{A} = \mathbf{P}_1\cdots\mathbf{P}_r$, $\mathbf{B} = \mathbf{Q}_1\cdots\mathbf{Q}_s$. Then $\mathbf{AB} = \mathbf{P}_1\cdots\mathbf{P}_r\mathbf{Q}_1\cdots\mathbf{Q}_s$ and hence

$$|\mathbf{AB}| = |\mathbf{P}_1|\cdots|\mathbf{P}_r|\cdot|\mathbf{Q}_1|\cdots|\mathbf{Q}_s| = |\mathbf{A}|\cdot|\mathbf{B}| \tag{5.13}$$

where we have used (5.12) first on $\mathbf{AB}$ and then on $\mathbf{A}, \mathbf{B}$ separately. If one of $\mathbf{A}, \mathbf{B}$ is not regular, its inner rank is less than n, hence the same is true of $\mathbf{AB}$, so in this case both sides of (5.13) are 0. Thus we have established the following:

Theorem 5.1
For any two square matrices $\mathbf{A}, \mathbf{B}$ *of the same order,*

$$|\mathbf{AB}| = |\mathbf{A}|\cdot|\mathbf{B}| \tag{5.14}$$

This formula is sometimes called the **multiplication theorem for determinants**. Later, in section 5.13, we shall meet another proof of (5.14).

The reduction after the pattern of Chapter 2 can be used to evaluate any determinant. We apply the operations α, β, γ, noting at each stage the change in the value of the determinant. In the first example of this section we have

$$\begin{vmatrix} 7 & 2 & 4 \\ 3 & -4 & 5 \\ 1 & 3 & -2 \end{vmatrix} = \begin{vmatrix} 0 & -19 & 18 \\ 0 & -13 & 11 \\ 1 & 3 & -2 \end{vmatrix} = \begin{vmatrix} -19 & 18 \\ -13 & 11 \end{vmatrix} = \begin{vmatrix} -1 & 18 \\ -2 & 11 \end{vmatrix} = \begin{vmatrix} -1 & 18 \\ 0 & -25 \end{vmatrix} = 25$$

The process of systematically reducing all but one of the entries of a given column (or row) to zero by adding multiples of a fixed row (or column) to the others is called **pivotal condensation**, the unreduced element being the pivot. Usually it is convenient, either by division or by taking a combination of rows (or columns), to reduce the pivot to 1. We give a further example, where the pivot at each stage is indicated by an asterisk:

$$\begin{vmatrix} 2^* & 6 & -2 & 4 \\ -1 & -2 & 1 & 0 \\ 7 & 13 & 5 & 9 \\ 4 & 8 & -6 & 11 \end{vmatrix} = \begin{vmatrix} 2 & 0 & 0 & 0 \\ -1 & 1 & 0 & 2 \\ 7 & -8 & 12 & -5 \\ 4 & -4 & -2 & 3 \end{vmatrix} = 2\begin{vmatrix} 1^* & 0 & 2 \\ -8 & 12 & -5 \\ -4 & -2 & 3 \end{vmatrix}$$

$$= 2\begin{vmatrix} 1 & 0 & 0 \\ -8 & 12 & 11 \\ -4 & -2 & 11 \end{vmatrix} = 2\begin{vmatrix} 12 & 11 \\ -2 & 11 \end{vmatrix} = 2.11\begin{vmatrix} 12 & 1 \\ -2 & 1 \end{vmatrix}$$

$$= 2.11.14 = 308$$

5.12 A FORMULA FOR THE INVERSE MATRIX

If **A** is a regular matrix, so that its inverse $\mathbf{A}^{-1}$ exists, we can express this inverse in terms of the determinant and cofactors of **A** as follows. We first form the transpose of the matrix of cofactors:

$$\begin{pmatrix} A_{11} & A_{21} & \cdots & A_{n1} \\ A_{12} & A_{22} & \cdots & A_{n2} \\ \cdots & \cdots & \cdots & \cdots \\ A_{1n} & A_{2n} & \cdots & A_{nn} \end{pmatrix}$$

This matrix is denoted by adj(**A**) and is called the **adjoint*** of **A** or the **adjugate matrix**. We now take any square matrix **A** and evaluate the product $\mathbf{A}\cdot\mathrm{adj}(\mathbf{A})$. The (1, 1)th entry is

$$a_{11}A_{11} + a_{12}A_{12} + \cdots + a_{1n}A_{1n}$$

which equals $|\mathbf{A}|$, by (5.8). Similarly, (5.8) shows that every element on the main diagonal has the value $|\mathbf{A}|$. Next consider the (2, 1)th entry:

$$a_{21}A_{11} + a_{22}A_{12} + \cdots + a_{2n}A_{1n}$$

*The term 'adjoint' is also used for the linear transformation corresponding to the transposed matrix; the intended meaning is usually clear from the context.

This represents the expansion by the first row of the determinant

$$\begin{vmatrix} a_{21} & a_{22} & a_{23} & \cdots & a_{2n} \\ a_{21} & a_{22} & a_{23} & \cdots & a_{2n} \\ a_{31} & a_{32} & a_{33} & \cdots & a_{3n} \\ \cdots & \cdots & \cdots & \cdots & \cdots \\ a_{n1} & a_{n2} & a_{n3} & \cdots & a_{nn} \end{vmatrix}$$

which has the same 2nd, 3rd,..., nth row as $|\mathbf{A}|$ (because the cofactors $A_{11}, A_{12}, \ldots, A_{1n}$ involve only elements from these rows), but its first row is equal to its second row and so its value is zero, by D.2^{T}. Generally, we have

$$a_{i1}A_{j1} + a_{i2}A_{j2} + \cdots + a_{in}A_{jn} = 0 \quad \text{for } i \neq j$$

because the left-hand side represents the expansion of a determinant which differs from $|\mathbf{A}|$ only in its jth row, which is equal to its ith row, so that two rows are the same (an 'expansion by alien cofactors'). Thus the entries on the main diagonal of $\mathbf{A}\cdot\mathrm{adj}(\mathbf{A})$ are $|\mathbf{A}|$ and all others are zero; in other words, it is a scalar multiple of the unit matrix:

$$\mathbf{A}\cdot\mathrm{adj}(\mathbf{A}) = |\mathbf{A}|\cdot\mathbf{I} \tag{5.15}$$

Similarly, using (5.9) for the diagonal entries and D.2 for the non-diagonal entries, we find that

$$\mathrm{adj}(\mathbf{A})\cdot\mathbf{A} = |\mathbf{A}|\cdot\mathbf{I} \tag{5.16}$$

These formulae hold for any square matrix $\mathbf{A}$. Now assume that $|\mathbf{A}| \neq 0$ and form the matrix

$$\mathbf{B} = \frac{1}{|\mathbf{A}|}\mathrm{adj}(\mathbf{A})$$

Equations (5.15) and (5.16) show that $\mathbf{AB} = \mathbf{BA} = \mathbf{I}$, hence $\mathbf{A}$ has the inverse $\mathbf{B}$ given by

$$\mathbf{A}^{-1} = \frac{1}{|\mathbf{A}|}\mathrm{adj}(\mathbf{A}) \tag{5.17}$$

This holds for any matrix $\mathbf{A}$ with non-zero determinant; conversely, as we saw in section 5.11, a regular matrix has a non-zero determinant. Hence we have

Theorem 5.2

A square matrix $\mathbf{A}$ is regular if and only if $|\mathbf{A}| \neq 0$. Hence a system of homogeneous linear equations with square matrix $\mathbf{A}$ has a non-trivial solution precisely when $|\mathbf{A}| = 0$.

Proof

The first part has just been proved, while the second part follows by combining the first part with Theorem 4.1.

In order to find the inverse of a matrix **A** by this method we go through the following steps:

1. Form the matrix of minors (α_{ij}).
2. Affix signs by the chessboard rule: $((-1)^{i+j}\alpha_{ij})$.
3. Take the transpose and so obtain the adjoint matrix $\text{adj}(\mathbf{A}) = (A_{ij})^{\mathrm{T}}$.
4. Form the product $\mathbf{A}\cdot\text{adj}(\mathbf{A})$.

Now there are two cases: Either $\mathbf{A}\cdot\text{adj}(\mathbf{A}) = \mathbf{0}$; then $|\mathbf{A}| = 0$ and **A** has no inverse. Or $\mathbf{A}\cdot\text{adj}(\mathbf{A}) = \delta\mathbf{I}$ for some $\delta \neq 0$; then $\mathbf{A}^{-1} = \delta^{-1}\cdot\text{adj}(\mathbf{A})$.

This method is self-checking, in that any mistake in the calculation will show up by $\mathbf{A}\cdot\text{adj}(\mathbf{A})$ not being a scalar matrix (unless our mistake gives a wrong value for the scalar, or is cancelled by a second mistake, a relatively rare occurrence). But the whole procedure is quite lengthy and it is usually quicker to find the inverse by the method of row reduction, as in section 3.10.

To illustrate the above method, let us take the simplest non-trivial case, a 2×2 matrix:

$$\mathbf{A} = \begin{pmatrix} a & b \\ c & d \end{pmatrix}$$

We follow the above steps:

1. The minor of each entry is found by omitting the row and column containing that entry. Thus for a we have d etc., giving as the matrix of minors

$$\begin{pmatrix} d & c \\ b & a \end{pmatrix}$$

2. The two non-diagonal entries are multiplied by -1:

$$\begin{pmatrix} d & -c \\ -b & a \end{pmatrix}$$

3. Now form the transpose to obtain the adjoint:

$$\text{adj}(\mathbf{A}) = \begin{pmatrix} d & -b \\ -c & a \end{pmatrix}$$

4. $$\mathbf{A}\cdot\text{adj}(\mathbf{A}) = \begin{pmatrix} a & b \\ c & d \end{pmatrix}\begin{pmatrix} d & -b \\ -c & a \end{pmatrix} = (ad - bc)\cdot\mathbf{I}$$

The result is a scalar matrix, as expected, with the scalar factor $ad - bc$, which must therefore be the determinant of **A**. Hence **A** has an inverse if and only if $ad - bc \neq 0$, and it is then given by

$$\mathbf{A}^{-1} = \delta^{-1}\begin{pmatrix} d & -b \\ -c & a \end{pmatrix},$$

where $\delta = ad - bc$.

5.13 CRAMER'S RULE

The formula (5.17) for the inverse matrix may also be used to obtain an explicit formula for the solution of a regular system. By (5.17), the solution of

$$\mathbf{Ax} = \mathbf{k} \tag{5.18}$$

is

$$\mathbf{x} = \mathbf{A}^{-1}\mathbf{k} = \frac{1}{|\mathbf{A}|}\operatorname{adj}(\mathbf{A})\mathbf{k}$$

Thus to obtain x_i we multiply the terms of the ith row of adj(**A**) by the components of **k** and sum and divide by $|\mathbf{A}|$:

$$x_i = \frac{1}{|\mathbf{A}|}(A_{1i}k_1 + A_{2i}k_2 + \cdots + A_{ni}k_n)$$

If we denote the jth column of **A** by $\mathbf{a}_j$, then the bracket on the right may be obtained as the expansion, by the ith column, of the determinant of the matrix

$$\mathbf{A}^{(i)} = (\mathbf{a}_1, \ldots, \mathbf{a}_{i-1}, \mathbf{k}, \mathbf{a}_{i+1}, \ldots, \mathbf{a}_n) \tag{5.19}$$

which is obtained when the ith column of **A** is replaced by **k**. Hence the solution of (5.18) may be written as the quotient of two determinants:

$$x_i = \frac{|\mathbf{A}^{(i)}|}{|\mathbf{A}|} \quad (i = 1, \ldots, n) \tag{5.20}$$

where the matrix $\mathbf{A}^{(i)}$ is defined by (5.19). This formula for the solution of a regular system of equations is known as **Cramer's rule** (it appeared in Cramer's work, *Introduction à l'analyse des lignes courbes*, 1750). It is mainly of theoretical interest, since for a practical solution the Gaussian elimination of Chapter 2 is considerably quicker than evaluating $n + 1$ determinants of order n.

With the help of Cramer's rule we can give another proof, due to Kronecker, of the multiplication theorem of determinants (Theorem 5.1):

$$|\mathbf{AB}| = |\mathbf{A}| \cdot |\mathbf{B}| \tag{5.14}$$

This equation may be regarded as an identity in the $2n^2$ entries of **A** and **B**. If we regard these entries as variables, the determinants $|\mathbf{A}|$, $|\mathbf{B}|$, $|\mathbf{AB}|$ are all non-zero, because they do not vanish identically – for example, they are

1 when $\mathbf{A} = \mathbf{B} = \mathbf{I}$. Thus we may assume $\mathbf{A}, \mathbf{B}$, and $\mathbf{AB}$ to be regular in what follows.

The system $\mathbf{Bx} = \mathbf{k}$ has a unique solution, which by Cramer's rule may be written

$$x_i = \frac{|\mathbf{B}^{(i)}|}{|\mathbf{B}|}$$

where

$$\mathbf{B}^{(i)} = (\mathbf{b}_1, \ldots, \mathbf{b}_{i-1}, \mathbf{k}, \mathbf{b}_{i+1}, \ldots, \mathbf{b}_n)$$

Here $\mathbf{b}_1, \ldots, \mathbf{b}_n$ are the columns of $\mathbf{B}$, so that $\mathbf{B}^{(i)}$ does not involve the ith column of $\mathbf{B}$. Now the solution of $\mathbf{Bx} = \mathbf{k}$ plainly also satisfies the equation

$$\mathbf{ABx} = \mathbf{Ak}$$

whose solution is, again by Cramer's rule,

$$x_i = \frac{|\mathbf{AB}^{(i)}|}{|\mathbf{AB}|}$$

where

$$\begin{aligned}\mathbf{AB}^{(i)} &= \mathbf{A}(\mathbf{b}_1, \ldots, \mathbf{b}_{i-1}, \mathbf{k}, \mathbf{b}_{i+1}, \ldots, \mathbf{b}_n)\\ &= (\mathbf{Ab}_1, \ldots, \mathbf{Ab}_{i-1}, \mathbf{Ak}, \mathbf{Ab}_{i+1}, \ldots, \mathbf{Ab}_n)\end{aligned}$$

Since the solution is unique, it is the same in each case, so

$$\frac{|\mathbf{B}^{(i)}|}{|\mathbf{B}|} = \frac{|\mathbf{AB}^{(i)}|}{|\mathbf{AB}|} \qquad i = 1, \ldots, n$$

Here neither numerator vanishes if we regard the components of $\mathbf{k}$ as variables, so we obtain

$$\frac{|\mathbf{AB}|}{|\mathbf{B}|} = \frac{|\mathbf{AB}^{(i)}|}{|\mathbf{B}^{(i)}|} \qquad (i = 1, \ldots, n) \tag{5.21}$$

In this equation the right-hand side does not involve the ith column of $\mathbf{B}$, therefore neither does the left-hand side. Since this holds for $i = 1, \ldots, n$, $|\mathbf{AB}|/|\mathbf{B}|$ is independent of $\mathbf{B}$ altogether and we may find its value by giving $\mathbf{B}$ any particular value, for example $\mathbf{B} = \mathbf{I}$:

$$\frac{|\mathbf{AB}|}{|\mathbf{B}|} = \frac{|\mathbf{A}|}{|\mathbf{I}|} = |\mathbf{A}|$$

Multiplying up, we obtain $|\mathbf{AB}| = |\mathbf{A}| \cdot |\mathbf{B}|$, that is (5.14).

5.14 A DETERMINANTAL CRITERION FOR LINEAR DEPENDENCE

If we recall the definition of a regular matrix given in section 3.9, we see from Theorem 5.2 that the vanishing of $|\mathbf{A}|$ is a necessary and sufficient condition for the linear dependence of the columns of **A**, so that the determinant provides a numerical criterion for the linear dependence of n vectors in n dimensions. In a similar way we can characterize the linear dependence of any number of vectors by means of determinants. We know already that a linearly independent set of n-vectors contains at most n vectors (Theorem 2.3), so we may suppose that we are dealing with r vectors, where $r \leqslant n$. From the components of these r vectors we can form a determinant of order r by choosing any r rows. In this way we obtain $\binom{n}{r} = n!/r!(n-r)!$ determinants of order r.

The vanishing of these $\binom{n}{r}$ determinants is necessary and sufficient for the linear dependence of the r vectors.

For example, the vectors $\mathbf{u} = (u_1, u_2, u_3)^{\mathrm{T}}$ and $\mathbf{v} = (v_1, v_2, v_3)^{\mathrm{T}}$ are linearly dependent if and only if

$$\begin{vmatrix} u_2 & v_2 \\ u_3 & v_3 \end{vmatrix} = \begin{vmatrix} u_1 & v_1 \\ u_2 & v_2 \end{vmatrix} = \begin{vmatrix} u_1 & v_1 \\ u_3 & v_3 \end{vmatrix} = 0$$

Geometrically this means that two vectors in space are linearly dependent if and only if their projections on the three coordinate planes are linearly dependent.

To prove the assertion, let $\mathbf{u}_1, \ldots, \mathbf{u}_r$ be linearly dependent vectors in n dimensions, say,

$$\mathbf{u}_1 = \lambda_2 \mathbf{u}_2 + \cdots + \lambda_r \mathbf{u}_r$$

Then in any determinant formed by taking r rows of the $n \times r$ matrix $(\mathbf{u}_1, \ldots, \mathbf{u}_r)$, we can make the first column (C1) zero by subtracting $\lambda_2 . \mathrm{C}2 + \cdots + \lambda_r \mathrm{C}r$. Hence all these determinants vanish.

For the converse assume that the r vectors $\mathbf{u}_1, \ldots, \mathbf{u}_r$ are linearly independent and consider again the $n \times r$ matrix $(\mathbf{u}_1, \ldots, \mathbf{u}_r)$. Let s be the maximum number of linearly independent rows; to complete the proof we need only show that $s = r$, for these r rows then form a regular matrix of order r, whose determinant is non-zero, by Theorem 5.2. We may suppose the rows so numbered that the first s are linearly independent, while the remaining $n - s$ depend on them. This means that the vector equation

$$\mathbf{u}_1 x_1 + \cdots + \mathbf{u}_r x_r = \mathbf{0} \tag{5.22}$$

is equivalent to the s scalar equations

$$\begin{aligned} u_{11} x_1 + \cdots + u_{1r} x_r &= 0 \\ \cdots \quad & \cdots \\ u_{s1} x_1 + \cdots + u_{sr} x_r &= 0 \end{aligned} \tag{5.23}$$

where $u_{1i}, u_{2i}, \ldots, u_{ni}$ are the components of $\mathbf{u}_i$. Since the rows have r components, a linearly independent set of rows can have at most r elements (Theorem 2.3) and so $s \leqslant r$. If we had $s < r$, then the system (5.23) would have more than s columns, and these columns would therefore be linearly dependent, again by Theorem 2.3. This means that (5.23) would have a non-trivial solution, and so would (5.22), because it is equivalent to (5.23). But this contradicts the linear independence of $\mathbf{u}_1, \ldots, \mathbf{u}_r$. Hence $s = r$, and now the determinant of the system (5.23) provides a non-zero determinant of order r.

5.15 A DETERMINANTAL EXPRESSION FOR THE RANK

To describe the rank of a matrix in terms of determinants we consider the determinants which can be formed from an $m \times n$ matrix $\mathbf{A}$ by omitting all but k rows and k columns. Such a determinant will be called a **minor** of order k of $\mathbf{A}$. In this terminology the minors of an $n \times n$ matrix introduced in section 5.10 would be 'minors of order $n-1$'. Thus for example, the matrix

$$\begin{pmatrix} a_1 & a_2 & a_3 \\ b_1 & b_2 & b_3 \end{pmatrix}$$

has six first-order minors: a_i, b_i $(i = 1, 2, 3)$ and three second-order minors, obtained by omitting each of the three columns in turn.

If $\mathbf{A}$ has rank r, then we can, by the result of section 5.14, choose a non-zero minor of order r. And if $\mathbf{A}$ contains a non-zero minor of order s, then the s columns which go to make up this minor are linearly independent, again by the result of section 5.1, so that the rank is at least s in this case. Hence the rank may be characterized in terms of determinants as follows:

Theorem 5.3
The rank of a matrix $\mathbf{A}$ *is the greatest number* r *such that* $\mathbf{A}$ *contains a non-vanishing minor of order* r.

This then gives a fourth characterization of the rank of a matrix, sometimes called the **determinantal rank** (in addition to row, column and inner rank). Since the determinant of a square matrix equals that of its transpose (rule D.4 of section 5.9), this again shows the equality of row and column rank. Applied to systems of linear equations it means that a system of rank r has just r (and no more) equations whose left-hand sides are linearly independent. This result is already implicit in the process of solving homogeneous systems of equations described in section 4.3, where we found that any system of rank r could be reduced to an equivalent system consisting of r equations, which were linearly independent, because each equation served for the determination of a different unknown in terms of the $n - r$ free variables.

EXERCISES

5.1. Evaluate: (i) $\varepsilon(14532)$; (ii) $\varepsilon(41532)$; (iii) $\varepsilon(14325)$; (iv) $\varepsilon(521436)$; (v) $\varepsilon(5314726)$; (vi) $\varepsilon(231)$; (vii) $\varepsilon(2341)$; (viii) $\varepsilon(23451)$; (ix) $\varepsilon(321)$; (x) $\varepsilon(4321)$; (xi) $\varepsilon(54321)$; (xii) $\varepsilon(654321)$.

5.2. Find the determinants of the matrices occurring in Exercise 2.1.

5.3. Test the linear dependence of the sets of vectors in Exercise 1.2 by finding non-zero determinants of maximal order formed from their components.

5.4. Determine the ranks of matrices in (a) Exercises 4.3 and (b) Exercise 4.4, by using determinants.

5.5. Evaluate

$$\text{(i)}\begin{vmatrix} 1 & 3 & 2 \\ 8 & 4 & 0 \\ 2 & 1 & 1 \end{vmatrix} \quad \text{(ii)}\begin{vmatrix} 6 & 5 & 2 \\ 3 & 0 & -1 \\ -7 & 2 & 4 \end{vmatrix} \quad \text{(iii)}\begin{vmatrix} 1 & -2 & -3 & 4 \\ -2 & 3 & 4 & -5 \\ 3 & -4 & -5 & 6 \\ -4 & 5 & 6 & -7 \end{vmatrix}$$

$$\text{(iv)}\begin{vmatrix} 1 & 1 & 1 & -1 \\ 1 & 1 & -1 & 1 \\ 1 & -1 & 1 & 1 \\ -1 & 1 & 1 & 1 \end{vmatrix} \quad \text{(v)}\begin{vmatrix} 3 & 9 & 27 & 81 \\ 1 & 1 & 1 & 1 \\ -2 & 4 & -8 & 16 \\ 2 & 4 & 8 & 16 \end{vmatrix}$$

5.6. Evaluate

$$\text{(i)}\begin{vmatrix} 1 & 1 & 1 & 1 \\ a & b & c & d \\ a^2 & b^2 & c^2 & d^2 \\ a^3 & b^3 & c^3 & d^3 \end{vmatrix} \quad \text{(ii)}\begin{vmatrix} a & b & c & d \\ a^2 & b^2 & c^2 & d^2 \\ a^3 & b^3 & c^3 & d^3 \\ a^4 & b^4 & c^4 & d^4 \end{vmatrix} \quad \text{(iii)}\begin{vmatrix} 1 & 1 & 1 & 1 \\ a & b & c & d \\ a^2 & b^2 & c^2 & d^2 \\ a^4 & b^4 & c^4 & d^4 \end{vmatrix}$$

$$\text{(iv)}\begin{vmatrix} 0 & a & b & c \\ -a & 0 & d & e \\ -b & -d & 0 & f \\ -c & -e & -f & 0 \end{vmatrix}$$

(Case (i) is known as a **Vandermonde determinant** (after Alexandre Théophile Vandermonde, 1735–96); we shall return to it in Chapter 8.)

5.7. Show that for any $n \times n$ matrix $\mathbf{A}$ and any scalar λ, $\det(\lambda\mathbf{A}) = \lambda^n.\det(\mathbf{A})$.

5.8. Use the multiplication theorem of determinants to give a direct proof of the fact that a matrix which has an inverse has a non-zero determinant.

5.9. Show that for a regular matrix $\mathbf{A}$ of order n, $|\text{adj}(\mathbf{A})| = |\mathbf{A}|^{n-1}$. (This result is true for any square matrix, regular or not.)

5.10. Show that an nth-order determinant with an $r \times s$ block of zeros vanishes whenever $r + s > n$. Give an example of a non-vanishing determinant with an $r \times (n-r)$ block of zeros.

5.11. Show that if a matrix has a non-vanishing rth-order minor but every $(r+1)$th-order minor containing the given rth-order minor is zero, then the matrix is of rank r.

5.12. Let $\mathbf{A}, \mathbf{B}$ be square matrices of order n and consider the matrix $\mathbf{AB} \oplus \mathbf{I}_n$ of order $2n$. Show that by elementary row and column operations which leave the determinant unchanged, it can be transformed to

$$\begin{pmatrix} \mathbf{0} & \mathbf{A} \\ -\mathbf{B} & \mathbf{I} \end{pmatrix}$$

By writing this matrix as a product

$$\begin{pmatrix} \mathbf{A} & \mathbf{0} \\ \mathbf{0} & \mathbf{I} \end{pmatrix}\begin{pmatrix} \mathbf{0} & \mathbf{I} \\ -\mathbf{I} & \mathbf{I} \end{pmatrix}\begin{pmatrix} \mathbf{B} & \mathbf{0} \\ \mathbf{0} & \mathbf{I} \end{pmatrix}$$

show that its determinant is $|\mathbf{A}|\cdot|\mathbf{B}|$ and so obtain another proof of the multiplication theorem of determinants.

6

Coordinate geometry

6.1 THE GEOMETRIC INTERPRETATION OF VECTORS

We have already seen in section 1.2 how points in the plane and in space can be described by 2- or 3-vectors, respectively, relative to a given coordinate system. It is useful to think of a vector $\mathbf{v}$, going from the origin O to a point A say, not as describing A, but as the **translation** from O to A. We shall also denote this vector by OA and call it the **position vector** of A relative to O. Given any two points P, P' in space, there exists just one translation of space which moves an object situated at P to P'; this translation displaces any object in the direction of the line PP' by an amount equal to the distance PP'. If P and P' are represented by the vectors $\mathbf{u}$ and $\mathbf{u}'$ (Fig. 6.1), and the translation PP' is represented by $\mathbf{v}$, then we have

$$\mathbf{u}' = \mathbf{u} + \mathbf{v}$$

This is just the geometric interpretation of vector addition. We also recall from section 1.5 that for a vector $\mathbf{x} = OP$ and a real number λ, the vector $\lambda\mathbf{x}$ represents a point on the line OP but at λ times the distance from O, on the same side as P if $\lambda > 0$ and on the opposite side if $\lambda < 0$; of course, for $\lambda = 0$ we just have the point O.

Our object in this chapter is to introduce two important operations on vectors, the scalar product and the vector product, and to see how they can be used to describe lines and planes in space. Of these operations the scalar product can be defined more generally in n-dimensional space and we also show how to do this since it often arises in applications.

6.2 COORDINATE SYSTEMS

To set up a coordinate system in space we can now proceed as follows. We choose a point O as origin and take three linearly independent vectors $\mathbf{e}_1, \mathbf{e}_2, \mathbf{e}_3$. If P is any point in space, its position vector relative to O may be expressed as a linear combination of the $\mathbf{e}$s:

$$OP = x_1\mathbf{e}_1 + x_2\mathbf{e}_2 + x_3\mathbf{e}_3$$

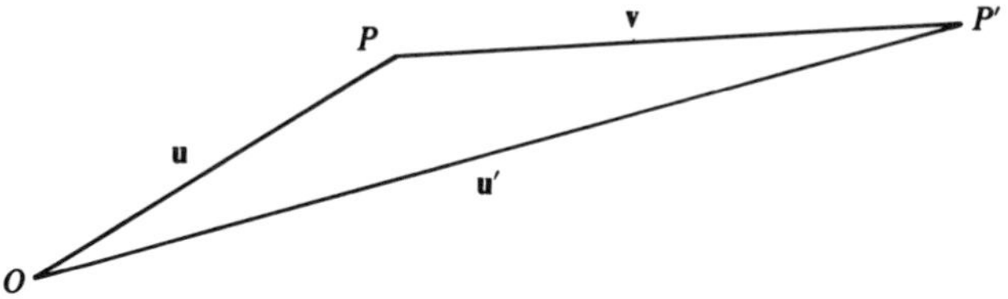

Fig. 6.1

The coefficients x_1, x_2, x_3 are called the **coordinates** of P relative to the basis vectors $\mathbf{e}_1, \mathbf{e}_2, \mathbf{e}_3$ and the origin O. In this way every point in space is described by a triple of real numbers. Conversely, any triple of real numbers (x_1, x_2, x_3) describes exactly one point in space: we form the linear combination $\mathbf{x} = x_1\mathbf{e}_1 + x_2\mathbf{e}_2 + x_3\mathbf{e}_3$ and apply the translation defined by $\mathbf{x}$ to the point O. We shall also express this correspondence by writing

$$\mathbf{x} \leftrightarrow (x_1, x_2, x_3)$$

where a fixed coordinate system is understood. The lines through O in the directions of the basis vectors are called the **coordinate axes**; the axes defined by $\mathbf{e}_1, \mathbf{e}_2, \mathbf{e}_3$ are the 1-axis, 2-axis and 3-axis, respectively. Any two axes define a **coordinate plane**; thus the 12-plane is the plane containing the 1- and 2-axes, and the 13-plane and 23-plane are defined similarly. The fact that the $\mathbf{e}_i$ were chosen to be linearly independent ensures that they span the whole of the space and not merely a plane or less.

In using coordinate systems to solve problems one naturally adopts a system best suited to the particular problem, but there is one type of system which is of general importance: a rectangular coordinate system. A coordinate system is said to be **rectangular** if its basis vectors are mutually orthogonal vectors of unit length; they are said to form an **orthonormal** basis. By contrast, a general coordinate system is often described as **oblique**.

In a rectangular system the endpoints of the vectors $x_1\mathbf{e}_1, x_2\mathbf{e}_2, x_3\mathbf{e}_3$ (all starting at O) are the feet of the perpendiculars from the endpoint of the vector $\mathbf{x}$ to the coordinate axes. Thus $x_1\mathbf{e}_1$ is the projection of $\mathbf{x}$ on the

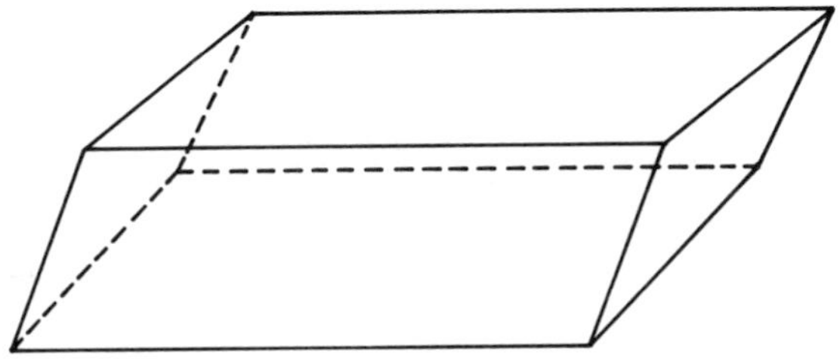

Fig. 6.2

1-axis. Since $\mathbf{e}_1$ is a vector of unit length, x_1 represents the ratio of this projection to 1, so x_1 is the projection of $\mathbf{x}$ on $\mathbf{e}_1$. We note in particular that x_1 is positive, negative or zero according as the angle between the directions of $\mathbf{x}$ and $\mathbf{e}_1$ is less than, greater than or equal to a right angle. In any coordinate system, the solid whose edges are $x_1\mathbf{e}_1, x_2\mathbf{e}_2, x_3\mathbf{e}_3$ has the vector $\mathbf{x}$ as a diagonal; it is called a **3-cell** or also a **parallelepiped** (Fig. 6.2). If the $\mathbf{e}_i$ are mutually orthogonal (so the solid is shaped like a brick), we speak of a **rectangular 3-cell**.

6.3 THE LENGTH OF A VECTOR

Let $\mathbf{a}$ be any vector in 3-space, corresponding to the position vector OA. The **length** of $\mathbf{a}$ is defined as the distance from O to A; it is a positive number or zero which we shall denote by $|\mathbf{a}|$, thus in writing $|\mathbf{a}|$ we ignore the direction of $\mathbf{a}$. This length function has the following properties:

L.1 $|\mathbf{a}| > 0$ *unless* $\mathbf{a} = \mathbf{0}$.
L.2 $|\lambda\mathbf{a}| = |\lambda|\cdot|\mathbf{a}|$.

Here $|\lambda|$ denotes the usual absolute value of the scalar λ. Since the square of any real number is positive or zero, we have $|\lambda|^2 = \lambda^2$ and so we have

$$|\lambda\mathbf{a}|^2 = \lambda^2|\mathbf{a}|^2 \tag{6.1}$$

A vector of unit length is called a **unit vector**. We can give an expression for the length of a vector in terms of its coordinates; for the moment we shall confine ourselves to a rectangular coordinate system, leaving the general case for section 6.7.

If $\mathbf{e}_1, \mathbf{e}_2, \mathbf{e}_3$ are the basis vectors of a rectangular coordinate system and

$$\mathbf{a} = a_1\mathbf{e}_1 + a_2\mathbf{e}_2 + a_3\mathbf{e}_3$$

then the segment OA representing $\mathbf{a}$ forms the diagonal of a rectangular 3-cell (Fig. 6.3) whose edges lie along $\mathbf{e}_1, \mathbf{e}_2$ and $\mathbf{e}_3$ and are of lengths a_1, a_2, a_3, respectively. The square of the length of this diagonal is found by a double application of Pythagoras's theorem, to be the sum of the squares of the

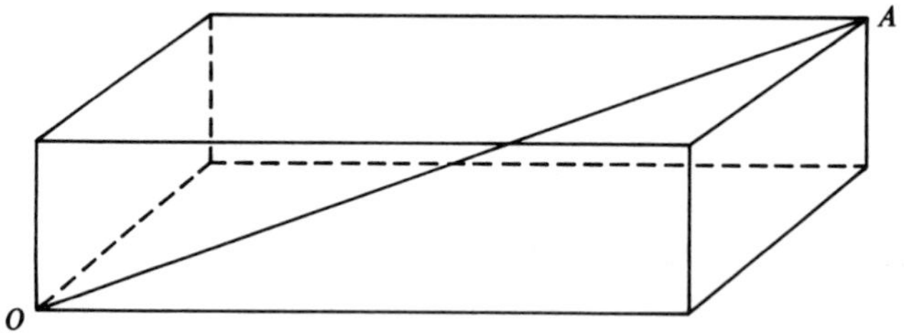

Fig. 6.3

edge lengths, thus

$$|\mathbf{a}|^2 = a_1^2 + a_2^2 + a_3^2 \tag{6.2}$$

6.4 THE SCALAR PRODUCT

With any pair of vectors $\mathbf{a}$ and $\mathbf{b}$ we associate a scalar $\mathbf{a}\cdot\mathbf{b}$ which is defined by the equation

$$\mathbf{a}\cdot\mathbf{b} = \tfrac{1}{2}\{|\mathbf{a}+\mathbf{b}|^2 - |\mathbf{a}|^2 - |\mathbf{b}|^2\} \tag{6.3}$$

This expression $\mathbf{a}\cdot\mathbf{b}$ is called the **scalar product** (or sometimes the **dot product**) of $\mathbf{a}$ and $\mathbf{b}$. It is linear in each factor ('bilinear'), and commutative, i.e.

$$(\lambda\mathbf{a} + \mu\mathbf{b})\cdot\mathbf{c} = \lambda\mathbf{a}\cdot\mathbf{c} + \mu\mathbf{b}\cdot\mathbf{c} \tag{6.4}$$

$$\mathbf{a}\cdot(\lambda\mathbf{b} + \mu\mathbf{c}) = \lambda\mathbf{a}\cdot\mathbf{b} + \mu\mathbf{a}\cdot\mathbf{c} \tag{6.5}$$

$$\mathbf{a}\cdot\mathbf{b} = \mathbf{b}\cdot\mathbf{a} \tag{6.6}$$

Moreover, the length of a vector may be expressed in terms of the scalar product by the equation

$$|\mathbf{a}| = (\mathbf{a}\cdot\mathbf{a})^{\frac{1}{2}} \tag{6.7}$$

The rules (6.6) and (6.7) are almost immediate consequences of the definition (6.3) and L.1–2. To prove (6.4) and (6.5) we first express $\mathbf{a}\cdot\mathbf{b}$ in terms of coordinates referred to a rectangular coordinate system. In such a system, if $\mathbf{a} = \sum a_i\mathbf{e}_i$ and $\mathbf{b} = \sum b_i\mathbf{e}_i$, then $\mathbf{a}+\mathbf{b} = \sum(a_i + b_i)\mathbf{e}_i$ and hence, by (6.2) and (6.3),

$$\mathbf{a}\cdot\mathbf{b} = \tfrac{1}{2}\left\{\sum (a_i + b_i)^2 - \sum a_i^2 - \sum b_i^2\right\}$$

that is,

$$\mathbf{a}\cdot\mathbf{b} = a_1b_1 + a_2b_2 + a_3b_3 \tag{6.8}$$

This equation defines the scalar product in terms of the coordinates in a rectangular coordinate system and from it we can now verify (6.4) and (6.5) without difficulty, simply by writing out both sides. This may be left to the reader.

Two vectors are said to be **orthogonal** if their scalar product is zero.

6.5 VECTORS IN *n* DIMENSIONS

So far we have assumed our vector space to be 3-dimensional, but it is clear how the scalar product can be defined in n dimensions. We fix a basis $\mathbf{u}_1, \ldots, \mathbf{u}_n$ and if $\mathbf{a} = \sum a_i\mathbf{u}_i$, we define the **length** $|\mathbf{a}|$ of $\mathbf{a}$ by the equation $|\mathbf{a}| = (\sum a_i^2)^{1/2}$. Clearly $|\mathbf{a}| \geqslant 0$ with equality if and only if $\mathbf{a} = \mathbf{0}$, because a sum of squares of real numbers, not all zero, is positive. Now the scalar product $\mathbf{a}\cdot\mathbf{b}$ can again be defined by (6.3). Explicitly we have

$$\mathbf{a}\cdot\mathbf{b}=\sum a_i b_i \tag{6.9}$$

and it again satisfies (6.4)–(6.6). We remark that the basis $\mathbf{u}_1,\ldots,\mathbf{u}_n$ used in the definition has the property

$$\mathbf{u}_i\cdot\mathbf{u}_j=\delta_{ij} \tag{6.10}$$

where δ_{ij} is the Kronecker delta. Thus the vectors $\mathbf{u}_1,\ldots,\mathbf{u}_n$ are pairwise orthogonal unit vectors. A basis with this property is again called **orthonormal**, giving rise to a rectangular system of coordinates in n dimensions. We remark that the above definition of the scalar product in n dimensions was based on declaring a given basis to be orthonormal. By contrast, in three dimensions we were able to make use of the intrinsic definition of length.

We note that any orthonormal set of vectors, or more generally, any set of pairwise orthogonal non-zero vectors, must be linearly independent. For if $\mathbf{v}_1,\ldots,\mathbf{v}_r$ are pairwise orthogonal, non-zero and there is a linear dependence

$$\lambda_1\mathbf{v}_1+\cdots+\lambda_r\mathbf{v}_r=\mathbf{0}$$

then on taking the scalar product with $\mathbf{v}_i$ we obtain $\lambda_i(\mathbf{v}_i\cdot\mathbf{v}_i)=0$, and since $\mathbf{v}_i\cdot\mathbf{v}_i\neq 0$ by hypothesis, it follow that $\lambda_i=0$ for $i=1,\ldots,r$, so the given relation was indeed trivial.

Orthonormal bases are particularly convenient to work with. Thus, if $\mathbf{u}_1,\ldots,\mathbf{u}_n$ is an orthonormal basis of our space, to express any vector $\mathbf{x}$ in terms of the **u**s, we know that $\mathbf{x}=\sum\alpha_i\mathbf{u}_i$ for some scalars α_i. To find α_i we take the scalar product with $\mathbf{u}_j$ and find

$$\mathbf{x}\cdot\mathbf{u}_j=\sum_i\alpha_i(\mathbf{u}_i\cdot\mathbf{u}_j)$$

and this sum reduces to the single term α_j because $\mathbf{u}_i\cdot\mathbf{u}_j=\delta_{ij}$. Hence $\alpha_j=\mathbf{x}\cdot\mathbf{u}_j$ and we obtain the following formula for any vector $\mathbf{x}$:

$$\mathbf{x}=\sum_i(\mathbf{x}\cdot\mathbf{u}_i)\mathbf{u}_i \tag{6.11}$$

Suppose more generally that we have an orthonormal set of vectors $\mathbf{u}_1,\ldots,\mathbf{u}_r$, not necessarily a basis, and that we are looking for the linear combination of **u**s closest to a given vector $\mathbf{x}$. If we denote the difference by $\mathbf{x}'$, then

$$\mathbf{x}'=\mathbf{x}-\sum\alpha_i\mathbf{u}_i \tag{6.12}$$

where the αs are to be determined. Now we have

$$\begin{aligned}0\leqslant|\mathbf{x}'|^2&=|\mathbf{x}|^2-2\sum\alpha_i(\mathbf{x}\cdot\mathbf{u}_i)+\sum\alpha_i^2\\&=|\mathbf{x}|^2-\sum(\mathbf{x}\cdot\mathbf{u}_i)^2+\sum[(\mathbf{x}\cdot\mathbf{u}_i)-\alpha_i]^2\end{aligned}$$

This relation shows first, that the vector $\mathbf{x}'$ given by (6.12) is shortest when the last sum is zero, and so $\alpha_i = \mathbf{x}\cdot\mathbf{u}_i$; second, on giving this value to α_i we obtain the inequality

$$\sum(\mathbf{x}\cdot\mathbf{u}_i)^2 \leqslant |\mathbf{x}|^2, \tag{6.13}$$

known as **Bessel's inequality** (after the astronomer Friedrich Wilhelm Bessel, 1784–1864), which becomes an equality whenever the $\mathbf{u}_i$ form a basis. For then the difference $\mathbf{x}'$ can be made zero; in any case this equality also follows by taking lengths in (6.11). Bessel's inequality plays a role mainly in infinite-dimensional spaces, where the corresponding equality can be used to recognize bases of the space.

Inserting the values of α_i found, we obtain

$$\mathbf{x} = \sum(\mathbf{x}\cdot\mathbf{u}_i)\mathbf{u}_i + \mathbf{x}' \tag{6.14}$$

and $\mathbf{x}'$ is orthogonal to all the $\mathbf{u}_i$, as is easily verified. This shows again that for an orthonormal basis $\mathbf{u}_1,\ldots,\mathbf{u}_n$ we have $\mathbf{x}' = \mathbf{0}$ and (6.14) reduces to (6.11).

6.6 THE CONSTRUCTION OF ORTHONORMAL BASES

In section 1.9 we saw that from any set of vectors spanning a subspace of our space V we can select a basis of V. By taking suitable linear combinations we can ensure that we get an orthonormal basis of V; the method is known as the **Gram-Schmidt process** (Jörgen Pedersen Gram, 1850–1916; Erhard Schmidt, 1876–1959).

Let $\mathbf{u}_1,\ldots,\mathbf{u}_n$ be a basis of V. Then $\mathbf{u}_1 \neq \mathbf{0}$ and we can define

$$\mathbf{v}_1 = \frac{1}{|\mathbf{u}_1|}\cdot\mathbf{u}_1$$

and so obtain a unit vector $\mathbf{v}_1$ in the direction of $\mathbf{u}_1$. Suppose we have constructed pairwise orthogonal unit vectors $\mathbf{v}_1,\ldots,\mathbf{v}_{k-1}$ as linear combinations of $\mathbf{u}_1,\ldots,\mathbf{u}_{k-1}$. To find $\mathbf{v}_k$ we put

$$\mathbf{u}'_k = \mathbf{u}_k - \sum_{i=1}^{k-1}(\mathbf{u}_k\cdot\mathbf{v}_i)\mathbf{v}_i \tag{6.15}$$

Then $\mathbf{u}'_k \neq \mathbf{0}$, for otherwise $\mathbf{u}_k$ would depend linearly on $\mathbf{v}_1,\ldots,\mathbf{v}_{k-1}$ and hence on $\mathbf{u}_1,\ldots,\mathbf{u}_{k-1}$, which contradicts the fact that the **u**s form a basis. Now put

$$\mathbf{v}_k = \frac{1}{|\mathbf{u}'_k|}\mathbf{u}'_k$$

Then $\mathbf{v}_k$ is a unit vector in the direction of $\mathbf{u}'_k$ and so is orthogonal to $\mathbf{v}_1,\ldots,\mathbf{v}_{k-1}$, by (6.15). Thus $\mathbf{v}_1,\ldots,\mathbf{v}_k$ are pairwise orthogonal unit vectors spanning the same subspace as $\mathbf{u}_1,\ldots,\mathbf{u}_k$. By induction on k this holds for all $k \leqslant n$, and for $k = n$ it provides us with the required orthonormal basis.

For example, if we are given the vectors $\mathbf{u}_1 = (1,1,1)$, $\mathbf{u}_2 = (1,2,0)$, $\mathbf{u}_3 =$

$(1,0,1)$, then $\mathbf{v}_1 = (1/\sqrt{3})\mathbf{u}_1, \mathbf{u}_2' = \mathbf{u}_2 - (\mathbf{u}_2 \cdot \mathbf{v}_1)\mathbf{v}_1 = (1,2,0) - (3/3)(1,1,1) = (0,1,-1)$, $\mathbf{v}_2 = (1/\sqrt{2})(0,1,-1)$, $\mathbf{u}_3' = (1,0,1) - (2/3)(1,1,1) + (1/2)(0,1,-1) = (1/6)(2,-1,-1)$. We now have to normalize $\mathbf{u}_3'$; for simplicity we take $6\mathbf{u}_3' = (2,-1,-1)$, which gives $\mathbf{v}_3 = (1/\sqrt{6})(2,-1,-1)$. Thus the required orthonormal basis is

$$(1/\sqrt{3})(1,1,1), \qquad (1/\sqrt{2})(0,1,-1), \qquad (1/\sqrt{6})(2,-1,-1).$$

If we are given a set of vectors $\mathbf{u}_1, \ldots, \mathbf{u}_r$ spanning V, not necessarily a basis, we can use the same procedure as before to construct an orthonormal basis, but now it may happen that some $\mathbf{u}_k' = \mathbf{0}$, namely whenever $\mathbf{u}_k$ is a linear combination of $\mathbf{u}_1, \ldots, \mathbf{u}_{k-1}$. In that case we discard $\mathbf{u}_k'$ and continue with $\mathbf{u}_{k+1}$ as before.

6.7 THE CAUCHY–SCHWARZ INEQUALITY

Equation (6.8) may also be used to give a geometrical interpretation of the scalar product in two and three dimensions. First, we note that (6.8) holds in any rectangular coordinate system, because this is true of (6.2), from which (6.8) was derived. Now given any two vectors $\mathbf{a}$ and $\mathbf{b}$, we can choose a rectangular coordinate system in which the first basis vector $\mathbf{e}_1$ lies in the same direction as $\mathbf{a}$. Then $\mathbf{a} = |\mathbf{a}| \cdot \mathbf{e}_1$ and so $\mathbf{a} \cdot \mathbf{b} = |\mathbf{a}|(\mathbf{e}_1 \cdot \mathbf{b}) = |\mathbf{a}| \cdot b_1$; here b_1 represents the projection of $\mathbf{b}$ on the direction of $\mathbf{a}$, that is, $b_1 = |\mathbf{b}| \cdot \cos\theta$, where θ is the angle between $\mathbf{a}$ and $\mathbf{b}$ (Fig. 6.4). Thus

$$\mathbf{a} \cdot \mathbf{b} = |\mathbf{a}| \cdot |\mathbf{b}| \cdot \cos\theta \tag{6.16}$$

We note that $\mathbf{a} \cdot \mathbf{b}$ is positive when the directions of $\mathbf{a}$ and $\mathbf{b}$ make an acute angle and negative when they make an obtuse angle. In particular: $\mathbf{a} \cdot \mathbf{b} = 0$ precisely when $\mathbf{a} = \mathbf{0}$ or $\mathbf{b} = \mathbf{0}$ or $\mathbf{a}$ and $\mathbf{b}$ are at right angles to each other. Thus it is possible for $\mathbf{a} \cdot \mathbf{b}$ to be zero even though neither $\mathbf{a}$ nor $\mathbf{b}$ is the zero vector. Another important difference from ordinary products is that in the product $\mathbf{a} \cdot \mathbf{b}$ the factors are *vectors*, but the result is a *scalar*.

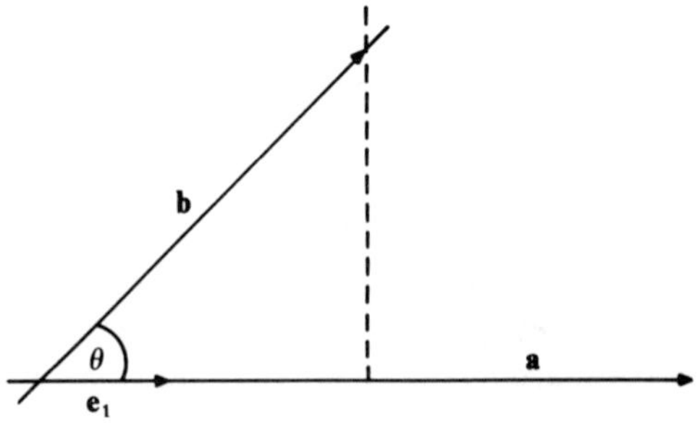

Fig. 6.4

If we recall that $|\cos\theta| \leqslant 1$ for any real θ, we obtain from (6.16) the inequality

$$(\mathbf{a}\cdot\mathbf{b})^2 \leqslant |\mathbf{a}|^2\cdot|\mathbf{b}|^2 \tag{6.17}$$

This is known as the **Cauchy–Schwarz inequality** (after Augustin Louis Cauchy, 1789–1857; and Hermann Amandus Schwarz, 1843–1921). But this inequality can also be derived without assuming any properties of the cosine. Expression (6.17) clearly holds (with equality) for $\mathbf{b} = \mathbf{0}$, so we may assume that $\mathbf{b} \neq \mathbf{0}$. Now take $\lambda \in \mathbf{R}$; by L.1 and (6.7) we have

$$(\mathbf{a} - \lambda\mathbf{b})\cdot(\mathbf{a} - \lambda\mathbf{b}) \geqslant 0$$

with equality precisely when $\mathbf{a} = \lambda\mathbf{b}$. Expanding the left-hand side and using (6.6), we obtain

$$|\mathbf{a}|^2 - 2\lambda(\mathbf{a}\cdot\mathbf{b}) + \lambda^2|\mathbf{b}|^2 \geqslant 0$$

If in this relation we put $\lambda = (\mathbf{a}\cdot\mathbf{b})/|\mathbf{b}|^2$ and multiply by $|\mathbf{b}|^2$, we find

$$|\mathbf{a}|^2\cdot|\mathbf{b}|^2 - 2(\mathbf{a}\cdot\mathbf{b})^2 + (\mathbf{a}\cdot\mathbf{b})^2 \geqslant 0$$

and now a rearrangement leads to (6.17). From this proof we see that equality holds in (6.17) precisely when $\mathbf{a} = \lambda\mathbf{b}$ for some λ or $\mathbf{b} = \mathbf{0}$, that is, if and only if $\mathbf{a}$ and $\mathbf{b}$ are linearly dependent.

6.8 THE LENGTH OF A VECTOR IN GENERAL COORDINATES

We can now give an expression for the length of a vector in a general (oblique) coordinate system, corresponding to the expression (6.2) for a rectangular system. Let $\mathbf{v}_1, \ldots, \mathbf{v}_n$ be a general basis and write

$$\gamma_{ij} = \mathbf{v}_i\cdot\mathbf{v}_j \qquad (i,j = 1,\ldots,n) \tag{6.18}$$

Of the n^2 constants γ_{ij} only $n(n+1)/2$ are independent, because $\gamma_{ij} = \gamma_{ji}$, by (6.6). The γs are determined by the lengths of the vectors $\mathbf{v}_i$ and the angles between any two of these vectors (cf. (6.16)). Using (6.4) and (6.5) we now find that for vectors $\mathbf{a}, \mathbf{b}$ with coordinates given by $\mathbf{a} = \sum a_i\mathbf{v}_i$, $\mathbf{b} = \sum b_i\mathbf{v}_i$, we have

$$\mathbf{a}\cdot\mathbf{b} = \sum_{ij} \gamma_{ij}a_ib_j \tag{6.19}$$

and in particular,

$$|\mathbf{a}|^2 = \mathbf{a}\cdot\mathbf{a} = \sum \gamma_{ij}a_ia_j$$

For $n = 3$ this equation, when written out in full, reads

$$|\mathbf{a}|^2 = \gamma_{11}a_1^2 + \gamma_{22}a_2^2 + \gamma_{33}a_3^2 + 2\gamma_{12}a_1a_2 + 2\gamma_{13}a_1a_3 + 2\gamma_{23}a_2a_3 \tag{6.20}$$

This is the generalization of (6.2) to general coordinate systems. In a rectangular system we have $\mathbf{v}_i \cdot \mathbf{v}_j = \delta_{ij}$ and with these values for γ_{ij}, (6.20) reduces to (6.2), while (6.19) reduces to (6.9). Because of these simple forms taken by (6.19) and (6.20) in rectangular coordinate systems we shall usually choose our system to be rectangular in any problem involving scalar products or the lengths of vectors.

6.9 THE VECTOR PRODUCT IN 3-SPACE

In three dimensions there is another type of product that is often used, the vector product. It arises when we determine the common perpendicular to two 3-vectors. Given two vectors **a** and **b**, the vector **x** is at right angles to both **a** and **b** precisely when $\mathbf{a} \cdot \mathbf{x} = \mathbf{b} \cdot \mathbf{x} = 0$. If in a rectangular coordinate system, $\mathbf{a} \leftrightarrow (a_1, a_2, a_3)$, $\mathbf{b} \leftrightarrow (b_1, b_2, b_3)$, then the coordinates (x_1, x_2, x_3) of **x** have to satisfy the equations

$$\begin{aligned} a_1x_1 + a_2x_2 + a_3x_3 &= 0 \\ b_1x_1 + b_2x_2 + b_3x_3 &= 0 \end{aligned} \tag{6.21}$$

Solving these equations for x_1, x_2, x_3, we obtain

$$\frac{x_1}{a_2b_3 - a_3b_2} = \frac{x_2}{a_3b_1 - a_1b_3} = \frac{x_3}{a_1b_2 - a_2b_1} \tag{6.22}$$

or equivalently,

$$\begin{aligned} x_1 &= \lambda(a_2b_3 - a_3b_2) \\ x_2 &= \lambda(a_3b_1 - a_1b_3) \\ x_3 &= \lambda(a_1b_2 - a_2b_1) \end{aligned} \tag{6.23}$$

where λ is an arbitrary scalar. This is the complete solution of the system (6.21), provided that the system is of rank 2 (by Theorem 4.1), which happens precisely when **a** and **b** are linearly independent. The form of the solutions (6.23) shows that this is so if at least one of the denominators in (6.22) is different from zero; they are, of course, just the second-order minors that can be formed from the matrix with columns **a**, **b**.

Let us denote by $\mathbf{a} \wedge \mathbf{b}$ the vector whose components are the terms multiplying λ in (6.23):

$$\mathbf{a} \wedge \mathbf{b} \leftrightarrow (a_2b_3 - a_3b_2, a_3b_1 - a_1b_3, a_1b_2 - a_2b_1) \tag{6.24}$$

Using the Kronecker ε function from section 5.6, we can write the result as follows, if $\mathbf{a} \wedge \mathbf{b} \leftrightarrow (c_1, c_2, c_3)$:

$$c_i = \sum \varepsilon(ijk) a_j b_k$$

We can state our result thus: For any two vectors **a** and **b** in three dimensions we can form the product defined by (6.24) (in a rectangular coordinate system), such that $\mathbf{a} \wedge \mathbf{b} = \mathbf{0}$ if and only if **a** and **b** are linearly dependent, and when they are independent, then every vector orthogonal to **a** and **b** is of the form $\lambda(\mathbf{a} \wedge \mathbf{b})$.

This result also has an immediate geometrical interpretation. To say that **a** and **b** are linearly independent just amounts to saying that **a** and **b** are not parallel, and so **a** and **b** define a plane in space. Now the above result shows that there is just one direction in space perpendicular to this plane.

The vector $\mathbf{a} \wedge \mathbf{b}$ is called the **vector product** of **a** and **b**; sometimes it is called the **cross product** (and is then denoted by $\mathbf{a} \times \mathbf{b}$). Its main properties are again bilinearity:

$$\begin{aligned}(\lambda\mathbf{a} + \mu\mathbf{b}) \wedge \mathbf{c} &= \lambda\mathbf{a} \wedge \mathbf{c} + \mu\mathbf{b} \wedge \mathbf{c} \\ \mathbf{a} \wedge (\lambda\mathbf{b} + \mu\mathbf{c}) &= \lambda\mathbf{a} \wedge \mathbf{b} + \mu\mathbf{a} \wedge \mathbf{c}\end{aligned} \tag{6.25}$$

and **anticommutativity**, expressed by the equation

$$\mathbf{a} \wedge \mathbf{b} = -\mathbf{b} \wedge \mathbf{a} \tag{6.26}$$

In particular, we also have the relation

$$\mathbf{a} \wedge \mathbf{a} = \mathbf{0} \tag{6.27}$$

These relations are an immediate consequence of the definition (6.24) and their verification may be left to the reader. To find the length of the vector $\mathbf{a} \wedge \mathbf{b}$, we have in a rectangular coordinate system,

$$\begin{aligned}|\mathbf{a} \wedge \mathbf{b}|^2 &= (a_2b_3 - a_3b_2)^2 + (a_3b_1 - a_1b_3)^2 + (a_1b_2 - a_2b_1)^2 \\ &= a_2^2b_3^2 + a_3^2b_2^2 + a_3^2b_1^2 + a_1^2b_3^2 + a_1^2b_2^2 + a_2^2b_1^2 \\ &\quad - 2(a_2b_2a_3b_3 + a_1b_1a_3b_3 + a_1b_1a_2b_2) \\ &= (a_1^2 + a_2^2 + a_3^2)(b_1^2 + b_2^2 + b_3^2) \\ &\quad - (a_1^2b_1^2 + a_2^2b_2^2 + a_3^2b_3^2 + 2a_1b_1a_2b_2 + 2a_1b_1a_3b_3 + 2a_2b_2a_3b_3) \\ &= |\mathbf{a}|^2 \cdot |\mathbf{b}|^2 - (\mathbf{a} \cdot \mathbf{b})^2\end{aligned}$$

Hence by (6.16),

$$\begin{aligned}|\mathbf{a} \wedge \mathbf{b}|^2 &= |\mathbf{a}|^2|\mathbf{b}|^2 - |\mathbf{a}|^2|\mathbf{b}|^2\cos^2\theta \\ &= |\mathbf{a}|^2|\mathbf{b}|^2\sin^2\theta\end{aligned}$$

and so

$$|\mathbf{a} \wedge \mathbf{b}| = |\mathbf{a}| \cdot |\mathbf{b}| \cdot |\sin\theta|, \tag{6.28}$$

where θ is the angle between **a** and **b**. The vectors **a** and **b** form a parallelogram whose base has length $|\mathbf{a}|$, while its height is $|\mathbf{b}|\sin\theta$; since its area is the length of the base times the height, we see that the length of the vector $\mathbf{a} \wedge \mathbf{b}$ is just the area of the parallelogram spanned by **a** and **b**. The sign of $\mathbf{a} \wedge \mathbf{b}$ also has a significance, which will be explained in section 7.4. Although the vector product was defined in terms of a particular coordinate system, we see from (6.28) that its length is independent of the choice of coordinate system (so long as it is a rectangular one). We already know that $\mathbf{a} \wedge \mathbf{b}$ is at right angles to **a** and **b**; therefore if (6.24) is applied in different rectangular coordinate systems, the resulting vectors can differ at most by a factor -1. In section 7.3 we shall restrict the coordinate systems (to right-handed rectangular systems) so as to avoid this ambiguity in sign. At the same time we shall show how to compute the vector product in general coordinate systems.

In n dimensions ($n > 3$) the vector product has to be modified; now $\mathbf{a} \wedge \mathbf{b}$ with components $a_i b_j - a_j b_i$ is no longer a vector, as a component count already shows, but it can be defined as a 'multi-vector' with $n(n-1)/2$ components. This leads to the exterior algebra on a vector space (cf. Cohn, 1991, section 3.4). This algebra and the related Grassmann algebra were introduced by Grassmann in his calculus of extension in 1844 but not taken up until much later. Independently, Hamilton for many years tried to find a multiplication of number triples analogous to the multiplication of number pairs provided by their interpretation as complex numbers. He finally succeeded in 1843 by his discovery of **quaternions**. These are defined as 4-vectors with the multiplication rule

$$\begin{aligned}(x_0, x_1, x_2, x_3)\cdot(y_0, y_1, y_2, y_3) \\ = (x_0y_0 - x_1y_1 - x_2y_2 - x_3y_3, x_0y_1 + x_1y_0 + x_2y_3 - x_3y_2, \\ x_0y_2 - x_1y_3 + x_2y_0 + x_3y_1, x_0y_3 + x_1y_2 - x_2y_1 + x_3y_0) \quad (6.29)\end{aligned}$$

The 0-component is called the **real part** and the other three form the **vector part**. Using the basis vectors $\mathbf{1} = (1, 0, 0, 0)$, $\mathbf{i} = (0, 1, 0, 0)$, $\mathbf{j} = (0, 0, 1, 0)$, $\mathbf{k} = (0, 0, 0, 1)$, we can also define the multiplication by the equations

$$\begin{aligned}\mathbf{ij} = -\mathbf{ji} = \mathbf{k} \\ \mathbf{jk} = -\mathbf{kj} = \mathbf{i} \\ \mathbf{ki} = -\mathbf{ik} = \mathbf{j}\end{aligned}$$

together with linearity. Thus he had to go into four dimensions as well as give up commutativity. The product (6.29) includes both the vector and scalar product. For if $\mathbf{x} = (x_1, x_2, x_3)$ and $\mathbf{y} = (y_1, y_2, y_3)$ are any 3-vectors, then by interpreting them as 4-vectors with real part zero, we find their product to have real part $-\mathbf{x}\cdot\mathbf{y}$ and vector part $\mathbf{x} \wedge \mathbf{y}$.

The vector product in its present form was introduced by Josiah Willard Gibbs (1839–1903).

6.10 LINES IN 3-SPACE

Let us now see how vectors can be used to study lines and planes in space. The different lines through a given point, O say, in 3-space differ only in their directions and may therefore be specified by a vector along them. A given line l through O is completely determined by laying a non-zero vector **a** along it; the vector **a** determines l, but **a** is determined only up to a non-zero scalar multiple by l. This means that l determines and is determined by the ratios $a_1:a_2:a_3$ of the components of $\mathbf{a}$ (in some coordinate system). For this reason the numbers a_1, a_2, a_3 are called the **direction ratios** of the line l in the given coordinate system.

Now, strictly speaking, there are two directions defined on each line l; if one of them is given by the vector $\mathbf{a}$, the other is given by $-\mathbf{a}$, and the vector $\lambda\mathbf{a}$ defines one or the other according as λ is positive or negative. For definiteness we shall single out one of these directions on l as the **positive direction** and denote the line with this direction by $\vec{l}$; we say that l has been **oriented**. In this way every line gives rise to two **oriented lines** which consist of the same line but with opposite directions. An oriented line may be thought of as a line with an arrow along it, pointing in the positive direction; this direction may also be defined by a unit vector along $\vec{l}$. If $\mathbf{a}$ is a non-zero vector along $\vec{l}$, then the unit-vector $\mathbf{u}$ describing $\vec{l}$ is given by

$$\mathbf{u} = \frac{1}{|\mathbf{a}|}\mathbf{a} \tag{6.30}$$

Let $\mathbf{u}$ be a unit vector corresponding to the oriented line $\vec{l}$ through O, and take a rectangular coordinate system with origin at O. By taking the scalar product of $\mathbf{u}$ with the basis vectors $\mathbf{e}_1, \mathbf{e}_2$ and $\mathbf{e}_3$ in turn, we find

$$u_i = \mathbf{u}\cdot\mathbf{e}_i = \cos\alpha_i \quad (i = 1, 2, 3) \tag{6.31}$$

where α_i is the angle which $\vec{l}$ makes with the direction of $\mathbf{e}_i$. The angles α_1, α_2, α_3 are called the **direction angles** and their cosines the **direction cosines** of the line $\vec{l}$. Each direction angle is uniquely determined by its cosine, because it must lie between 0° and 180° and this corresponds to a variation of its cosine from $+1$ to -1. To sum up, an oriented line through O may be determined by its direction angles, or by its direction cosines or, ignoring orientation, by its direction ratios.

Any three numbers, not all zero, represent the direction ratios of some line, because every non-zero vector determines a line. If $\mathbf{a}$ is a non-zero vector and $\vec{l}$ the oriented line through O determined by $\mathbf{a}$, then we obtain the direction cosines of $\vec{l}$ by **normalizing a**, that is to say, by dividing it by the scalar $|\mathbf{a}|$ (as in (6.30)) so as to obtain a unit vector along l. Explicitly, the components of this unit vector $\mathbf{u}$ (in a rectangular coordinate system) are

$$u_i = a_i(a_1^2 + a_2^2 + a_3^2)^{-1/2} \quad (i = 1, 2, 3)$$

For example, the line with the direction ratios $1:2:-3$ has direction cosines $1/\sqrt{14}$, $2/\sqrt{14}$, $-3/\sqrt{14}$ and angles 74.5°, 57.7° and 143.3°. The same line with the other orientation has direction cosines $-1/\sqrt{14}$, $-2/\sqrt{14}$, $3/\sqrt{14}$ and direction angles 105.5°, 122.3° and 36.7°.

Any three numbers u_1, u_2, u_3 such that $u_1^2 + u_2^2 + u_3^2 = 1$ may be taken as the direction cosines of some line, namely the oriented line in the direction of the vector $\mathbf{u} = \sum u_i \mathbf{e}_i$. From the interpretation of the u_i as cosines we see that the angles $\alpha_1, \alpha_2, \alpha_3$ which a line makes with the axes of a rectangular coordinate system satisfy the relation

$$\cos^2 \alpha_1 + \cos^2 \alpha_2 + \cos^2 \alpha_3 = 1 \tag{6.32}$$

Conversely, any three angles whose cosines satisfy this relation form the direction angles of an oriented line.

6.11 THE EQUATION OF A LINE IN SPACE

Next we wish to describe the different points on a line l. We suppose first that l passes through the origin of the coordinate system (which need not be rectangular). If $\mathbf{a}$ is a non-zero vector along l, then the points represented by the vectors $\lambda\mathbf{a}$, where λ runs over the scalars, all lie on l and we obtain all the points of the line l by varying λ. We also say that l is **spanned** by the vector $\mathbf{a}$. Thus the equation

$$\mathbf{x} = \lambda\mathbf{a} \tag{6.33}$$

may be regarded as an equation for the line l. Here $\mathbf{a}$ is any non-zero vector along l; the scalar λ is called the **parameter** and (6.33) is the **parametric form** of the equation of the line. For example, the coordinate axes have the equations $\mathbf{x} = \lambda\mathbf{e}_1$, $\mathbf{x} = \lambda\mathbf{e}_2$ and $\mathbf{x} = \lambda\mathbf{e}_3$, where $\mathbf{e}_1, \mathbf{e}_2, \mathbf{e}_3$ is the standard basis of vectors.

To represent a line l which does not pass through O (Fig. 6.5) we choose any point P on l; let $\mathbf{p}$ be the vector OP and let $\mathbf{a}$ be a non-zero vector in the direction of l. Then for any scalar λ, $\mathbf{p} + \lambda\mathbf{a}$ represents a point on l, and any point on l may be represented in this way.

Thus an equation for the line is

$$\mathbf{x} = \mathbf{p} + \lambda\mathbf{a} \tag{6.34}$$

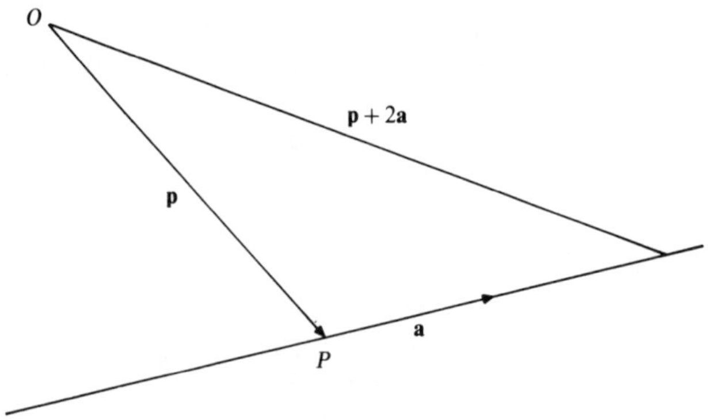

Fig. 6.5

Written out in terms of coordinates, this equation becomes

$$\begin{aligned} x_1 &= p_1 + \lambda a_1 \\ x_2 &= p_2 + \lambda a_2 \\ x_3 &= p_3 + \lambda a_3 \end{aligned} \tag{6.35}$$

which on elimination of the parameter λ leads to the form

$$\frac{x_1 - p_1}{a_1} = \frac{x_2 - p_2}{a_2} = \frac{x_3 - p_3}{a_3} (= \lambda) \tag{6.36}$$

for the scalar equations of the straight line through the point P with the coordinates (p_1, p_2, p_3) and with the direction ratios (a_1, a_2, a_3). Here the p_i and a_i are given constants to determine the line, while x_1, x_2, x_3 are the coordinates of a general point on the line. It is then convenient to abbreviate equations (6.35) by

$$\mathbf{x} \leftrightarrow (p_1, p_2, p_3) + \lambda(a_1, a_2, a_3) \tag{6.37}$$

in the notation of section 6.2, where, of course, a fixed coordinate system is understood. For example, the line through the point $(-2, 0, 5)$ with the direction ratios $1:2:-3$ is given by

$$\mathbf{x} \leftrightarrow (-2, 0, 5) + \lambda(1, 2, -3)$$

or, in scalar form,

$$\frac{x_1 + 2}{1} = \frac{x_2}{2} = \frac{x_3 - 5}{-3}$$

A line may also be specified by two of its points and it is not difficult to obtain an equation for it from these data. Thus let P and Q be two distinct points on the line l, whose position vectors relative to O are $\mathbf{p}$ and $\mathbf{q}$, respectively. Since $P \neq Q$, we have $\mathbf{p} \neq \mathbf{q}$ and $\mathbf{q} - \mathbf{p}$ is a vector in the direction of l. Hence an equation for l is

$$\mathbf{x} = \mathbf{p} + \lambda(\mathbf{q} - \mathbf{p}) \tag{6.38}$$

We note that P and Q are given by the values 0 and 1, respectively, of the parameter λ. As an example, the line through the points $(4, -1, 2)$ and $(3, 1, 0)$ is given by

$$\mathbf{x} \leftrightarrow (4, -1, 2) + \lambda(-1, 2, -2)$$

Equation (6.38) may also be written as

$$\mathbf{x} = \lambda\mathbf{p} + \mu\mathbf{q},$$

where $\lambda + \mu = 1$. For given values of λ and μ it represents a point R on the line PQ dividing the segment PQ in the ratio $\mu:\lambda$. For example, if $\lambda = \frac{1}{4}$, $\mu = \frac{3}{4}$, then R lies between P and Q and is 3 times as far from P as from Q.

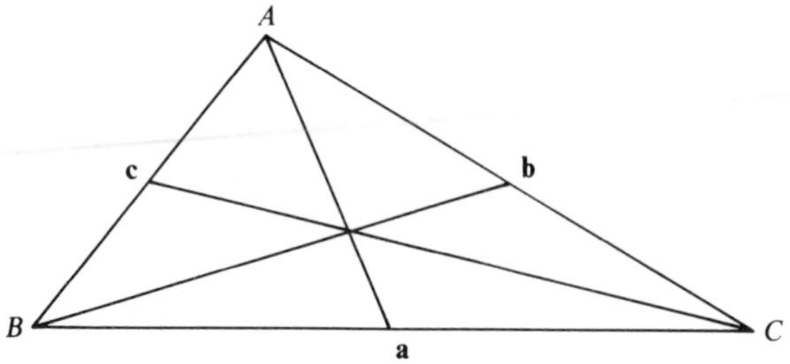

Fig. 6.6

If $\lambda < 0$, then R lies beyond P, so that P is between R and Q, while for $\lambda > 1$, $\mu < 0$ and Q lies between P and R.

As an application, let us use vectors to show that the three medians of a triangle meet in a point (Fig. 6.6). Let ABC be any triangle, with position vectors $\mathbf{a}, \mathbf{b}, \mathbf{c}$ for its vertices. The mid-points of AB, BC, CA are $\frac{1}{2}(\mathbf{a}+\mathbf{b})$, $\frac{1}{2}(\mathbf{b}+\mathbf{c})$, $\frac{1}{2}(\mathbf{c}+\mathbf{a})$. The median through A is the line through the points with position vectors $\mathbf{a}$, $\frac{1}{2}(\mathbf{b}+\mathbf{c})$, so the general point on the median has position vector $\lambda\mathbf{a} + \frac{1-\lambda}{2}(\mathbf{b}+\mathbf{c})$. We note that for $\lambda = 1/3$ this is the point M whose position vector has the symmetric form $\frac{1}{3}(\mathbf{a}+\mathbf{b}+\mathbf{c})$. The median through B is given by $\mu\mathbf{b} + \frac{1-\mu}{2}(\mathbf{c}+\mathbf{a})$, which again represents M for $\mu = 1/3$, and similarly for the median through C. Thus the three medians of the triangle meet in M, with coordinates $\frac{1}{3}(\mathbf{a}+\mathbf{b}+\mathbf{c})$. This form also shows that M divides the segment from A to the mid-point of BC in the ratio 2:1, and likewise for the other medians.

6.12 THE EQUATION OF A PLANE IN SPACE

To represent a plane in space, taking first the case of a plane through the origin O, we pick two linearly independent vectors in the plane, **a** and **b** say, and form all linear combinations of **a** and **b**. In this way we obtain as an equation for the plane

$$\mathbf{x} = \lambda\mathbf{a} + \mu\mathbf{b}, \tag{6.39}$$

where λ and μ are scalar parameters, the 'coordinates' of the plane relative to **a** and **b** as basis vectors.

To represent a general plane (not necessarily passing through O), we take any two linearly independent vectors **a** and **b** defining directions in the plane and the vector **p** of any point P of the plane, relative to O. The general point of the plane is then

$$\mathbf{x} = \mathbf{p} + \lambda\mathbf{a} + \mu\mathbf{b} \tag{6.40}$$

Conversely, any equation of the form (6.40), where **a** and **b** are linearly independent vectors, represents a plane. We shall refer to it as the plane through the point P with position vector **p** and **spanned** by the vectors **a** and **b**.

If P, Q, R are three non-collinear points with vectors $\mathbf{p}, \mathbf{q}, \mathbf{r}$, respectively, then $\mathbf{q} - \mathbf{p}$ and $\mathbf{r} - \mathbf{p}$ are linearly independent vectors; therefore an equation for the plane through P, Q, R is

$$\mathbf{x} = \mathbf{p} + \lambda(\mathbf{q} - \mathbf{p}) + \mu(\mathbf{r} - \mathbf{p})$$

For this is satisfied by $\mathbf{p}, \mathbf{q}, \mathbf{r}$ for the values $(0,0)$, $(1,0)$, $(0,1)$ of (λ, μ), respectively.

Equation (6.40) for the plane may be given in a more convenient form by eliminating the parameters λ and μ. Let **u** be a unit vector perpendicular to the plane. Taking the scalar product of (6.40) with **u** we find

$$\mathbf{u} \cdot \mathbf{x} = \mathbf{u} \cdot \mathbf{p} \tag{6.41}$$

the other terms vanishing because **a** and **b** lie in the plane and so are orthogonal to **u**. It follows that every vector **x** satisfying (6.40) must satisfy (6.41). Conversely, if **x** satisfies (6.41), then $\mathbf{x} - \mathbf{p}$ is orthogonal to **u**, hence it lies in the plane and so it is a linear combination of **a** and **b**, that is, it can be written in the form (6.40). We have thus shown that (6.40) is equivalent to (6.41), so that (6.41) may also be regarded as an equation for the plane. To obtain a unit vector perpendicular to the plane we may form $\mathbf{a} \wedge \mathbf{b}$ and divide by $|\mathbf{a} \wedge \mathbf{b}|$, which does not vanish because **a** and **b** are linearly independent.

Assume now that we have a rectangular coordinate system. Using coordinates, we can write (6.41) as

$$u_1 x_1 + u_2 x_2 + u_3 x_3 = k \tag{6.42}$$

where $k = \mathbf{u} \cdot \mathbf{p}$. The coefficients in (6.42) may be interpreted as follows (cf. Fig. 6.7): since **u** is a unit-vector, perpendicular to the plane, its coordinates

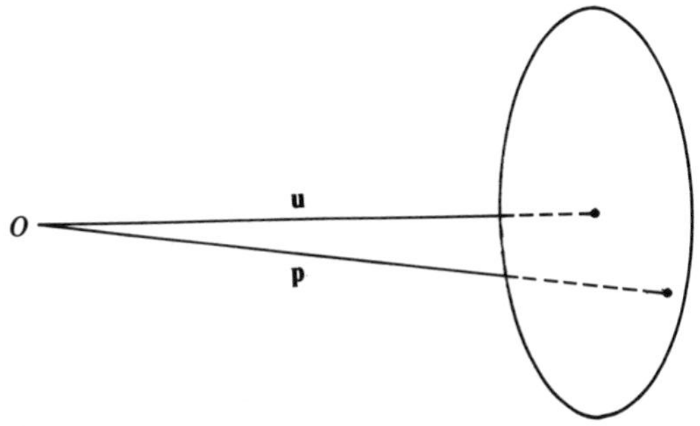

Fig. 6.7

u_1, u_2, u_3 are the direction cosines of the perpendicular. Further, $\mathbf{u}\cdot\mathbf{p}$ represents the projection of the vector $\mathbf{p}$ on $\mathbf{u}$, that is, the distance of the origin from the plane. The unique direction perpendicular to the plane is also called the direction **normal to the plane**.

The only restriction on the coefficients in (6.42) is that $u_1^2 + u_2^2 + u_3^2 = 1$. This shows that any equation

$$c_1x_1 + c_2x_2 + c_3x_3 = c \tag{6.43}$$

where c_1, c_2, c_3 are not all zero, represents a plane. For on dividing the whole equation by $(c_1^2 + c_2^2 + c_3^2)^{1/2}$ we can reduce it to the form (6.42). From the point of view of the theory of linear equations of Chapter 4 the condition that not all of c_1, c_2, c_3 vanish just represents the condition that the system (6.43) has rank 1, and hence has a solution with $3 - 1 = 2$ parameters.

As an example, the equation

$$2x_1 + x_2 - 5x_3 = 3$$

may be put in the form (6.42) by dividing by $(2^2 + 1^2 + 5^2)^{1/2} = \sqrt{30}$. It represents the plane whose normal has direction cosines $2/\sqrt{30}$, $1/\sqrt{30}$, $-5/\sqrt{30}$ and which is at a distance $3/\sqrt{30} = 0.548$ from the origin.

If in (6.43), $c \neq 0$, which happens precisely when the plane represented does not pass through the origin, then we can divide by c and obtain the form

$$a_1x_1 + a_2x_2 + a_3x_3 = 1 \tag{6.44}$$

This is called the **intercept form** of the equation, because the coefficients are the reciprocals of the intercepts of the plane with the coordinate axes. Thus the intersection of the plane with the 1-axis is given by putting $x_2 = x_3 = 0$ in (6.44), which yields

$$a_1x_1 = 1 \tag{6.45}$$

If $a_1 \neq 0$, this has the solution $x_1 = 1/a_1$, thus the plane meets the 1-axis in the point $(1/a_1, 0, 0)$. If $a_1 = 0$, (6.45) has no solution and the plane does not meet the 1-axis; in this case the plane is parallel to the 1-axis. The same argument shows that the plane meets the 2- and 3-axes, if at all, in the points $(0, 1/a_2, 0)$, $(0, 0, 1/a_3)$ respectively. For example, the plane through $(1,0,0)$, $(0,2,0)$, $(0,0,3)$ is given by

$$\frac{x_1}{1} + \frac{x_2}{2} + \frac{x_3}{3} = 1$$

or

$$6x_1 + 3x_2 + 2x_3 = 6 \tag{6.46}$$

Of course, if in (6.43) $c = 0$, the equation cannot be put in the form (6.44). In this case the plane passes through O, so its intercept with each axis is O, and therefore is not enough to determine the plane. For example, the plane through O parallel to (6.46) is given by

$$6x_1 + 3x_2 + 2x_3 = 0$$

The coordinate planes are a special case (of planes through O) – for example, the 12-plane has the equation $x_3 = 0$, and similarly for the others.

6.13 GEOMETRICAL INTERPRETATION OF THE SOLUTION OF LINEAR EQUATIONS

The description of planes in space may be used to give an interpretation to the solution of systems of linear equations in three dimensions, similar to the illustrations in the plane given in the Introduction. Let us take a 3×3 system

$$\mathbf{A}\mathbf{x} = \mathbf{k}, \tag{6.47}$$

where none of the rows of $\mathbf{A}$ is zero, so that the three equations of the system (6.47) represent three planes in space. These three planes may intersect in a point, a line, a plane or not at all, and the coordinates of all the points of intersection are plainly just the solutions $(x_1, x_2, x_3)^{\mathrm{T}}$ of (6.47). We shall indicate the possibilities in terms of the rank of $\mathbf{A}$ and of the augmented matrix $(\mathbf{A}, \mathbf{k})$. The rank of $\mathbf{A}$ can be 1, 2 or 3 (it cannot be 0, because $\mathbf{A}$ is not the zero matrix). The rank of $(\mathbf{A}, \mathbf{k})$ is either equal to the rank of $\mathbf{A}$ or it exceeds this value by 1; it cannot exceed the rank of $\mathbf{A}$ by more than 1, because we have only one more column at our disposal in $(\mathbf{A}, \mathbf{k})$.

(i) $\mathrm{rk}(\mathbf{A}) = \mathrm{rk}(\mathbf{A}, \mathbf{k}) = 3$. The equations have a unique solution and hence the planes intersect in a single point (Fig. 6.8).

(ii) (a) $\mathrm{rk}(\mathbf{A}) = \mathrm{rk}(\mathbf{A}, \mathbf{k}) = 2$. The equations have a solution and the general

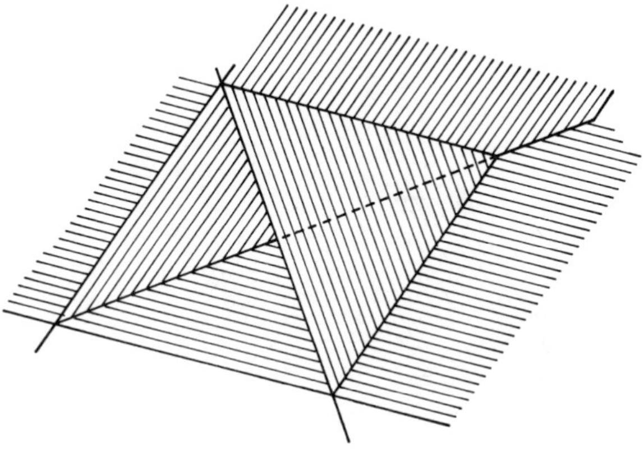

Fig. 6.8

solution depends on a single parameter. The three planes intersect in a line (Fig. 6.9).

(ii) (b) $\mathrm{rk}(\mathbf{A}) = 2$, $\mathrm{rk}(\mathbf{A}, \mathbf{k}) = 3$. The equations have no solution. The planes are parallel to the same straight line and intersect in pairs, but are not parallel (Fig. 6.10).

(iii) (a) $\mathrm{rk}(\mathbf{A}) = \mathrm{rk}(\mathbf{A}, \mathbf{k}) = 1$. Only one equation remains after elimination, hence the three equations are proportional in this case, and so represent the same plane. Thus the three planes coincide.

(iii) (b) $\mathrm{rk}(\mathbf{A}) = 1$, $\mathrm{rk}(\mathbf{A}, \mathbf{k}) = 2$. The three planes are parallel but do not coincide.

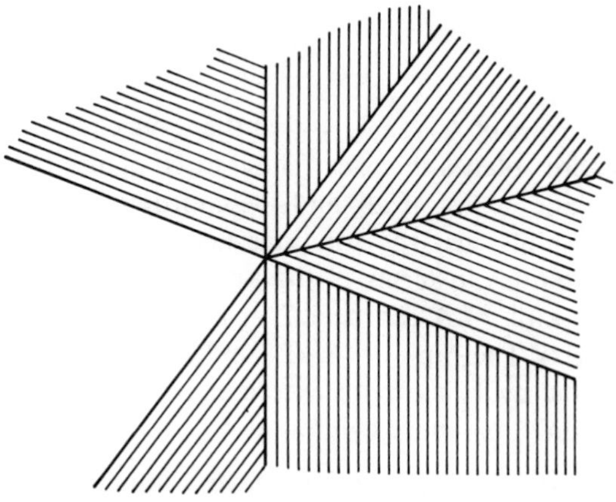

Fig. 6.9

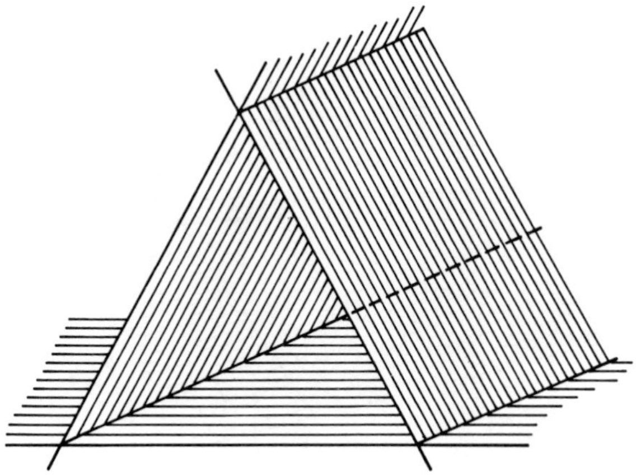

Fig. 6.10

6.14 PAIRS OF LINES

Two lines in space which intersect lie in a plane, and one can therefore speak of the angle between the lines. However, in general, two lines in space need not intersect. They may be parallel, in which case they still lie in a plane, or they may neither intersect nor be parallel; in the latter case they are said to be **skew**. For example, any one of the 12 edges of a cube meets four other edges and is parallel to three others and skew to the remaining four.

If we have two general lines l, m in space, we define the angle between them as the angle between the lines l', m' which pass through O (and hence intersect) and are parallel to l and m, respectively. This angle is uniquely determined if we restrict it to lie between 0° and 90°. If on the other hand we are dealing with oriented lines, the angle lies between 0° and 180° and is unique in this range. To calculate this angle, let the equations of l and m be

$$\mathbf{x} = \mathbf{p} + \lambda\mathbf{u},$$
$$\mathbf{x} = \mathbf{q} + \mu\mathbf{v},$$

where $\mathbf{u}$ and $\mathbf{v}$ are unit vectors.

If we take the positive directions along l and m to be defined by $\mathbf{u}$ and $\mathbf{v}$, respectively, we obtain two oriented lines, $\vec{l}$ and $\vec{m}$, say, whose direction cosines (in a rectangular coordinate system) are given by the components of $\mathbf{u}$ and $\mathbf{v}$, respectively. The angle between $\vec{l}$ and $\vec{m}$ is by definition the angle between $\mathbf{u}$ and $\mathbf{v}$. Denoting it by θ, we thus have $\cos\theta = \mathbf{u}\cdot\mathbf{v}$. Of course, the angle θ_0, say, between the unoriented lines l and m is θ or $180° - \theta$, whichever does not exceed 90°. Explicitly, we have

$$\cos\theta = u_1v_1 + u_2v_2 + u_3v_3$$

or, if $\alpha_1, \alpha_2, \alpha_3$ and $\beta_1, \beta_2, \beta_3$ are the direction angles of the lines,

$$\cos\theta = \cos\alpha_1\cos\beta_1 + \cos\alpha_2\cos\beta_2 + \cos\alpha_3\cos\beta_3$$

In particular, if $\mathbf{u}\cdot\mathbf{v} = 0$, the lines are at right angles, though they may not meet, and if $\mathbf{u}\cdot\mathbf{v} = \pm 1$, the lines are parallel, and hence lie in the same plane.

For example, consider the lines

$$\mathbf{x} \leftrightarrow (2,3,5) + \lambda(1,2,-1) \quad \text{and } \mathbf{x} \leftrightarrow (4,-8,0) + \mu(3,-1,4).$$

The cosine of the angle between the lines is

$$\frac{-3}{\sqrt{6}.\sqrt{26}} \sim -0.240 = -\cos 76.1°$$

it follows that the angle between the (unoriented) lines is 76.1°.

6.15 LINE AND PLANE

Suppose we are given a line and a plane in space. If we visualize this situation we see that either

(i) the line and the plane meet in a single point, or
(ii) (a) the line lies entirely in the plane, or
(ii) (b) the line lies parallel to, but not in the plane.

We shall now derive this result algebraically, using the equations of the line and plane obtained in sections 6.11 and 6.12. The algebraic method has the advantage that it applies even in more than three dimensions; on the other hand, the general method is illuminated by the geometrical interpretation, as we saw in section 6.12.

We take the equation of the line in the form

$$\mathbf{x} = \mathbf{p} + \lambda\mathbf{a} \quad (\mathbf{a} \neq \mathbf{0}) \tag{6.48}$$

and the plane

$$\mathbf{x}\cdot\mathbf{c} = k \quad (\mathbf{c} \neq \mathbf{0}) \tag{6.49}$$

Their intersection is given by those vectors $\mathbf{x}$ which satisfy both (6.48) and (6.49). Now the vector $\mathbf{x}$ given by (6.48) satisfies (6.49) if and only if $(\mathbf{p} + \lambda\mathbf{a})\cdot\mathbf{c} = k$, that is,

$$\mathbf{p}\cdot\mathbf{c} + \lambda\mathbf{a}\cdot\mathbf{c} = k \tag{6.50}$$

This is a linear equation for λ and its solution, if one exists, gives the intersection.

(i) $\mathbf{a}\cdot\mathbf{c} \neq 0$. This means that the direction of the line, namely $\mathbf{a}$, is not perpendicular to the normal (direction $\mathbf{c}$) of the plane; in other words, the line is not parallel to the plane. In this case (6.50) has the unique solution

$$\lambda = (k - \mathbf{p}\cdot\mathbf{c})/(\mathbf{a}\cdot\mathbf{c})$$

Inserting this value for λ in (6.48), we obtain the unique point of intersection.

(ii) $\mathbf{a}\cdot\mathbf{c} = 0$. Now the line is parallel to the plane. Either (ii) (a) $\mathbf{p}\cdot\mathbf{c} = k$; then every value of λ satisfies (6.50), so the line lies entirely in the plane; or (ii) (b) $\mathbf{p}\cdot\mathbf{c} \neq k$; then no value of λ satisfies (6.50), so the line and plane have no common point.

6.16 TWO PLANES IN SPACE

In a similar way we can deal with two planes in space. On geometrical grounds we expect two planes in space either (i) to intersect in a line or (ii) (a) to coincide, or (ii) (b) to be parallel but not coincident. Let us verify this algebraically. We take the equations for the planes in the form

$$\mathbf{x}\cdot\mathbf{a} = h, \quad \mathbf{x}\cdot\mathbf{b} = k \quad (\mathbf{a} \neq \mathbf{0}, \mathbf{b} \neq \mathbf{0}) \tag{6.51}$$

In case (i) the planes are not parallel, that is, their normals are not parallel, so that $\mathbf{a}$ and $\mathbf{b}$ are linearly independent. We show that there is then a point common to the two planes whose position vector is of the special form

$\alpha\mathbf{a} + \beta\mathbf{b}$. This amounts to showing that the equations

$$(\xi\mathbf{a} + \eta\mathbf{b})\cdot\mathbf{a} = h$$
$$(\xi\mathbf{a} + \eta\mathbf{b})\cdot\mathbf{b} = k \tag{6.52}$$

can be solved for ξ, η. These equations may be written

$$\xi\mathbf{a}\cdot\mathbf{a} + \eta\mathbf{b}\cdot\mathbf{a} = h$$
$$\xi\mathbf{a}\cdot\mathbf{b} + \eta\mathbf{b}\cdot\mathbf{b} = k \tag{6.53}$$

their determinant is

$$\begin{aligned}(\mathbf{a}\cdot\mathbf{a})(\mathbf{b}\cdot\mathbf{b}) - (\mathbf{a}\cdot\mathbf{b})^2 &= |\mathbf{a}|^2\cdot|\mathbf{b}|^2(1 - \cos^2\theta),\\ &= |\mathbf{a}|^2\cdot|\mathbf{b}|^2\sin^2\theta,\end{aligned}$$

where θ is the angle between **a** and **b**. Since **a** and **b** are linearly independent, they are not zero and θ is neither 0° nor 180°; hence the determinant does not vanish and so the equations (6.53) have a unique solution, say

$$\mathbf{p} = \alpha\mathbf{a} + \beta\mathbf{b}. \tag{6.54}$$

Let **c** be any vector orthogonal to **a** and **b**, for instance $\mathbf{c} = \mathbf{a} \wedge \mathbf{b}$. We shall show that the intersection of the planes consists of the line

$$\mathbf{x} = \mathbf{p} + \lambda\mathbf{c}, \tag{6.55}$$

where **p** is given by (6.54). By construction, $\mathbf{p}\cdot\mathbf{a} = h$, $\mathbf{p}\cdot\mathbf{b} = k$ and $\mathbf{c}\cdot\mathbf{a} = \mathbf{c}\cdot\mathbf{b} = 0$, whence we find

$$(\mathbf{p} + \lambda\mathbf{c})\cdot\mathbf{a} = \mathbf{p}\cdot\mathbf{a} = h$$
$$(\mathbf{p} + \lambda\mathbf{c})\cdot\mathbf{b} = \mathbf{p}\cdot\mathbf{b} = k$$

for all λ, which shows that every point of the line (6.55) lies in the intersection of the two planes. Conversely, if $\mathbf{x} = \mathbf{q}$ is any point of the intersection, then $\mathbf{q}\cdot\mathbf{a} = h$, $\mathbf{q}\cdot\mathbf{b} = k$, hence

$$(\mathbf{q} - \mathbf{p})\cdot\mathbf{a} = (\mathbf{q} - \mathbf{p})\cdot\mathbf{b} = 0$$

thus $\mathbf{q} - \mathbf{p}$ is orthogonal to both **a** and **b**, so $\mathbf{q} - \mathbf{p}$ is perpendicular to the plane spanned by **a** and **b**. Now the definition of **c** shows that $\mathbf{q} - \mathbf{p}$ must be linearly dependent on **c**, say $\mathbf{q} - \mathbf{p} = \lambda\mathbf{c}$. Thus $\mathbf{q} = \mathbf{p} + \lambda\mathbf{c}$, and this shows that **q** lies on the line (6.55). This line is therefore the intersection of the two planes.

We note that from the algebraic point of view, $\mathbf{x} = \mathbf{p}$ is a particular solution of the system (6.51) and $\mathbf{x} = \mathbf{c}$ is a solution of the associated homogeneous system (cf. section 4.2). When expressed in terms of coordinates, (6.51) is a system of two equations in three unknowns (the components of **x**) and it is of rank 2 provided that **a** and **b** are linearly independent.

When the planes are parallel, **a** and **b** are linearly dependent, say $\mathbf{b}=\gamma\mathbf{a}$. The equations (6.51) then read

$$\mathbf{x}\cdot\mathbf{a}=h$$
$$\gamma\mathbf{x}\cdot\mathbf{a}=k$$

Now either

(a) $k=\gamma h$ and the equations reduce to the single equation $\mathbf{x}\cdot\mathbf{a}=h$, or
(b) $k\neq\gamma h$ and the equations have no solution.

Accordingly, the planes either (a) coincide, or (b) are parallel without a common point.

As an illustration of the general case, consider the planes

$$\mathbf{x}\cdot\mathbf{a}=96$$
$$\mathbf{x}\cdot\mathbf{b}=18$$

where $\mathbf{a}\leftrightarrow(-2,7,-4)$ and $\mathbf{b}\leftrightarrow(1,1,-1)$ in rectangular coordinates. Equations (6.53) read

$$\begin{aligned}69\xi+9\eta&=96\\ 9\xi+3\eta&=18\end{aligned} \tag{6.56}$$

They have a unique solution, namely $\xi=1$, $\eta=3$ and hence

$$\mathbf{p}\leftrightarrow(-2,7,-4)+3(1,1,-1)=(1,10,-7).$$

is a common point in the form (6.56). To obtain a vector **c** orthogonal to **a** and **b** we form $\mathbf{a}\wedge\mathbf{b}$; its coordinates are $(-3,-6,-9)$ and, dividing by -3 (which simplifies the coordinates without affecting the direction), we get

$$\mathbf{c}\leftrightarrow(1,2,3).$$

Thus the direction ratios of the line of intersection are 1:2:3 and the intersection is

$$\mathbf{x}\leftrightarrow(1,10,-7)+\lambda(1,2,3)$$

6.17 PAIRS OF SKEW LINES

As we saw in section 6.14, two lines in space need not intersect, nor even lie in the same plane. But we shall now show that there always exists a line which meets both the given lines and is perpendicular to both. Moreover, if the given lines are not parallel, then there is exactly one such line.

Let l and m be the given lines, with equations

$$\mathbf{x}=\mathbf{p}+\lambda\mathbf{a}\quad(\mathbf{a}\neq\mathbf{0}) \tag{6.57}$$

$$\mathbf{x}=\mathbf{q}+\mu\mathbf{b}\quad(\mathbf{b}\neq\mathbf{0}) \tag{6.58}$$

Let R be the point on l with position vector given by (6.57) and S the point on m with position vector given by (6.58). Then the line segment RS is represented by the vector $\mathbf{q}+\mu\mathbf{b}-\mathbf{p}-\lambda\mathbf{a}$, and the condition for RS to be perpendicular to both l and m is expressed by the equations

$$(\mathbf{q}-\mathbf{p}+\mu\mathbf{b}-\lambda\mathbf{a})\cdot\mathbf{a}=0$$
$$(\mathbf{q}-\mathbf{p}+\mu\mathbf{b}-\lambda\mathbf{a})\cdot\mathbf{b}=0$$

These equations may be written

$$\begin{aligned}\lambda(\mathbf{a}\cdot\mathbf{a})-\mu(\mathbf{b}\cdot\mathbf{a})&=\mathbf{q}\cdot\mathbf{a}-\mathbf{p}\cdot\mathbf{a}\\ \lambda(\mathbf{a}\cdot\mathbf{b})-\mu(\mathbf{b}\cdot\mathbf{b})&=\mathbf{q}\cdot\mathbf{b}-\mathbf{p}\cdot\mathbf{b},\end{aligned} \qquad (6.59)$$

and, as we saw in section 6.16 (cf. (6.53)), these equations have a unique solution if $\mathbf{a}$ and $\mathbf{b}$ are linearly independent, that is to say, provided that l and m are not parallel. The solution gives the values of the parameters for the points in which the common perpendicular meets the two lines (Fig. 6.11).

As an example, let l and m be given by

$$\mathbf{x}\leftrightarrow(-3,3,6)+\lambda(3,0,-1)$$

and

$$\mathbf{x}\leftrightarrow(6,-4,3)+\mu(4,-5,2)$$

respectively (in rectangular coordinates). Then

$\mathbf{q}-\mathbf{p}+\mu\mathbf{b}-\lambda\mathbf{a}\leftrightarrow(9,-7,-3)+\mu(4,-5,2)-\lambda(3,0,-1)$, and equations (6.59)

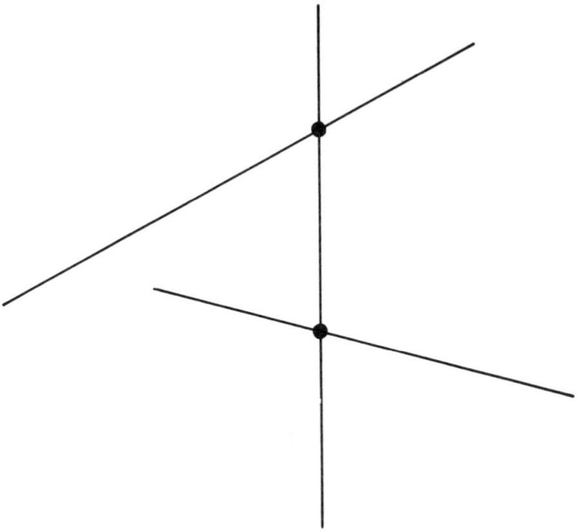

Fig. 6.11

take the form

$$10\lambda - 10\mu = 30$$
$$10\lambda - 45\mu = 65$$

The unique solution is $\lambda = 2$, $\mu = -1$. Hence the points in which the common perpendicular meets l and m respectively are

$$(-3, 3, 6) + 2(3, 0, -1) = (3, 3, 4)$$

and

$$(6, -4, 3) + (-1)(4, -5, 2) = (2, 1, 1)$$

An equation for the common perpendicular is obtained by writing down an expression for the line through these two points:

$$\mathbf{x} \leftrightarrow (2, 1, 1) + \lambda(1, 2, 3)$$

When the lines are parallel they still have a common perpendicular, but it is no longer unique. The lines may then be taken to be

$$\mathbf{x} = \mathbf{p} + \lambda\mathbf{a} \quad \text{and} \quad \mathbf{x} = \mathbf{q} + \mu\mathbf{a} \quad (\mathbf{a} \neq \mathbf{0})$$

Equations (6.59) now reduce to a single equation

$$(\mathbf{q} - \mathbf{p} + (\mu - \lambda)\mathbf{a})\cdot\mathbf{a} = 0$$

with the solution

$$\lambda - \mu = (\mathbf{q} - \mathbf{p})\cdot\mathbf{a}/|\mathbf{a}|^2 \tag{6.60}$$

We may fix μ arbitrarily, say $\mu = \mu_0$. Then λ is uniquely determined by (6.60), corresponding to the fact that there is one common perpendicular through each point of m (and of course through each point of l). For example, if the lines are given by

$$\mathbf{x} \leftrightarrow (2, 1, 4) + \lambda(3, -1, 2) \quad \text{and} \quad \mathbf{x} \leftrightarrow (3, 4, 2) + \mu(3, -1, 2)$$

then on solving (6.59) we find

$$\lambda - \mu = -4/14 = -2/7.$$

Thus the perpendicular through the point $\mu = \mu_0$ of m meets m and l in the points $(3, 4, 2) + \mu_0(3, -1, 2)$ and $(2, 1, 4) + (\mu_0 - 2/7)(3, -1, 2)$, respectively, and is given by

$$\mathbf{x} \leftrightarrow (3, 4, 2) + \mu_0(3, -1, 2) + \kappa\{(1, 3, -2) + 2/7(3, -1, 2)\},$$

where κ is the parameter.

Let us again consider two lines in space which are not parallel. If the lines intersect, then their intersection is a single point and their common perpendicular clearly passes through this point. Thus we have a method of determining the point of intersection of two lines, when there is such an intersection. If the lines are skew, the common perpendicular represents the shortest distance between them. This may be seen as follows: Let the two lines l and m be given by equations (6.57) and (6.58), and let the common perpendicular be

$$\mathbf{x} = \mathbf{r} + \nu\mathbf{c} \quad (\mathbf{c} \neq \mathbf{0})$$

Then $\mathbf{c}$ is a vector orthogonal to both $\mathbf{a}$ and $\mathbf{b}$. The lines l and m then lie in the planes

$$\mathbf{x}\cdot\mathbf{c} = \mathbf{p}\cdot\mathbf{c} \tag{6.61}$$

and

$$\mathbf{x}\cdot\mathbf{c} = \mathbf{q}\cdot\mathbf{c} \tag{6.62}$$

respectively, as follows from the fact that $\mathbf{a}$ and $\mathbf{b}$ are orthogonal to $\mathbf{c}$. These planes (6.61) and (6.62) are parallel, so the distance from any point R in the plane (6.61) to any point S in the plane (6.62) is least precisely when RS is perpendicular to the two planes, that is, when RS is parallel to $\mathbf{c}$. But this is just the case when R and S are the feet of a common perpendicular to the two planes.

EXERCISES

Any coordinates written out in these exercises are taken to refer to a rectangular coordinate system.

6.1. For any vectors $\mathbf{a}$ and $\mathbf{b}$ show that $(\mathbf{a}+\mathbf{b})^2 + (\mathbf{a}-\mathbf{b})^2 = 2(\mathbf{a}^2 + \mathbf{b}^2)$, where $\mathbf{a}^2 = \mathbf{a}\cdot\mathbf{a}$.

6.2. In a triangle with sides a, b, c and corresponding angles α, β, γ, show that $a^2 + b^2 - c^2 = 2ab\cos\gamma$. (*Hint*: Use (6.3) and (6.16).)

6.3 Show that for an orthonormal basis $\mathbf{u}_1, \ldots, \mathbf{u}_n$ and any vectors $\mathbf{a}, \mathbf{b}, \mathbf{a}\cdot\mathbf{b} = \sum(\mathbf{a}\cdot\mathbf{u}_i)\cdot(\mathbf{b}\cdot\mathbf{u}_i)$. (**Parseval's identity**.)

6.4. Show that any three vectors in 3-space satisfy the **Jacobi identity**: $(\mathbf{u}\wedge\mathbf{v})\wedge\mathbf{w} + (\mathbf{v}\wedge\mathbf{w})\wedge\mathbf{u} + (\mathbf{w}\wedge\mathbf{u})\wedge\mathbf{v} = \mathbf{0}$.

6.5. Prove **Lagrange's identity**:

$$(\mathbf{u}_1 \wedge \mathbf{u}_2)\cdot(\mathbf{v}_1 \wedge \mathbf{v}_2) = \begin{vmatrix} \mathbf{u}_1\cdot\mathbf{v}_1 & \mathbf{u}_1\cdot\mathbf{v}_2 \\ \mathbf{u}_2\cdot\mathbf{v}_1 & \mathbf{u}_2\cdot\mathbf{v}_2 \end{vmatrix}$$

6.6. Show that the points $(5,-5,4)$, $(3,-4,0)$, $(5,4,1)$ form a right-angled triangle in space and find the fourth point to complete the rectangle.

6.7. Show that two lines in space are coplanar if and only if they either meet or are parallel.

6.8. Find the circumcircle of the triangle with the vertices $\mathbf{0}$, $(3,1,2)$, $(1,2,-3)$. Find also the radius of the circumcircle. (The circumcircle of a triangle is the (unique) circle through its three vertices.)

6.9. Show that in a tetrahedron the joins of the mid-points of opposite edges meet in a point which bisects each of them.

6.10. A cube is placed with its sides parallel to the axis, one vertex at the origin and the opposite vertex at $P \leftrightarrow (1,1,1)$. Find the equations for

(i) the lines containing the edges through P; (ii) the lines containing the diagonals (through P) of the faces through P; (iii) the lines containing the diagonals (not through P) of the faces through P.

6.11. Find the lines through the pairs of points with the following coordinates: (i) $(1,1,1)$ and $(4,2,3)$; (ii) $(2,0,1)$ and $(7,4,2)$; (iii) $(1,4,2)$ and $(0,5,3)$. Find the angles between any two of these lines.

6.12. In a regular tetrahedron $ABCD$ find the cosine of the angle between the edge AD and the face ABC. Find also the angle between two faces (the angle between a line and a plane is defined as the complement of the acute angle between the line and the normal to the plane; the angle between two planes is defined as the acute angle between their normals).

6.13. Find the planes through the following sets of points: (i) $(1,2,3)$, $(3,5,7)$, $(3,-1,-3)$; (ii) $(4,-5,8)$, $(1,2,3)$, $(2,9,-4)$; (iii) $(0,12,2)$, $(2,-2,14)$, $(1,2,3)$. Find the intersection of any two of these planes.

6.14. Find the points where the common perpendicular to each of the following pairs of lines meets these lines: (i) $(1,4,0)+\lambda(0,1,1)$ and $(1,-1,4)+\mu(2,1,2)$; (ii) $(0,1,1)+\lambda(1,1,1)$ and $(1,0,0)+\mu(1,2,-3)$.

6.15. For any vectors $\mathbf{a},\mathbf{b},\mathbf{c}$ show that the four points with position vectors $\mathbf{a},\mathbf{b},\mathbf{c},\frac{1}{3}(\mathbf{a}+\mathbf{b}+\mathbf{c})$ are coplanar. What are the conditions on λ, μ and ν for $\lambda\mathbf{a}+\mu\mathbf{b}+\nu\mathbf{c}$ to represent a point coplanar with the points represented by $\mathbf{a},\mathbf{b},\mathbf{c}$?

6.16. The position vectors of three vertices of a tetrahedron relative to the fourth as origin are $\mathbf{a},\mathbf{b}$ and $\mathbf{c}$. Through each vertex a plane is drawn parallel to the opposite face. Express the position vectors of the vertices of the tetrahedron formed by these planes in terms of $\mathbf{a},\mathbf{b}$ and $\mathbf{c}$.

6.17. Show that a parallelogram whose sides, referred to a vertex as origin, are given by the vectors $\mathbf{u},\mathbf{v}$, has sides of length $|\mathbf{u}|$, $|\mathbf{v}|$. If the angle between these sides is α, then the altitude of the parallelogram with $\mathbf{u}$ as base is $|\mathbf{v}|\sin\alpha$. Find the expression obtained in section 5.2 for the area μ in terms of the components of $\mathbf{u},\mathbf{v}$ by evaluating μ^2.

6.18. Show that (6.26) is a consequence of (6.27) and (6.25).

6.19. Let U, U' be subspaces of a vector space V, write $U\cap U'$ for the intersection of U and U' and $U+U'$ for the space of vectors $\mathbf{u}+\mathbf{u}'$ (where $\mathbf{u}\in U$, $\mathbf{u}'\in U'$) as in Exercises 1.12 and 1.14. Take any basis $\mathbf{u}_1,\ldots,\mathbf{u}_r$ of $U\cap U'$ and complete it to a basis $\mathbf{u}_1,\ldots,\mathbf{u}_r,\mathbf{v}_1,\ldots,\mathbf{v}_s$ of U and to a basis $\mathbf{u}_1,\ldots,\mathbf{u}_r,\mathbf{w}_1,\ldots,\mathbf{w}_t$ of U'. Show that $\mathbf{u}_1,\ldots,\mathbf{u}_r,\mathbf{v}_1,\ldots,\mathbf{v}_s$, $\mathbf{w}_1,\ldots,\mathbf{w}_t$ is a basis of $U+U'$ and deduce the formula $\dim(U+U')=\dim U+\dim U'-\dim(U\cap U')$.

6.20. Let V be the vector space of continuous functions in the interval $[0,2\pi]$. Show that

$$f\cdot g=\frac{1}{\pi}\int_0^{2\pi} fg\,\mathrm{d}x$$

defines a scalar product on V. Show further that $\{1/\sqrt{2}, \sin x, \cos x, \sin 2x, \cos 2x, \ldots\}$ is an orthonormal set in V.

6.21. Show that the vector space of all polynomials has a scalar product, defined by

$$f \cdot g = \int_{-1}^{1} f g \, \mathrm{d}x$$

By applying the Gram–Schmidt process to the set $\{1, x, x^2, x^3, x^4\}$, find an orthonormal basis for the polynomials of degree at most 4. (These polynomials, with a scalar factor adjusted to give the value 1 for $x = 1$, are called **Legendre polynomials**.)

6.22. Prove the **triangle inequality** for vectors: $|\mathbf{x} + \mathbf{y}| \leqslant |\mathbf{x}| + |\mathbf{y}|$. (If $OP \leftrightarrow \mathbf{x}$, $OQ \leftrightarrow \mathbf{y}$, this inequality states that the side PQ does not exceed the sum of the sides OP and OQ in length. (*Hint*: Expand $|\mathbf{x} + \mathbf{y}|^2$ and apply (6.17).)

7

Coordinate transformations and linear mappings

7.1 THE MATRIX OF A TRANSFORMATION

We have seen that points in space may be described as linear combinations of three linearly independent vectors, referred to a given point as origin, the scalar coefficients being the coordinates relative to these basis vectors (section 6.2). Our next task is to compare the coordinates of a given point in different coordinate systems; for simplicity we shall at first keep the same point O as origin. Let $\mathbf{u}_1, \mathbf{u}_2, \mathbf{u}_3$ and $\mathbf{v}_1, \mathbf{v}_2, \mathbf{v}_3$ be any two sets of basis vectors and consider the two coordinate systems defined by these sets of vectors, relative to the origin O. For a given point P we have

$$OP = \sum \mathbf{u}_i x_i \tag{7.1}$$

and

$$OP = \sum \mathbf{v}_j y_j \tag{7.2}$$

say. To obtain a relation between the xs and ys let us express the vectors $\mathbf{v}_j$ in terms of the $\mathbf{u}_i$. We have linear expressions

$$\mathbf{v}_j = \sum_i \mathbf{u}_i a_{ij} \quad (j = 1, 2, 3), \tag{7.3}$$

and inserting these expressions in (7.2), we find

$$OP = \sum \mathbf{u}_i a_{ij} y_j \tag{7.4}$$

where the summation is over i and j. Now the coordinates of OP relative to the $\mathbf{u}_i$ are uniquely determined; we may therefore equate the coefficients of $\mathbf{u}_1$ in (7.1) and (7.4) and obtain

$$x_1 = \sum a_{1j} y_j$$

and similarly with 1 replaced by 2 and 3. Thus we have

$$x_i = \sum_j a_{ij} y_j \quad (i = 1, 2, 3) \tag{7.5}$$

In matrix notation this may be written (cf. section 3.1)

$$\mathbf{x} = \mathbf{A}\mathbf{y} \tag{7.6}$$

where $\mathbf{x}$ and $\mathbf{y}$ are columns consisting of x_1, x_2, x_3 and y_1, y_2, y_3, respectively, and $\mathbf{A}$ is the 3×3 matrix with (i, j)th entry a_{ij}. From (7.3) we see that the columns of $\mathbf{A}$ represent the coordinates of $\mathbf{v}_1$, $\mathbf{v}_2, \mathbf{v}_3$ relative to $\mathbf{u}_1, \mathbf{u}_2, \mathbf{u}_3$ as basis vectors.

Of course, the same argument may be used with the roles of the **u**s and **v**s interchanged. We have to express the **u**s in terms of the **v**s:

$$\mathbf{u}_j = \sum \mathbf{v}_i b_{ij} \quad (j = 1, 2\,3) \tag{7.7}$$

for some coefficients b_{ij} and from these equations together with (7.1) and (7.2) we find

$$y_i = \sum b_{ij} x_j$$

or in matrix form, writing $\mathbf{B} = (b_{ij})$,

$$\mathbf{y} = \mathbf{B}\mathbf{x} \tag{7.8}$$

Equations (7.6) and (7.8) express the transformation of coordinates from $\mathbf{x}$ to $\mathbf{y}$ and vice versa. If we substitute from (7.8) into (7.6), we find

$$\mathbf{x} = \mathbf{A}\mathbf{B}\mathbf{x},$$

and since this holds for all columns $\mathbf{x}$, we find as in section 3.3 that $\mathbf{AB} = \mathbf{I}$. Similarly, if we substitute from (7.6) into (7.8) we find that $\mathbf{y} = \mathbf{BA}\mathbf{y}$ for all columns $\mathbf{y}$, and hence $\mathbf{BA} = \mathbf{I}$. This shows that $\mathbf{B}$ is just the inverse of $\mathbf{A}$:

$$\mathbf{B} = \mathbf{A}^{-1} \tag{7.9}$$

In particular, this shows that the matrix $\mathbf{A}$ of the transformation is regular (section 3.9).

Conversely, given any coordinate system with basis vectors $\mathbf{u}_1$, $\mathbf{u}_2, \mathbf{u}_3$, say, any regular matrix $\mathbf{A}$ defines a coordinate transformation. We need only define the vectors $\mathbf{v}_1, \mathbf{v}_2, \mathbf{v}_3$ by the rule

$$\mathbf{v}_j \leftrightarrow (a_{1j}, a_{2j}, a_{3j})^{\mathrm{T}} \tag{7.10}$$

Since the columns of $\mathbf{A}$ are linearly independent, so are the vectors $\mathbf{v}_1, \mathbf{v}_2, \mathbf{v}_3$ and hence they can again be used as basis vectors for a coordinate system. Writing the columns of $\mathbf{A}$ as $\mathbf{a}_1, \mathbf{a}_2, \mathbf{a}_3$, we can express (7.10) as

$$\mathbf{v}_j \leftrightarrow \mathbf{a}_j \quad (j = 1, 2, 3)$$

In practical situations we may have to find $\mathbf{A}$ when the **u**s and **v**s are given in terms of the standard basis $\mathbf{e}_1, \mathbf{e}_2, \mathbf{e}_3$. The matrix $\mathbf{P}$ of transformation from the **e**s to the **u**s has for its columns the coordinates of $\mathbf{u}_1, \mathbf{u}_2, \mathbf{u}_3$ (relative to the standard basis), so if $\mathbf{z}$ denotes the coordinates of the general point relative to the **e**s, then we have

$$\mathbf{x} = \mathbf{P}\mathbf{z}$$

Similarly, the matrix **Q** whose columns are the coordinates of $\mathbf{v}_1, \mathbf{v}_2, \mathbf{v}_3$ relative to the **e**s gives the transformation from the **e**s to the **v**s, thus we have

$$\mathbf{y} = \mathbf{Q}\mathbf{z}$$

Hence $\mathbf{x} = \mathbf{P}\mathbf{Q}^{-1}\mathbf{y}$ and this shows that

$$\mathbf{A} = \mathbf{P}\mathbf{Q}^{-1}$$

To give an example, let the **u**s be (0, 1, 1), (1, 0, 1), (1, 1, 0) and the **v**s (1, 0, 0), (1, 2, 3), (1, 1, 2); then we have

$$\mathbf{P} = \begin{pmatrix} 0 & 1 & 1 \\ 1 & 0 & 1 \\ 1 & 1 & 0 \end{pmatrix}$$

$$\mathbf{Q} = \begin{pmatrix} 1 & 1 & 1 \\ 0 & 2 & 1 \\ 0 & 3 & 2 \end{pmatrix} \quad \mathbf{Q}^{-1} = \begin{pmatrix} 1 & 1 & -1 \\ 0 & 2 & -1 \\ 0 & -3 & 2 \end{pmatrix}$$

$$\mathbf{P}\mathbf{Q}^{-1} = \begin{pmatrix} 0 & -1 & 1 \\ 1 & -2 & 1 \\ 1 & 3 & -2 \end{pmatrix}$$

More generally, we can change the origin as well by a transformation of the form

$$\mathbf{x} = \mathbf{A}\mathbf{y} + \mathbf{b} \tag{7.11}$$

This is called an **affine** transformation. It is no longer linear, but has the following property. If $\mathbf{x}' = \mathbf{A}\mathbf{y}' + \mathbf{b}$, $\mathbf{x}'' = \mathbf{A}\mathbf{y}'' + \mathbf{b}$ and λ is any scalar, then

$$\lambda\mathbf{x}' + (1 - \lambda)\mathbf{x}'' = \mathbf{A}[\lambda\mathbf{y}' + (1 - \lambda)\mathbf{y}''] + \mathbf{b}$$

Geometrically this means that all points on the straight line determined by the points with coordinates $\mathbf{x}', \mathbf{x}''$ are transformed in the same way; it follows that the transformation (7.11) preserves straight lines.

Finally, we remark that although we have confined ourselves here to three dimensions, everything that has been said applies equally well in *n*-dimensional space.

7.2 ORTHOGONAL MATRICES

We now take two *rectangular* coordinate systems (with the same origin) and ask for the conditions which the matrix of transformation has to satisfy. If this matrix is again denoted by **A**, we have to express the fact that the columns of **A** represent pairwise orthogonal unit vectors in a rectangular system. Thus we must have

$$\mathbf{a}_i^{\mathrm{T}}\mathbf{a}_j = \delta_{ij} \quad (i, j = 1, 2, 3)$$

where $\mathbf{a}_i$ are the columns of **A** and $\mathbf{a}_i^{\mathrm{T}}$ denotes the row which is the transpose of $\mathbf{a}_i$, while δ_{ij} is the Kronecker delta (section 3.6). These equations may be summed up in a single matrix equation

$$\mathbf{A}^{\mathrm{T}}\mathbf{A} = \mathbf{I} \tag{7.12}$$

Conversely, the transformation defined by a matrix **A** satisfying (7.12), when applied to a rectangular system, leads again to a rectangular system, for it ensures that the new basis vectors are pairwise orthogonal unit vectors.

A square matrix **A** satisfying (7.12) is said to be **orthogonal** and the corresponding transformation is called an **orthogonal transformation**. For any orthogonal matrix **A** we have

$$1 = |\mathbf{A}^{\mathrm{T}}\mathbf{A}| = |\mathbf{A}^{\mathrm{T}}|\cdot|\mathbf{A}| = |\mathbf{A}|^2$$

and hence

$$|\mathbf{A}| = \pm 1 \tag{7.13}$$

(Here the vertical bars on the left-hand side denote the determinant and must not be confused with the absolute value of a real number.) The orthogonal matrix **A** (as well as the transformation defined by it) is called **proper** or **improper orthogonal** according as its determinant is $+1$ or -1; the geometrical interpretation of these cases will be given later, in section 7.3.

An orthogonal matrix **A** must be regular, since its determinant is non-zero, by (7.13), so it has an inverse, and as (7.12) shows, this is given by the transpose:

$$\mathbf{A}^{-1} = \mathbf{A}^{\mathrm{T}} \tag{7.14}$$

It follows that the inverse of an orthogonal matrix is again orthogonal, for this inverse is $\mathbf{A}^{\mathrm{T}}$ and we have $\mathbf{A}^{\mathrm{TT}}\mathbf{A}^{\mathrm{T}} = \mathbf{A}\mathbf{A}^{\mathrm{T}} = \mathbf{A}\mathbf{A}^{-1} = \mathbf{I}$. The product of two orthogonal matrices is also orthogonal, for if $\mathbf{A}^{\mathrm{T}}\mathbf{A} = \mathbf{B}^{\mathrm{T}}\mathbf{B} = \mathbf{I}$, then

$$(\mathbf{AB})^{\mathrm{T}}\mathbf{AB} = \mathbf{B}^{\mathrm{T}}\mathbf{A}^{\mathrm{T}}\mathbf{AB} = \mathbf{B}^{\mathrm{T}}\mathbf{B} = \mathbf{I}$$

Moreover, if **A** and **B** are proper orthogonal, then so is their product, because $|\mathbf{AB}| = |\mathbf{A}|\cdot|\mathbf{B}| = 1$. Similarly the product of two improper orthogonal matrices is proper orthogonal: $|\mathbf{AB}| = |\mathbf{A}|\cdot|\mathbf{B}| = (-1)^2 = 1$. On the other hand, the product of one proper and one improper orthogonal matrix is improper orthogonal.

Equation (7.14) greatly simplifies the discussion of orthogonal transformations. Thus the transformation from one rectangular system to another was described by equation (7.6):

$$\mathbf{x} = \mathbf{A}\mathbf{y}$$

where **A** is orthogonal. If we want to express the new coordinates in terms of the old, we need the inverse of **A**; but this is $\mathbf{A}^T$, as we saw in (7.14). Hence the inverse transformation is

$$\mathbf{y} = \mathbf{A}^T\mathbf{x} \tag{7.15}$$

For example, if the equation of a transformation is

$$\mathbf{x} = \begin{pmatrix} 1/\sqrt{3} & 1/\sqrt{2} & -1/\sqrt{6} \\ 1/\sqrt{3} & -1/\sqrt{2} & -1/\sqrt{6} \\ -1/\sqrt{3} & 0 & -2/\sqrt{6} \end{pmatrix} \mathbf{y}$$

then the inverse transformation is given by

$$\mathbf{y} = \begin{pmatrix} 1/\sqrt{3} & 1/\sqrt{3} & -1/\sqrt{3} \\ 1/\sqrt{2} & -1/\sqrt{2} & 0 \\ -1/\sqrt{6} & -1/\sqrt{6} & -2/\sqrt{6} \end{pmatrix} \mathbf{x}$$

If we have a third coordinate system, again rectangular and with the same origin as the other two, in which the coordinates of the general point are denoted by **z**, then this is related to the second system by an orthogonal transformation, say

$$\mathbf{y} = \mathbf{B}\mathbf{z} \tag{7.16}$$

The transformation from the original to the final system is obtained by eliminating **y** from equations (7.6) and (7.16):

$$\mathbf{x} = \mathbf{A}\mathbf{B}\mathbf{z} \tag{7.17}$$

Since (7.17) and (7.15) are transformations between rectangular coordinate systems, these equations provide another proof of the fact that products and inverses of orthogonal matrices are again orthogonal.

Once again we can use (7.12) to define orthogonal matrices in n dimensions and everything that has been said also applies in the general case.

7.3 ROTATIONS AND REFLEXIONS

Let us now consider some examples of orthogonal transformations in two and three dimensions. If we rotate the axes of a rectangular system like a rigid body about the origin, we shall again obtain a rectangular system, and the resulting transformation must therefore be orthogonal. In order to describe the matrix we need to clarify the relation between different rectangular systems.

Suppose we have two rectangular coordinate systems with the same origin O. If we think of each system as rigidly movable about O, we can rotate the second system so that its 1-axis coincides (as an oriented line) with the 1-axis

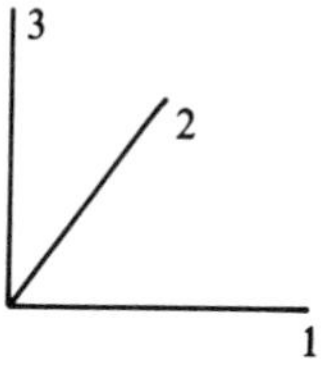

Fig. 7.1

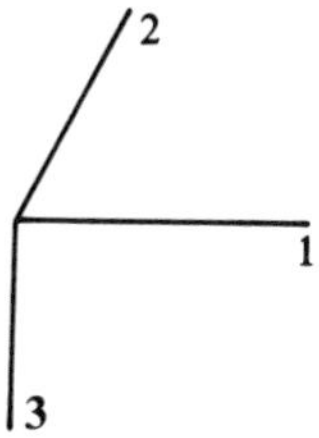

Fig. 7.2

of the first system. Now the second system may be rotated about the common 1-axis so that their 2-axes coincide. Then the 3-axes of the two systems lie along the same line but they have either the same orientation or opposite orientations. Accordingly, we can distinguish two possible **orientations** of a rectangular coordinate system: if the rotation through 90° from the 1-axis to the 2-axis is related to the positive direction of the 3-axis as are rotation and motion of a right-handed screw, the system is called **right-handed** (Fig. 7.1); in the opposite case it is **left-handed** (Fig. 7.2). The distinction may be easily remembered by the fact that thumb, index and middle finger of the right hand, when extended, form a right-handed system, while the left hand exemplifies a left-handed system. In this way the orientation may be defined for any system, not necessarily rectangular. It is clear that a rotation can never change a right-handed system into a left-handed one or vice versa. In other words, the orientation of a coordinate system is preserved by rotations.

Consider now the matrix of transformation when a right-handed rectangular system is rotated about the 3-axis. Denote the angle of rotation by α, the sense of rotation being anticlockwise when looking along the 3-axis from the positive direction. Then the basis vectors of the new system, expressed in terms of the old, have the coordinates

$$(\cos\alpha, \sin\alpha, 0), (-\sin\alpha, \cos\alpha, 0), (0, 0, 1)$$

These are therefore the columns of the required matrix, and so the transformation is given by

$$\mathbf{x} = \begin{pmatrix} \cos\alpha & -\sin\alpha & 0 \\ \sin\alpha & \cos\alpha & 0 \\ 0 & 0 & 1 \end{pmatrix} \mathbf{y}$$

with a proper orthogonal matrix. Written out in terms of scalar equations, the transformation has the form

$$x_1 = y_1 \cos\alpha - y_2 \sin\alpha$$
$$x_2 = y_1 \sin\alpha + y_2 \cos\alpha$$
$$x_3 = y_3$$

In a similar way it may be shown that any rotation is represented by a proper orthogonal matrix. The remarks at the beginning of this section show that we can get from any right-handed rectangular system to any other by a series of rotations (in fact a single rotation will do, as we shall see in a moment); so the transformation from one to the other is proper orthogonal. On the other hand, the transformation from a right-handed to a left-handed system must be improper; for if $\mathbf{x}$ and $\mathbf{y}$ denote the coordinates of a given point in the two systems, we introduce a third system whose 1- and 2-axes agree with those of the second (left-handed) system, but whose 3-axis has the opposite orientation. This system is therefore right-handed; if the same point has coordinates $\mathbf{z}$ in this system, then

$$\mathbf{x} = \mathbf{Az}$$

where $\mathbf{A}$, as matrix of transformation between right-handed systems, is proper orthogonal. Now the relation between $\mathbf{y}$ and $\mathbf{z}$ is given by $z_1 = y_1$, $z_2 = y_2$, $z_3 = -y_3$, or in matrix form

$$\mathbf{z} = \mathbf{Ty},$$

where $\mathbf{T}$, given by the equation

$$\mathbf{T} = \begin{pmatrix} 1 & 0 & 0 \\ 0 & 1 & 0 \\ 0 & 0 & -1 \end{pmatrix} \tag{7.18}$$

is an improper orthogonal matrix. The transformation from $\mathbf{x}$ to $\mathbf{y}$ is therefore

$$\mathbf{x} = \mathbf{ATy}$$

and $|\mathbf{AT}| = |\mathbf{A}|\cdot|\mathbf{T}| = -1$, so the transformation is improper. The transformation given by (7.18) consists in reflecting every point in the 12-plane, as in a mirror. Such a transformation is called a **reflexion**, and it can be shown that any two coordinate systems related by a reflexion have opposite orientations.

We have seen that any rotation is represented by a proper orthogonal transformation. Conversely, we now show that every proper orthogonal transformation (in 3-space) represents a rotation.

Theorem 7.1
Any 3×3 proper orthogonal matrix represents a rotation about some axis through the origin.

Proof
Let $\mathbf{A}$ be a proper orthogonal matrix. Suppose first that $\mathbf{A}$ is a 2×2 matrix, say

$$\mathbf{A} = \begin{pmatrix} a_1 & b_1 \\ a_2 & b_2 \end{pmatrix}$$

Since the columns represent unit vectors, we have $a_1^2 + a_2^2 = b_1^2 + b_2^2 = 1$. If we regard (a_1, a_2) and (b_1, b_2) as plane coordinates, then in terms of polar coordinates we can express $\mathbf{A}$ as

$$\mathbf{A} = \begin{pmatrix} \cos\alpha & \cos\beta \\ \sin\alpha & \sin\beta \end{pmatrix} \tag{7.19}$$

Now $|\mathbf{A}| = 1$, therefore $\sin(\beta - \alpha) = \sin\beta \cdot \cos\alpha - \cos\beta \cdot \sin\alpha = 1$, and so $\beta = \alpha + 90^\circ + n \cdot 360^\circ$, where n is an integer. Inserting this value in (7.19), we obtain

$$\mathbf{A} = \begin{pmatrix} \cos\alpha & -\sin\alpha \\ \sin\alpha & \cos\alpha \end{pmatrix}$$

which represents a rotation through an angle α about the origin in the plane.

We now take $\mathbf{A}$ to be a 3×3 proper orthogonal matrix. If this is to be a rotation, it must leave a line through the origin fixed, namely the axis of rotation. If $\mathbf{v}$ is a non-zero vector along this line, then we must have

$$\mathbf{Av} = \mathbf{v} \tag{7.20}$$

and such a vector $\mathbf{v}$ $(\neq \mathbf{0})$ exists precisely when $\mathbf{A} - \mathbf{I}$ is singular. Now $\mathbf{A}$ is orthogonal, so we have

$$\mathbf{A} - \mathbf{I} = \mathbf{A} - \mathbf{A}^{\mathrm{T}}\mathbf{A} = (\mathbf{I} - \mathbf{A}^{\mathrm{T}})\mathbf{A}$$

Taking determinants, we find that $|\mathbf{A} - \mathbf{I}| = |\mathbf{I} - \mathbf{A}^{\mathrm{T}}| \cdot |\mathbf{A}|$. Now $|\mathbf{A}| = 1$, by hypothesis, and $|\mathbf{A} - \mathbf{I}| = |\mathbf{I} - \mathbf{A}^{\mathrm{T}}| = |\mathbf{I} - \mathbf{A}|$, by transposition. But we also have

$$|\mathbf{I} - \mathbf{A}| = (-1)^3 |\mathbf{A} - \mathbf{I}| = -|\mathbf{A} - \mathbf{I}|$$

therefore $|\mathbf{A} - \mathbf{I}| = -|\mathbf{A} - \mathbf{I}|$, that is,

$$|\mathbf{A} - \mathbf{I}| = 0 \tag{7.21}$$

as we wished to show. Thus $\mathbf{A} - \mathbf{I}$ is singular and as a consequence the homogeneous system (7.20) has a non-zero solution. We shall complete the proof by showing that $\mathbf{A}$ represents a rotation about the line $\mathbf{x} = \mathbf{v}$ as axis.

If $\mathbf{A}$ represents a transformation between two right-handed rectangular coordinate systems, we have an equation

$$\mathbf{x} = \mathbf{Ay}$$

relating the coordinates of a given point in the two systems. Now equation (7.20) means that there is a line through O, namely $\mathbf{x} = \lambda\mathbf{v}$, whose points have the same coordinates in both systems. We take this line as the 3-axis of a right-handed rectangular coordinate system. If the coordinates of its basis vectors, referred to the x-system, are $\mathbf{u}_1, \mathbf{u}_2, \mathbf{u}_3$, respectively, then $\mathbf{u}_3$ is a solution of (7.20), so that $\mathbf{Au}_3 = \mathbf{u}_3$. Moreover, since $\mathbf{A}$ is proper orthogonal, $\mathbf{Au}_1, \mathbf{Au}_2, \mathbf{Au}_3$ again form a right-handed rectangular system. The two sets $\mathbf{u}_1, \mathbf{u}_2, \mathbf{u}_3$ and $\mathbf{Au}_1, \mathbf{Au}_2, \mathbf{Au}_3$ have the third vector in common, and hence

$\mathbf{Au}_1, \mathbf{Au}_2$ are related to $\mathbf{u}_1, \mathbf{u}_2$ by a 2×2 proper orthogonal matrix. In other words, $\mathbf{A}$ represents a rotation about the line $\mathbf{x} \leftrightarrow \lambda \mathbf{u}_3$, as we had to show.

The theorem just proved deals completely with the case of a 3×3 proper orthogonal matrix. Suppose now that $\mathbf{A}$ is an improper orthogonal matrix. We take any matrix $\mathbf{T}$ representing a reflexion (for example (7.18)) and form $\mathbf{R} = \mathbf{AT}$; this matrix is proper orthogonal and we have a rotation, so $\mathbf{A} = \mathbf{RT}$ has been expressed as a rotation followed by a reflexion. Moreover, the reflexion could be chosen at will, and the rotation is then uniquely determined.

What is more interesting is that any orthogonal matrix in 3-space can be expressed as a product of at most three reflexions; this will follow if we show how to express a rotation as a product of two reflexions.

Any improper orthogonal 2×2 matrix $\mathbf{A}$ has the form

$$\mathbf{A} = \begin{pmatrix} \cos\alpha & \sin\beta \\ \sin\alpha & \cos\beta \end{pmatrix}$$

where $\cos(\alpha + \beta) = \cos\alpha\cos\beta - \sin\alpha\sin\beta = -1$. Hence $\alpha + \beta = 180^\circ$ and we have $\sin\beta = \sin\alpha$, $\cos\beta = -\cos\alpha$; thus the general improper orthogonal 2×2 matrix is given by

$$\mathbf{A} = \begin{pmatrix} \cos\alpha & \sin\alpha \\ \sin\alpha & -\cos\alpha \end{pmatrix}$$

This matrix represents the reflexion in the line making an angle $\alpha/2$ with the positive 1-axis. For the matrix $\mathbf{A}$, applied to the general unit-vector $(\cos\theta, \sin\theta)^{\mathrm{T}}$ gives $(\cos(\alpha - \theta), \sin(\alpha - \theta))^{\mathrm{T}}$. We now form the product of two reflexions:

$$\begin{pmatrix} \cos\alpha & \sin\alpha \\ \sin\alpha & -\cos\alpha \end{pmatrix}\begin{pmatrix} \cos\beta & \sin\beta \\ \sin\beta & -\cos\beta \end{pmatrix} = \begin{pmatrix} \cos(\alpha-\beta) & -\sin(\alpha-\beta) \\ \sin(\alpha-\beta) & \cos(\alpha-\beta) \end{pmatrix}.$$

Thus the result is a rotation through an angle $\alpha - \beta$. This shows that the general rotation in the plane, through an angle θ, can be obtained by first reflecting in the line making an angle $\alpha/2$ and then reflecting in the line making an angle $(\alpha - \theta)/2$ with the positive 1-axis. Here α can be chosen arbitrarily.

We have already seen that any proper orthogonal matrix in space represents a rotation, and taking the axis of rotation as the 3-axis, we have a rotation in the 12-plane, which we know is a product of two reflexions. Summing up, we have:

Theorem 7.2

Any orthogonal 3×3 matrix can be written as a product of at most three reflexions.

By comparing determinants we see that a proper orthogonal matrix is either $\mathbf{I}$ or a product of two reflexions, while an improper orthogonal matrix is a product of either one or three reflexions. For some improper orthogonal matrices all three reflexions are indeed needed; for example the matrix

$$\begin{pmatrix} -1 & 0 & 0 \\ 0 & -1 & 0 \\ 0 & 0 & -1 \end{pmatrix}$$

is improper orthogonal, but is not itself a reflexion.

The results of this section can all be carried over to n dimensions, but as they are not needed later, we simply state the results without proof. As in the plane and 3-space one can distinguish left-handed and right-handed coordinate systems in n dimensions, and while proper orthogonal transformations (those with determinant 1) preserve the orientation, improper transformations reverse it. A rotation in n-dimensional space leaves an $(n-2)$-dimensional subspace fixed; this is harder to visualize than a rotation in 3-space and one tends to express everything in terms of reflexions leaving a hyperplane $(=(n-1)$-space) fixed, which acts as a 'mirror'. There is then an analogue to Theorem 7.2, stating that every orthogonal transformation in n-space can be written as a product of at most n reflexions – see, for example, Artin (1957), and for the general background, Cohn (1989, section 6.3).

7.4 THE TRIPLE PRODUCT

With any triad of vectors $\mathbf{a}_1, \mathbf{a}_2, \mathbf{a}_3$ in 3-space a scalar called their **triple product** may be associated as follows. Let

$$\mathbf{a}_i \leftrightarrow \mathbf{x}_i \tag{7.22}$$

in some right-handed rectangular coordinate system. Then the triple product of $\mathbf{a}_1, \mathbf{a}_2, \mathbf{a}_3$ is defined as the determinant of the matrix with $\mathbf{x}_1, \mathbf{x}_2, \mathbf{x}_3$ as columns:

$$[\mathbf{a}_1, \mathbf{a}_2, \mathbf{a}_3] = |\mathbf{x}_1, \mathbf{x}_2, \mathbf{x}_3| \tag{7.23}$$

If in a second coordinate system, not necessarily rectangular, we have $\mathbf{a}_i \leftrightarrow \mathbf{y}_i$, then

$$\mathbf{x}_i = \mathbf{A}\mathbf{y}_i \quad (i = 1, 2, 3)$$

with a regular matrix $\mathbf{A}$ of transformation. Hence

$$\begin{aligned} |\mathbf{x}_1, \mathbf{x}_2, \mathbf{x}_3| &= |\mathbf{A}\mathbf{y}_1, \mathbf{A}\mathbf{y}_2, \mathbf{A}\mathbf{y}_3| \\ &= |\mathbf{A}| \cdot |\mathbf{y}_1, \mathbf{y}_2, \mathbf{y}_3| \end{aligned}$$

thus

$$|\mathbf{x}_1, \mathbf{x}_2, \mathbf{x}_3| = |\mathbf{A}| \cdot |\mathbf{y}_1, \mathbf{y}_2, \mathbf{y}_3| \tag{7.24}$$

We consider especially the case where the y-system is again rectangular. Then $\mathbf{A}$ is an orthogonal matrix, proper or improper according as the y-system is right- or left-handed. According to (7.23) we therefore have

$$[\mathbf{a}_1, \mathbf{a}_2, \mathbf{a}_3] = |\mathbf{y}_1, \mathbf{y}_2, \mathbf{y}_3| \text{ in a right-handed system} \tag{7.25}$$

$$[\mathbf{a}_1, \mathbf{a}_2, \mathbf{a}_3] = -|\mathbf{y}_1, \mathbf{y}_2, \mathbf{y}_3| \text{ in a left-handed system} \tag{7.26}$$

Of course, (7.23) may also be used to express the triple product in oblique coordinates, as in (7.24), but this will not be needed.

The triple product can also be used to give an invariant definition of the vector product. Let $\mathbf{a}$ and $\mathbf{b}$ be vectors and $\mathbf{a} \wedge \mathbf{b}$ their vector product in a fixed right-handed rectangular system, as defined by (6.24) in section 6.9. Then for any vector $\mathbf{c}$ we have, by the expansion rule for determinants ((5.7) in section 5.7),

$$[\mathbf{a}, \mathbf{b}, \mathbf{c}] = (\mathbf{a} \wedge \mathbf{b}) \cdot \mathbf{c} \tag{7.27}$$

Together with (7.25) this shows that the same rule (6.24) may be used to obtain the coordinates of $\mathbf{a} \wedge \mathbf{b}$ in *any* right-handed rectangular coordinate system, while in a left-handed system we have, by (7.26),

$$\mathbf{a} \wedge \mathbf{b} \leftrightarrow -(a_2b_3 - a_3b_2, a_3b_1 - a_1b_3, a_1b_2 - a_2b_1) \tag{7.28}$$

in terms of the coordinates (a_1, a_2, a_3) of $\mathbf{a}$ and (b_1, b_2, b_3) of $\mathbf{b}$.

Thus the general definition of $\mathbf{a} \wedge \mathbf{b}$, in any rectangular coordinate system – right- or left-handed – with basis vectors $\mathbf{e}_1, \mathbf{e}_2, \mathbf{e}_3$ may be written

$$\mathbf{a} \wedge \mathbf{b} = [\mathbf{a}, \mathbf{b}, \mathbf{e}_1]\mathbf{e}_1 + [\mathbf{a}, \mathbf{b}, \mathbf{e}_2]\mathbf{e}_2 + [\mathbf{a}, \mathbf{b}, \mathbf{e}_3]\mathbf{e}_3 \tag{7.29}$$

To obtain a geometrical interpretation of the triple product we use (7.27) and note that $\mathbf{a} \wedge \mathbf{b}$ is a vector at right angles to both $\mathbf{a}$ and $\mathbf{b}$, and of length equal to the area of the parallelogram spanned by $\mathbf{a}$ and $\mathbf{b}$. Now $[\mathbf{a}, \mathbf{b}, \mathbf{c}]$ is the product of this area by the projection of $\mathbf{c}$ on the perpendicular to $\mathbf{a}$ and $\mathbf{b}$; in other words, $[\mathbf{a}, \mathbf{b}, \mathbf{c}]$ is just the volume of the 3-cell spanned by $\mathbf{a}, \mathbf{b}, \mathbf{c}$, this volume being base area times height. This shows also that $[\mathbf{a}, \mathbf{b}, \mathbf{c}]$ vanishes if and only if $\mathbf{a}, \mathbf{b}$ and $\mathbf{c}$ lie in a plane, that is to say, are linearly dependent; of course, this also follows from the determinantal criterion for linear dependence established in section 5.14. Finally, to determine the sign, we take a rectangular system with basis vectors $\mathbf{e}_1, \mathbf{e}_2, \mathbf{e}_3$ chosen so that $\mathbf{e}_1$ is a positive scalar multiple of $\mathbf{a}, \mathbf{e}_2$ is in the plane of $\mathbf{a}$ and $\mathbf{b}$ and such that $\mathbf{b} \cdot \mathbf{e}_2 > 0$, and $\mathbf{e}_3$ is such that $\mathbf{c} \cdot \mathbf{e}_3 > 0$. This is always possible if $\mathbf{a}, \mathbf{b}, \mathbf{c}$ are linearly independent and by construction the triple $\mathbf{e}_1, \mathbf{e}_2, \mathbf{e}_3$ has the same orientation as $\mathbf{a}, \mathbf{b}, \mathbf{c}$, that is, both are right-handed or both are left-handed. Moreover, in this coordinate system $\mathbf{a} \leftrightarrow (a_1, 0, 0)$, $\mathbf{b} \leftrightarrow (b_1, b_2, 0)$, $\mathbf{c} \leftrightarrow (c_1, c_2, c_3)$ and a_1, b_2, c_3 are all positive. Hence the determinant of the coordinates of $\mathbf{a}, \mathbf{b}, \mathbf{c}$, namely $a_1b_2c_3$, is positive and by (7.25) and (7.26) $[\mathbf{a}, \mathbf{b}, \mathbf{c}]$ is positive or negative or zero according as $\mathbf{a}, \mathbf{b}, \mathbf{c}$ is a right-handed set or a left-handed set or is linearly dependent. If the angle between $\mathbf{a}$ and $\mathbf{b}$ is α, then

$|\mathbf{a} \wedge \mathbf{b}| = |\mathbf{a}||\mathbf{b}||\sin\alpha|$; if, further, $\mathbf{c}$ makes an angle γ with the plane of $\mathbf{a}$ and $\mathbf{b}$, then it makes an angle $90^\circ - \gamma$ with the normal to this plane and $|(\mathbf{a} \wedge \mathbf{b})\cdot\mathbf{c}| = |\mathbf{a}||\mathbf{b}||\mathbf{c}||\sin\alpha||\cos(90^\circ - \gamma)|$, hence the volume is $|\mathbf{a}|\,|\mathbf{b}|\,|\mathbf{c}|\,|\sin\alpha|\,|\sin\gamma|$.

If in (7.27) we take $\mathbf{c} = \mathbf{a} \wedge \mathbf{b}$, we see from the last result that $\mathbf{a} \wedge \mathbf{b}$ is oriented so that $\mathbf{a}, \mathbf{b}, \mathbf{a} \wedge \mathbf{b}$ form a right-handed set, unless $\mathbf{a}$ and $\mathbf{b}$ are linearly dependent.

As an example of the application of vector products and triple products let us calculate the shortest distance between two skew lines. We take two lines l and m, given by the equations

$$\mathbf{x} = \mathbf{p} + \lambda\mathbf{a} \quad (\mathbf{a} \neq \mathbf{0})$$
$$\mathbf{x} = \mathbf{q} + \mu\mathbf{b} \quad (\mathbf{b} \neq \mathbf{0})$$

The shortest distance between these lines is the length of their common perpendicular, as we saw in section 6.17. It may be obtained by taking any vector from a point on l to a point on m and projecting it on the common perpendicular to the two lines. Hence the shortest distance is $\mathbf{u}\cdot(\mathbf{q} - \mathbf{p})$, where $\mathbf{u}$ is a unit vector perpendicular to both $\mathbf{a}$ and $\mathbf{b}$. Such a unit vector may be obtained by normalizing $\mathbf{a} \wedge \mathbf{b}$, and so we obtain the expression

$$[\mathbf{a}, \mathbf{b}, \mathbf{q} - \mathbf{p}]/|\mathbf{a} \wedge \mathbf{b}|$$

for the shortest distance. In particular, the lines l and m intersect if and only if

$$[\mathbf{a}, \mathbf{b}, \mathbf{q} - \mathbf{p}] = 0 \tag{7.30}$$

In the example in section 6.17, $\mathbf{a} \wedge \mathbf{b} \leftrightarrow (5, 10, 15)$ and $\mathbf{q} - \mathbf{p} \leftrightarrow (9, -7, -3)$, hence the shortest distance is $14/\sqrt{14} = \sqrt{14} \sim 3.74$.

In n-space one can in a similar way form a determinant from the components of any n-tuple of vectors and this will represent the volume of the n-cell spanned by these vectors (with an appropriate definition of volume in n dimensions).

7.5 LINEAR MAPPINGS

We note that when we transform from the basis vectors $\mathbf{u}_i$ to $\mathbf{v}_j = \sum \mathbf{u}_i a_{ij}$, the coordinates transform from $\mathbf{x}$ to $\mathbf{y} = \mathbf{A}^{-1}\mathbf{x}$, so when the basis vectors are transformed by $\mathbf{A}$, the coordinates are transformed by its inverse $\mathbf{A}^{-1}$. This is expressed by saying that the coordinates transform **contragrediently** to the basis. It is a familiar phenomenon: if the currency **halves** in value through inflation, then prices will be **twice** as high. To give a geometrical example, taking the plane for simplicity, suppose that the coordinate axes are rotated through 30° (Fig. 7.3), so that

$$\mathbf{v}_1 = \tfrac{1}{2}(\sqrt{3}\mathbf{u}_1 + \mathbf{u}_2)$$
$$\mathbf{v}_2 = \tfrac{1}{2}(-\mathbf{u}_1 + \sqrt{3}\mathbf{u}_2)$$

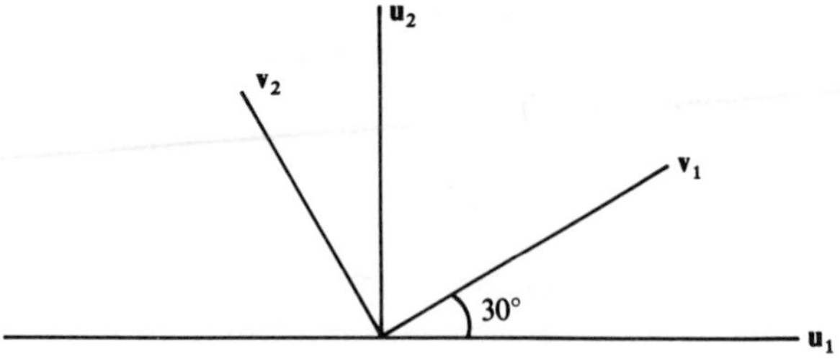

Fig. 7.3

The old coordinates **x** and the new coordinates **y** are related by the equations

$$\begin{aligned} \mathbf{y}_1 &= \tfrac{1}{2}(\sqrt{3}\mathbf{x}_1 - \mathbf{x}_2) \\ \mathbf{y}_2 &= \tfrac{1}{2}(\mathbf{x}_1 + \sqrt{3}\mathbf{x}_2) \end{aligned} \qquad (7.31)$$

Here the inverse matrix is the transpose, because the matrix is orthogonal. In geometric terms, we can achieve the transformation either by rotating the coordinate axes through $+30°$, or we can leave the axes fixed but rotate all points through $-30°$.

This provides us with another way of regarding (7.31); we may look upon it as an operation on all points of the plane, passing from $\mathbf{x} = (x_1, x_2)$ to $\mathbf{y} = (y_1, y_2)$ given by (7.31). Such an operation is called a **mapping**; this is nothing other than a function on our vector space, assigning to the vector $\mathbf{x}$ another vector $\mathbf{y} = f(\mathbf{x})$.

More generally, we may consider mappings from one vector space V to another, W. Such a mapping f is often written as $f: V \to W$. With each vector $\mathbf{x}$ in V it associates a vector $f(\mathbf{x})$ in W, a correspondence which is also expressed as $f: \mathbf{x} \mapsto f(\mathbf{x})$.

All our mappings are **linear**, meaning that they preserve linear combinations:

$$f(\lambda\mathbf{x}' + \mu\mathbf{x}'') = \lambda f(\mathbf{x}') + \mu f(\mathbf{x}'') \qquad (7.32)$$

Let us take bases $\mathbf{v}_1, \ldots, \mathbf{v}_n$ of V and $\mathbf{w}_1, \ldots, \mathbf{w}_m$ of W. Then any vector $\mathbf{x}$ in V may be written $\mathbf{x} = \sum \mathbf{v}_j x_j$ and by induction we obtain from (7.32),

$$f(\mathbf{x}) = \sum f(\mathbf{v}_j) x_j \qquad (7.33)$$

Here $f(\mathbf{v}_j)$ is a vector in W and so it may be written as a linear combination of the **w**s: $f(\mathbf{v}_j) = \sum \mathbf{w}_i a_{ij}$. Inserting this value in (7.33), we find

$$f(\mathbf{x}) = \sum \mathbf{w}_i a_{ij} x_j$$

Thus if $f(\mathbf{x}) = \mathbf{y} = \sum \mathbf{w}_i y_i$ say, then we have $y_i = \sum a_{ij} x_j$ or, in matrix form

$$\mathbf{y} = \mathbf{A}\mathbf{x} \qquad (7.34)$$

where $\mathbf{A} = (a_{ij})$. This shows that every linear mapping from V to W can be represented by an $m \times n$ matrix, operating on column vectors from V to produce column vectors in W.

Conversely, every mapping of the form (7.34) is linear, for as we saw in Chapter 3,

$$\mathbf{A}(\lambda\mathbf{x}' + \mu\mathbf{x}'') = \lambda\mathbf{A}\mathbf{x}' + \mu\mathbf{A}\mathbf{x}'' \tag{7.35}$$

where λ and μ range over all scalars. In the more general case where the mapping is of the form

$$\mathbf{y} = \mathbf{A}\mathbf{x} + \mathbf{b}$$

it is **affine**, and it is characterized by (7.35) holding for all λ, μ such that $\lambda + \mu = 1$.

Let us now look at some examples, taking for simplicity the case where $V = W$ and $n = 2$; thus we are dealing with linear mappings of a 2-dimensional space into itself. From (7.34) it is clear that the origin remains fixed under a linear mapping. We have already seen that the **rotation** through an angle α,

$$\mathbf{R}_\alpha = \begin{pmatrix} \cos\alpha & -\sin\alpha \\ \sin\alpha & \cos\alpha \end{pmatrix}$$

is a linear mapping. Another example is given by

$$\mathbf{D} = \begin{pmatrix} a & 0 \\ 0 & b \end{pmatrix}$$

which stretches vectors by a factor a in the 1-direction and by a factor b in the 2-direction; $\mathbf{D}$ is called a **dilation**. We remark that $\mathbf{D}$ may be singular; this will be so precisely when $ab = 0$, that is, when $a = 0$ or $b = 0$. Suppose that $a = 0$; then $\mathbf{D}$ represents a projection on the 2-axis and subsequent stretch by a factor b. The line $x_2 = h$ is mapped to the point $(0, bh)$. Here a 'stretch by a factor a' is of course understood to mean a contraction if $|a| < 1$ and a reversal of direction if $a < 0$.

A third type of mapping is given by the matrix

$$\mathbf{T} = \begin{pmatrix} 1 & c \\ 0 & 1 \end{pmatrix}$$

It transforms (x_1, x_2) to $(x_1 + cx_2, x_2)$; the effect is to shear each line $x_2 = h$ along the 1-axis by an amount ch, and $\mathbf{T}$ is called a **shear**.

These three types may be regarded as basic, in that every linear mapping can be built up from them. More precisely we can say that every 2×2 matrix can be written as a product of a rotation, a dilation and a shear. Intuitively this is clear, at least for regular matrices, for if the basis vectors $\mathbf{u}_1, \mathbf{u}_2$ become $f(\mathbf{u}_1)$, $f(\mathbf{u}_2)$, we rotate the plane to bring $f(\mathbf{u}_1)$ to lie along $\mathbf{u}_1$, then apply a shear to bring $f(\mathbf{u}_2)$ to lie along $\mathbf{u}_2$ and finally apply a dilation to bring $f(\mathbf{u}_1)$, $f(\mathbf{u}_2)$ to coincidence with $\mathbf{u}_1, \mathbf{u}_2$. Moreover, it does not matter whether we perform a shear first or a dilation, since we have

$$\begin{pmatrix} a & 0 \\ 0 & b \end{pmatrix}\begin{pmatrix} 1 & c \\ 0 & 1 \end{pmatrix} = \begin{pmatrix} a & ac \\ 0 & b \end{pmatrix} = \begin{pmatrix} 1 & ac/b \\ 0 & 1 \end{pmatrix}\begin{pmatrix} a & 0 \\ 0 & b \end{pmatrix}.$$

Here we have assumed $f(\mathbf{u}_1)$, $f(\mathbf{u}_2)$ linearly independent, which is the case just when the given matrix is regular, but the result holds quite generally. To prove it, we take a 2×2 matrix $\mathbf{A}$ and first assume that the first column is non-zero. We can express this column in polar coordinates as $(a\cos\alpha, a\sin\alpha)$. If we now multiply $\mathbf{A}$ by $\mathbf{R}_{-\alpha}$ on the left we reduce the (2, 1)th entry to zero:

$$\mathbf{R}_{-\alpha}\mathbf{A} = \begin{pmatrix} a & b \\ 0 & d \end{pmatrix}$$

This also covers the case previously excluded, where the first column is zero. If a, d are both non-zero (the regular case), we apply the inverse of the dilation $\mathbf{D} = \operatorname{diag}(a, d)$:

$$\mathbf{D}^{-1}\mathbf{R}_{-\alpha}\mathbf{A} = \begin{pmatrix} 1 & b/a \\ 0 & 1 \end{pmatrix}$$

This is a shear $\mathbf{T}$, say, and so we have expressed $\mathbf{A}$ in the desired form:

$$\mathbf{A} = \mathbf{R}_\alpha \mathbf{D}\mathbf{T} \tag{7.36}$$

The same formula holds even if $d = 0$. If $a = 0$ but $\mathbf{A} \neq \mathbf{0}$, we have

$$\mathbf{A} = \begin{pmatrix} 0 & b \\ 0 & d \end{pmatrix}$$

Either $d \neq 0$; then

$$\mathbf{A} = \begin{pmatrix} 1 & b/d \\ 0 & 1 \end{pmatrix}\begin{pmatrix} 0 & 0 \\ 0 & d \end{pmatrix} = \mathbf{T}\mathbf{D}$$

where $\mathbf{T}$ is a shear and $\mathbf{D}$ a dilation. Or $d = 0$ but $b \neq 0$; then, by a suitable rotation, we can transform the second column to a vector with non-zero second component and so reach the form

$$\mathbf{A} = \mathbf{R}_\alpha \mathbf{T}\mathbf{D}.$$

This only leaves the case $\mathbf{A} = \mathbf{0}$, which can be regarded as a special case of a dilation.

We remark that the rotation and shear leave areas unchanged, while the dilation $\mathbf{D} = \operatorname{diag}(a, b)$ multiplies areas by ab. In general any area is multiplied by $|\mathbf{A}|$, when a linear mapping with matrix $\mathbf{A}$ is applied; this follows from (7.36) and the multiplication theorem of determinants.

It is also possible to express every linear mapping as a product of shears and a dilation. Given any matrix

$$\mathbf{A} = \begin{pmatrix} a & b \\ c & d \end{pmatrix}$$

let us apply a shear $\begin{pmatrix} 1 & h \\ 0 & 1 \end{pmatrix}$. We obtain

$$\begin{pmatrix} a & b \\ c & d \end{pmatrix}\begin{pmatrix} 1 & h \\ 0 & 1 \end{pmatrix} = \begin{pmatrix} a & ah+b \\ c & ch+d \end{pmatrix}$$

Suppose that $a \neq 0$; then on taking $h = -b/a$ and writing $d' = d - bc/a$, we obtain

$$\begin{pmatrix} a & 0 \\ c & d' \end{pmatrix} = \begin{pmatrix} 1 & 0 \\ c/a & 1 \end{pmatrix}\begin{pmatrix} a & 0 \\ 0 & d' \end{pmatrix}$$

hence

$$\mathbf{A} = \begin{pmatrix} 1 & 0 \\ c/a & 1 \end{pmatrix}\begin{pmatrix} a & 0 \\ 0 & d' \end{pmatrix}\begin{pmatrix} 1 & b/a \\ 0 & 1 \end{pmatrix}$$

so **A** has been expressed as a product of a shear along the 2-axis, a dilation and a shear along the 1-axis. If $a = 0$, we multiply **A** by

$$\begin{pmatrix} 1 & 1 \\ 0 & 1 \end{pmatrix}\begin{pmatrix} 1 & 0 \\ -1 & 1 \end{pmatrix}\begin{pmatrix} 1 & 1 \\ 0 & 1 \end{pmatrix} = \begin{pmatrix} 0 & 1 \\ -1 & 0 \end{pmatrix}$$

on the right or left, according as $b \neq 0$ or $c \neq 0$, to obtain a matrix with non-zero (1, 1)th entry. If $a = b = c = 0$, **A** already has the form of a dilation.

In the general case of n dimensions it can again be shown that every linear mapping is a product of shears and a dilation – cf. for example, Cohn (1991, section 4.7).

As another illustration of linear mappings, and one with important applications, we consider the multiplication of complex numbers, where the set **C** of complex numbers is regarded as a 2-dimensional real vector space. The multiplication by $c = a + ib$ defines the linear mapping $z \mapsto cz$; writing $z = x + iy$, where x, y are real, we have

$$x + iy \mapsto (a + ib)(x + iy) = ax - by + i(ay + bx) \tag{7.37}$$

The matrix corresponding to this linear mapping is

$$\begin{pmatrix} a & -b \\ b & a \end{pmatrix}$$

as we see by comparing components, and this interpretation can in fact be used to introduce complex numbers. We define the matrix **J** as

$$\mathbf{J} = \begin{pmatrix} 0 & -1 \\ 1 & 0 \end{pmatrix}$$

and consider the linear mappings given by $a\mathbf{I} + b\mathbf{J}$ in a 2-dimensional real vector space. Since $\mathbf{IJ} = \mathbf{JI} = \mathbf{J}$, $\mathbf{J}^2 = -\mathbf{I}$, we obtain in this way all the properties of complex numbers. By using a model such as this one can prove the consistency of the complex numbers (which is more tedious to check when complex numbers are defined abstractly as number pairs with the multiplication given by (7.37)).

7.6 A NORMAL FORM FOR THE MATRIX OF A LINEAR MAPPING

In the last section we saw that every linear mapping f from an n-dimensional vector space V to an m-dimensional vector space W can be represented by an $m \times n$ matrix, relative to given bases in V and W. A natural question at this point asks how this matrix changes under a change of basis in V and W. Let us take bases $\mathbf{v}_1, \ldots, \mathbf{v}_n$ of V and $\mathbf{w}_1, \ldots, \mathbf{w}_m$ of W and suppose that the matrix of f relative to these bases is $\mathbf{A}$; so if $\mathbf{r} = \sum \mathbf{v}_i x_i$ is mapped to $\mathbf{s} = \sum \mathbf{w}_j y_j$, we have, on writing $\mathbf{x} = (x_1, \ldots, x_n)^{\mathrm{T}}$, $\mathbf{y} = (y_1, \ldots, y_m)^{\mathrm{T}}$,

$$\mathbf{y} = \mathbf{A}\mathbf{x} \tag{7.38}$$

If $\mathbf{v}'_1, \ldots, \mathbf{v}'_n$ is a second basis of V, related to the first by equations

$$\mathbf{v}'_j = \sum \mathbf{v}_i p_{ij}$$

where $\mathbf{P} = (p_{ij})$, the matrix of transformation from the $\mathbf{v}_i$ to the $\mathbf{v}'_i$, is regular of order n, then we have for the general vector $\mathbf{r}$ in V,

$$\mathbf{r} = \sum \mathbf{v}_i x_i = \sum \mathbf{v}'_j x'_j = \sum \mathbf{v}_i p_{ij} x'_j$$

and hence $x_i = \sum p_{ij} x'_j$, or, in matrix form,

$$\mathbf{x} = \mathbf{P}\mathbf{x}' \tag{7.39}$$

Similarly, let $\mathbf{w}'_1, \ldots, \mathbf{w}'_m$ be a second basis of W, related to the first by

$$\mathbf{w}'_l = \sum \mathbf{w}_k q_{kl}$$

where $\mathbf{Q} = (q_{kl})$ is a regular $m \times m$ matrix; then the general vector $\mathbf{s}$ in W is given by

$$\mathbf{s} = \sum \mathbf{w}_k y_k = \sum \mathbf{w}'_l y'_l = \sum \mathbf{w}_k q_{kl} y'_l$$

so we have

$$\mathbf{y} = \mathbf{Q}\mathbf{y}' \tag{7.40}$$

If the matrix of f relative to the new bases $\mathbf{v}'_i$ of V and $\mathbf{w}'_k$ of W is $\mathbf{A}'$, then corresponding to (7.38) we have

$$\mathbf{y}' = \mathbf{A}'\mathbf{x}'$$

Now we substitute from (7.39) and (7.40) into (7.38) and find $\mathbf{Q}\mathbf{y}' = \mathbf{A}\mathbf{P}\mathbf{x}'$, hence $\mathbf{y}' = \mathbf{Q}^{-1}\mathbf{A}\mathbf{P}\mathbf{x}'$ and so we obtain

$$\mathbf{A}' = \mathbf{Q}^{-1}\mathbf{A}\mathbf{P} \tag{7.41}$$

This formula expresses the matrix in the new coordinates in terms of the matrix in the old coordinates and the matrices **P**, **Q** describing the transformations of coordinates.

From (7.41) it follows that **A** and **A**′ have the same rank. This is implicit in the results of Chapter 4 (which showed that multiplication by a regular matrix does not change the rank), but it may also be seen as follows. We have $\mathrm{rk}(\mathbf{AP}) \leqslant \mathrm{rk}(\mathbf{A})$, because the columns of **AP** are linear combinations of those of **A**. Similarly, using row rank, we see that $\mathrm{rk}(\mathbf{Q}^{-1}\mathbf{AP}) \leqslant \mathrm{rk}(\mathbf{AP})$, hence $\mathrm{rk}(\mathbf{A}') \leqslant \mathrm{rk}(\mathbf{A})$. By symmetry, writing $\mathbf{A} = \mathbf{QA}'\mathbf{P}^{-1}$, we have $\mathrm{rk}(\mathbf{A}) \leqslant \mathrm{rk}(\mathbf{A}')$, and hence

$$\mathrm{rk}(\mathbf{A}) = \mathrm{rk}(\mathbf{A}').$$

This number is called the **rank** of the linear mapping f, written $\mathrm{rk}\, f$. Soon we shall meet a geometric interpretation of this number.

As an example let us take the mapping $f:(x_1, x_2, x_3) \mapsto (x_2, x_1 + x_3)$ from $\mathbf{R}^3$ to $\mathbf{R}^2$ with the standard bases, and introduce new bases (1, 1, 0), (1, 0, 1), (0, 1, 1) and (2, 3), (3, 5). Then

$$\mathbf{A} = \begin{pmatrix} 0 & 1 & 0 \\ 1 & 0 & 1 \end{pmatrix}$$

$$\mathbf{P} = \begin{pmatrix} 1 & 1 & 0 \\ 1 & 0 & 1 \\ 0 & 1 & 1 \end{pmatrix}$$

$$\mathbf{Q} = \begin{pmatrix} 2 & 3 \\ 3 & 5 \end{pmatrix} \quad \mathbf{Q}^{-1} = \begin{pmatrix} 5 & -3 \\ -3 & 2 \end{pmatrix}$$

hence

$$\mathbf{A}' = \begin{pmatrix} 5 & -3 \\ -3 & 2 \end{pmatrix} \begin{pmatrix} 0 & 1 & 0 \\ 1 & 0 & 1 \end{pmatrix} \begin{pmatrix} 1 & 1 & 0 \\ 1 & 0 & 1 \\ 0 & 1 & 1 \end{pmatrix} = \begin{pmatrix} 5 & -3 \\ -3 & 2 \end{pmatrix} \begin{pmatrix} 1 & 0 & 1 \\ 1 & 2 & 1 \end{pmatrix}$$

$$= \begin{pmatrix} 2 & -6 & 2 \\ -1 & 4 & -1 \end{pmatrix}$$

In general one will choose the bases in V and W so as to give the matrix of f a particularly simple form. As an illustration we shall use Theorem 4.3 to prove a result on the normal form of the matrix for f:

Theorem 7.3

Let $f: V \to W$ be any linear mapping. Then we can find bases $\mathbf{v}_1, \ldots, \mathbf{v}_n$ of V and $\mathbf{w}_1, \ldots, \mathbf{w}_m$ of W such that f is given by

$$\mathbf{v}_i \mapsto \begin{cases} \mathbf{w}_i & i = 1, \ldots, r \\ 0 & i = r+1, \ldots, n \end{cases} \tag{7.42}$$

where r is the rank of f. Thus the matrix of f relative to these bases is $\mathbf{I}_r \oplus \mathbf{0}$.

Proof

Choose any bases in V and W and let $\mathbf{A}$ be the matrix of f relative to these bases. By Theorem 4.3, if $\text{rk}(\mathbf{A}) = r$, there exist invertible matrices $\mathbf{P}, \mathbf{Q}$ such that

$$\mathbf{Q}^{-1}\mathbf{A}\mathbf{P} = \begin{pmatrix} \mathbf{I}_r & \mathbf{0} \\ \mathbf{0} & \mathbf{0} \end{pmatrix} \tag{7.43}$$

If we now transform the bases in V and W by $\mathbf{P}$ and $\mathbf{Q}$, respectively, the matrix of f in the new coordinates will be $\mathbf{A}' = \mathbf{I}_r \oplus \mathbf{0}$, by (7.41) and (7.43); in detail, we have $\mathbf{A}' = (a'_{ki})$, where $a'_{ki} = 1$ for $k = i \leqslant r$ and 0 otherwise. If the bases are $\mathbf{v}_1, \ldots, \mathbf{v}_n$ and $\mathbf{w}_1, \ldots, \mathbf{w}_m$, we have $f: \mathbf{v}_i \mapsto \sum \mathbf{w}_k a'_{ki}$ which is $\mathbf{w}_i$ if $i \leqslant r$ and 0 otherwise, and this is just (7.42).

7.7 THE IMAGE AND KERNEL OF A LINEAR MAPPING

Theorem 7.3 of the last section can also be proved more directly, in a way which will help us to understand the nature of linear mappings. A linear mapping $f: V \to W$ assigns to each vector $\mathbf{x}$ in V a vector in W. Thus we can say that for each $\mathbf{x} \in V$ we have $f(\mathbf{x}) \in W$. The set of all these images is also called the **image** of f, written $\text{im}\, f$. This set is itself a subspace of W; for $\mathbf{0} \in \text{im}\, f$, and if $\mathbf{y}', \mathbf{y}'' \in \text{im}\, f$, say $\mathbf{y}' = f(\mathbf{x}')$, $\mathbf{y}'' = f(\mathbf{x}'')$, then $\lambda\mathbf{y}' + \mu\mathbf{y}'' = \lambda f(\mathbf{x}') + \mu f(\mathbf{x}'') = f(\lambda\mathbf{x}' + \mu\mathbf{x}'')$ by linearity, hence $\lambda\mathbf{y}' + \mu\mathbf{y}'' \in \text{im}\, f$, and this shows $\text{im}\, f$ to be a subspace of W. If this subspace is the whole of W, thus $\text{im}\, f = W$, then f is said to map V **onto** W and is also called **surjective**.

A second subspace associated with f, the **kernel** of f, denoted by $\ker f$, is defined as the set of all vectors in V mapped to $\mathbf{0}$ by f. To verify that it is a subspace, we have $\mathbf{0} \in \ker f$, and if $\mathbf{x}', \mathbf{x}'' \in \ker f$, then $f(\mathbf{x}') = f(\mathbf{x}'') = \mathbf{0}$, hence for any scalars λ, μ, $f(\lambda\mathbf{x}' + \mu\mathbf{x}'') = \lambda f(\mathbf{x}') + \mu f(\mathbf{x}'') = \mathbf{0}$, which shows that $\lambda\mathbf{x}' + \mu\mathbf{x}'' \in \ker f$, so $\ker f$ is indeed a subspace. The mapping f is said to be **one-one** or **injective** if distinct vectors have distinct images. We note that a necessary and sufficient condition for f to be injective is that $\ker f = \mathbf{0}$. For if f is injective, then no vector other than zero can map to zero; conversely, if $\ker f = \mathbf{0}$, and $f(\mathbf{u}) = f(\mathbf{u}')$, then $f(\mathbf{u} - \mathbf{u}') = f(\mathbf{u}) - f(\mathbf{u}') = \mathbf{0}$, so $\mathbf{u} - \mathbf{u}' \in \ker f$ and hence $\mathbf{u} = \mathbf{u}'$, so f is injective. If f is injective and surjective, it is called an **isomorphism** between V and W, with inverse f^{-1}.

Given a linear mapping $f: V \to W$, let us take a basis of $\ker f$, say $\mathbf{v}_1, \ldots, \mathbf{v}_s$ and complete this set to a basis of V (as in section 1.10), say $\mathbf{v}_1, \ldots, \mathbf{v}_s, \mathbf{v}_{s+1}, \ldots, \mathbf{v}_n$. Then $\text{im}\, f$ is spanned by the vectors $f(\mathbf{v}_i)$, $i = 1, \ldots, n$. Of these, $f(\mathbf{v}_1) = \cdots = f(\mathbf{v}_s) = \mathbf{0}$ by construction, since $\mathbf{v}_i \in \ker f$ for $i \leqslant s$. We claim that $f(\mathbf{v}_{s+1}), \ldots, f(\mathbf{v}_n)$ are linearly independent and so form a basis of $\text{im}\, f$. For suppose that

$$\alpha_{s+1} f(\mathbf{v}_{s+1}) + \cdots + \alpha_n f(\mathbf{v}_n) = \mathbf{0}. \tag{7.44}$$

Then $f(\sum \alpha_i \mathbf{v}_i) = \sum \alpha_i f(\mathbf{v}_i) = \mathbf{0}$, hence $\sum_{s+1}^n \alpha_i \mathbf{v}_i \in \ker f$, and since $\mathbf{v}_1, \ldots, \mathbf{v}_s$ is a basis for ker f, we have

$$\sum_{s+1}^{n} \alpha_i \mathbf{v}_i = \sum_{1}^{s} \lambda_j \mathbf{v}_j$$

for some scalars λ_j. But $\mathbf{v}_1, \ldots, \mathbf{v}_n$ are linearly independent, so all the coefficients in this relation must vanish, and this shows (7.44) to be the trivial relation. Hence $f(\mathbf{v}_{s+1}), \ldots, f(\mathbf{v}_n)$ are linearly independent and so form a basis of im f.

Since the vectors $\mathbf{w}_j = f(\mathbf{v}_{s+j})$, $j = 1, \ldots, n-s$ are linearly independent, we can complete them to a basis $\mathbf{w}_1, \ldots, \mathbf{w}_m$ of W and the linear mapping f is then completely defined by the rule

$$\mathbf{v}_i \mapsto \begin{cases} \mathbf{0} & i = 1, \ldots, s \\ \mathbf{w}_{i-s} & i = s+1, \ldots, n \end{cases}$$

and this is just of the form (7.42), except for the numbering. We have thus obtained another proof of formula (7.43), which incidentally provides an interpretation for the rank of f: rk(f) is the dimension of im f. Further, we note the relation between the dimensions of ker f and im f: the construction of the bases shows that $n - s = r = \text{rk}(f)$, so we have proved the dimension formula for linear mappings:

Theorem 7.4
For any linear mapping $f: V \to W$ the dimensions of the image and the kernel of f are related by the formula

$$\dim \ker f + \dim \operatorname{im} f = \dim V \tag{7.45}$$

Finally, we note the special case when $W = V$. If we have a linear mapping $f: V \to V$, whose matrix relative to one basis is $\mathbf{A}$ and relative to another $\mathbf{A}'$, then

$$\mathbf{A}' = \mathbf{P}^{-1}\mathbf{A}\mathbf{P}, \tag{7.46}$$

where $\mathbf{P}$ is the matrix of transformation between these bases. Now we can no longer achieve the form (7.43); the normal form under such transformations (7.46) will be discussed in Chapter 8.

EXERCISES

7.1. A right-handed coordinate system is rotated through an angle of 120° about the line $\mathbf{x} \leftrightarrow \lambda(1, 1, 1)$; find the matrix $\mathbf{A}$ of transformation. Verify that $\mathbf{A}$ is proper orthogonal and show also that $\mathbf{A}^3 = \mathbf{I}$.

7.2. Let $\mathbf{a}$ be a non-zero vector. Verify that the transformation defined by

$$\mathbf{x} = \mathbf{y} - 2\frac{\mathbf{a} \cdot \mathbf{y}}{\mathbf{a} \cdot \mathbf{a}} \cdot \mathbf{a}$$

is orthogonal, and give a geometrical interpretation. (*Hint*: Find the points fixed under this transformation.)

7.3. Show that in n dimensions $-\mathbf{I}$ is orthogonal and is proper if n is even, and improper if n is odd.

7.4. Show that if $\mathbf{A}$ is a 3×3 orthogonal matrix such that $\mathbf{I} + \mathbf{A}$ is singular, then either $\mathbf{A}$ is improper or $\mathbf{A}$ represents a rotation through 180°.

7.5. A matrix $\mathbf{S}$ is said to be **skew-symmetric** if it equals its negative transpose, that is, if $\mathbf{S} = -\mathbf{S}^T$. Show that any orthogonal matrix $\mathbf{A}$ such that $\mathbf{I} + \mathbf{A}$ is regular can be expressed in the form $\mathbf{A} = (\mathbf{I} - \mathbf{S})(\mathbf{I} + \mathbf{S})^{-1}$, where $\mathbf{S}$ is a skew-symmetric matrix. (*Hint*: Find $\mathbf{S}$ to satisfy $(\mathbf{I} + \mathbf{A})(\mathbf{I} + \mathbf{S}) = (\mathbf{I} + \mathbf{S})\cdot(\mathbf{I} + \mathbf{A}) = 2\mathbf{I}$.)

7.6. Show that three points A, B, C with position vectors $\mathbf{a}, \mathbf{b}, \mathbf{c}$, respectively, are coplanar with the origin if and only if $[\mathbf{a}, \mathbf{b}, \mathbf{c}] = 0$.

7.7. Show that the equation of the plane through three non-collinear points with position vectors $\mathbf{a}, \mathbf{b}, \mathbf{c}$ may be written

$$[\mathbf{x} - \mathbf{a}, \mathbf{x} - \mathbf{b}, \mathbf{x} - \mathbf{c}] = 0$$

(*Hint*: Use Exercise 7.6 to express the condition for four points to be collinear.)

7.8. Show that for any vectors $\mathbf{a}, \mathbf{b}, \mathbf{c}$,

$$(\mathbf{a} \wedge \mathbf{b}) \wedge \mathbf{c} = (\mathbf{a} \cdot \mathbf{c}) \cdot \mathbf{b} - (\mathbf{b} \cdot \mathbf{c}) \cdot \mathbf{a}$$

(*Hint*: Show first that the left-hand side is linearly dependent on $\mathbf{a}$ and $\mathbf{b}$ and determine the ratio of the coefficients by taking scalar products with $\mathbf{c}$. This determines the right-hand side up to a constant factor, which may be found, for example, by comparing coefficients of $a_1 b_2 c_1$ in a rectangular coordinate system.)

7.9. In any coordinate system with basis vectors $\mathbf{a}_1, \mathbf{a}_2, \mathbf{a}_3$, let $\mathbf{b}_1, \mathbf{b}_2, \mathbf{b}_3$ be such that $\mathbf{a}_i \cdot \mathbf{b}_j = \delta_{ij}$. Show that $\mathbf{b}_1, \mathbf{b}_2, \mathbf{b}_3$ again form a basis (it is called the basis **dual** to the **a**s). Show also that the coordinates of $\mathbf{x} \wedge \mathbf{y}$ in this system are $[\mathbf{x}, \mathbf{y}, \mathbf{b}_1]$, $[\mathbf{x}, \mathbf{y}, \mathbf{b}_2]$, $[\mathbf{x}, \mathbf{y}, \mathbf{b}_3]$; thus

$$\mathbf{x} \wedge \mathbf{y} = \sum_i [\mathbf{x}, \mathbf{y}, \mathbf{b}_i]\mathbf{a}_i$$

7.10. Show that for any vectors $\mathbf{a}, \mathbf{b}, \mathbf{c}, \mathbf{d}$,

$$(\mathbf{a} \wedge \mathbf{b}) \cdot (\mathbf{c} \wedge \mathbf{d}) = (\mathbf{a} \cdot \mathbf{c})(\mathbf{b} \cdot \mathbf{d}) - (\mathbf{a} \cdot \mathbf{d})(\mathbf{b} \cdot \mathbf{c})$$

and apply the result to prove (6.28) in section 6.9.

7.11. Show that if $\mathbf{a}$ is any non-zero vector and $\mathbf{u}$ is a unit vector orthogonal to $\mathbf{a}$, then the equation

$$\mathbf{u} \wedge \mathbf{x} = \mathbf{a}$$

represents the line in the direction $\mathbf{u}$ through the point $\mathbf{u} \wedge \mathbf{a}$.

7.12. Find the shortest distance between pairs of opposite edges (produced if necessary) of the tetrahedron $ABCD$, where $A\leftrightarrow(0,0,0)$, $B\leftrightarrow(3,1,2)$, $C\leftrightarrow(2,-1,4)$, $D\leftrightarrow(-1,3,2)$, in a rectangular coordinate system.

7.13. Show that a linear mapping of a finite-dimensional vector space into itself is injective if and only if it is surjective (and hence it is then an isomorphism).

7.14. Show that complex conjugation: $a\mapsto\bar{a}$ is a linear mapping of **C** considered as a vector space over **R** but not when considered as a space over **C**.

7.15. Show that on the vector space of all polynomials in x the operator $\mathbf{T}: f(x)\mapsto f(x+1)$ is a linear mapping, which is invertible, and find its inverse.

7.16. Let V be the vector space of all polynomials in x. Show that the rule associating with each polynomial $f(x)$ its derivative $f'(x)$ is a linear mapping of V into itself. Verify that this mapping is surjective but not injective. Find an example of a linear mapping of V into itself which is injective but not surjective.

8

Normal forms of matrices

This chapter is devoted to the classification of square matrices. In section 8.5 we shall see that quadratic forms are described by symmetric matrices and the object will be to show that each quadratic form can be transformed to a diagonal form which is essentially unique. In terms of the matrices this provides a normal form for symmetric matrices under congruence transformations. Secondly, there is the more general problem of classifying square matrices under similarity transformations; this amounts to finding a coordinate system in which a given linear mapping has a simple description. This is superficially a different process, but it includes an important special case of the former. Here the complete answer (as well as the proof) is more complicated, so we begin by treating a number of special cases of this reduction in sections 8.1–8.4, then deal with quadratic forms in sections 8.5 and 8.6, and in section 8.7 complete the classification by presenting the Jordan normal form.

8.1 SIMILARITY OF MATRICES

Let V be a vector space of finite dimension n, say, and $f: V \to V$ a linear mapping. Relative to a basis of V with coordinates $\mathbf{x} = (x_1, \ldots, x_n)^T$ we can describe f by an $n \times n$ matrix $\mathbf{A}: \mathbf{x} \mapsto \mathbf{y}$, where

$$\mathbf{y} = \mathbf{A}\mathbf{x} \tag{8.1}$$

If the coordinates in a second basis of V are $\mathbf{x}'$, related to $\mathbf{x}$ by

$$\mathbf{x} = \mathbf{P}\mathbf{x}' \tag{8.2}$$

with a regular matrix $\mathbf{P}$, then in the new coordinates f takes the form $\mathbf{A}': \mathbf{x}' \mapsto \mathbf{y}'$, where $\mathbf{y}' = \mathbf{A}'\mathbf{x}'$, and hence

$$\mathbf{A}'\mathbf{x}' = \mathbf{y}' = \mathbf{P}^{-1}\mathbf{y} = \mathbf{P}^{-1}\mathbf{A}\mathbf{x} = \mathbf{P}^{-1}\mathbf{A}\mathbf{P}\mathbf{x}'$$

so

$$\mathbf{A}' = \mathbf{P}^{-1}\mathbf{A}\mathbf{P} \tag{8.3}$$

as we already saw in (7.46) in section 7.7. Two matrices $\mathbf{A}$, $\mathbf{A}'$ are said to be **similar** if there exists an invertible matrix $\mathbf{P}$ such that (8.3) holds. What has been said shows that the matrices of a given linear mapping (of a vector space into itself) in different coordinates are similar. Conversely, if $\mathbf{A}$ is a matrix similar to $\mathbf{A}'$, related by (8.3), and $\mathbf{A}$ represents the linear mapping f relative to coordinates $\mathbf{x}$ in V, then we have $\mathbf{y} = \mathbf{Ax}$, and, by (8.3), $\mathbf{P}^{-1}\mathbf{y} = \mathbf{P}^{-1}\mathbf{AP}\cdot\mathbf{P}^{-1}\mathbf{x}$, so on defining new coordinates $\mathbf{x}'$ in V by (8.2), we see that $\mathbf{A}'$ is the matrix of f in the new coordinates. Thus we have:

Theorem 8.1

The matrices of a given linear mapping f of a vector space V in different coordinate systems are similar, and conversely, if f is represented by $\mathbf{A}$ in one system, then any matrix similar to $\mathbf{A}$ represents f in a suitable coordinate system in V.

For example, the identity mapping in V, leaving all vectors in V fixed, has the matrix $\mathbf{I}$ in any coordinate system, for $\mathbf{I}$ does not change under similarity transformation: $\mathbf{P}^{-1}\mathbf{IP} = \mathbf{I}$. The same holds for the zero mapping, $\mathbf{x} \mapsto \mathbf{0}$, or more generally, for the scalar mapping $\mathbf{x} \mapsto \alpha\mathbf{x} (\alpha \in \mathbf{R})$. In all other cases the matrix for f changes under similarity transformation and this fact can be exploited to choose the coordinates so as to give the matrix of f a particularly simple form.

As an example, let us take the matrix

$$\mathbf{A} = \begin{pmatrix} 15 & -21 \\ 10 & -14 \end{pmatrix}$$

We cannot hope to transform $\mathbf{A}$ to scalar form, because, as we saw, any matrix similar to a scalar matrix is itself scalar; but it may be possible to transform $\mathbf{A}$ to diagonal form. Thus we must find a regular matrix $\mathbf{P}$ such that $\mathbf{P}^{-1}\mathbf{AP}$ is diagonal. If $\mathbf{P} = \begin{pmatrix} p & q \\ r & s \end{pmatrix}$, we need to solve

$$\begin{pmatrix} 15 & -21 \\ 10 & -14 \end{pmatrix}\begin{pmatrix} p & q \\ r & s \end{pmatrix} = \begin{pmatrix} p & q \\ r & s \end{pmatrix}\begin{pmatrix} \alpha & 0 \\ 0 & \beta \end{pmatrix}$$

for suitable α, β. This gives the four equations

$$\begin{aligned} 15p - 21r &= \alpha p \\ 10p - 14r &= \alpha r \\ 15q - 21s &= \beta q \\ 10q - 14s &= \beta s \end{aligned}$$

Hence $(15 - \alpha)p = 21r$, $10p = (14 + \alpha)r$, and so

$$\frac{15 - \alpha}{10} = \frac{21}{14 + \alpha} \tag{8.4}$$

which is satisfied by $\alpha = 0$, $p = 7$, $r = 5$. The same equations hold for q and

s, leading again to (8.4) with α replaced by β. If we again take $\beta = 0$, the values of q and s would make $\mathbf{P}$ singular, but in fact (8.4) is equivalent to a quadratic equation for α:

$$(15 - x)(14 + x) - 210 = 0$$

which simplifies to $x^2 - x = 0$, with roots $x = 0, 1$. If we use the root $\beta = 1$, we obtain $14q = 21s$, or $2q = 3s$, which is satisfied by $q = 3$, $s = 2$, and we obtain

$$\mathbf{P} = \begin{pmatrix} 7 & 3 \\ 5 & 2 \end{pmatrix} \tag{8.5}$$

This is a regular matrix with inverse

$$\mathbf{P}^{-1} = \begin{pmatrix} -2 & 3 \\ 5 & -7 \end{pmatrix}$$

and so we find

$$\mathbf{P}^{-1}\mathbf{A}\mathbf{P} = \begin{pmatrix} -2 & 3 \\ 5 & -7 \end{pmatrix}\begin{pmatrix} 15 & -21 \\ 10 & -14 \end{pmatrix}\begin{pmatrix} 7 & 3 \\ 5 & 2 \end{pmatrix} = \begin{pmatrix} -2 & 3 \\ 5 & -7 \end{pmatrix}\begin{pmatrix} 0 & 3 \\ 0 & 2 \end{pmatrix}$$
$$= \begin{pmatrix} 0 & 0 \\ 0 & 1 \end{pmatrix}$$

Thus we have accomplished the similarity transformation of $\mathbf{A}$ to diagonal form.

Secondly, let us take

$$\mathbf{A} = \begin{pmatrix} 35 & -49 \\ 25 & -35 \end{pmatrix} \tag{8.6}$$

We have to solve

$$\begin{pmatrix} 35 & -49 \\ 25 & -35 \end{pmatrix}\begin{pmatrix} p & q \\ r & s \end{pmatrix} = \begin{pmatrix} p & q \\ r & s \end{pmatrix}\begin{pmatrix} \alpha & 0 \\ 0 & \beta \end{pmatrix},$$

which is equivalent to the equations

$$35p - 49r = \alpha p$$
$$25p - 35r = \alpha r$$
$$35q - 49s = \beta q$$
$$25q - 35s = \beta s$$

We have $(35 - \alpha)p = 49r$, $25p = (35 + \alpha)r$, and so

$$\frac{35 - \alpha}{49} = \frac{25}{35 + \alpha}$$

The quadratic for α is now $(35 - x)(35 + x) - 1225 = 0$, which simplifies to $x^2 = 0$. So there is only one possible value for α or β, namely $\alpha = \beta = 0$. This already shows the diagonal form to be impossible; for if $\alpha = \beta = 0$, we have a scalar matrix on the right and we saw that this cannot happen. In fact, we again have $p = 7$, $r = 5$, but now q, s are necessarily proportional to p, r, so we cannot find a regular matrix **P** to transform **A** to diagonal form. Thus not every square matrix is similar to a diagonal matrix; if we transform **A** by the matrix **P** in (8.5), used in the last example, we find

$$\begin{pmatrix} -2 & 3 \\ 5 & -7 \end{pmatrix}\begin{pmatrix} 35 & -49 \\ 25 & -35 \end{pmatrix}\begin{pmatrix} 7 & 3 \\ 5 & 2 \end{pmatrix} = \begin{pmatrix} 5 & -7 \\ 0 & 0 \end{pmatrix}\begin{pmatrix} 7 & 3 \\ 5 & 2 \end{pmatrix} = \begin{pmatrix} 0 & 1 \\ 0 & 0 \end{pmatrix}$$

It is easily checked directly that the matrix on the right is not similar to a diagonal matrix (Exercise 8.2).

In the course of this chapter we shall describe a normal form for matrices under similarity, but we begin by dealing with the special case of matrices similar to a matrix in diagonal form.

8.2 THE CHARACTERISTIC EQUATION; EIGENVALUES

Let us consider the general problem of transforming a matrix to diagonal form by similarity transformation. If the square matrix **A** is similar to a diagonal matrix **D**, then for some regular matrix **P** we have $\mathbf{P}^{-1}\mathbf{A}\mathbf{P} = \mathbf{D}$ or

$$\mathbf{AP} = \mathbf{PD} \tag{8.7}$$

Let **D** have diagonal elements $d_1, \ldots, d_n$, thus $\mathbf{D} = \mathrm{diag}(d_1, \ldots, d_n)$, and put $\mathbf{A} = (a_{ij})$, $\mathbf{P} = (p_{ij})$. Then we can write this equation as

$$\sum_j a_{ij} p_{jk} = p_{ik} d_k.$$

If the columns of **P** are denoted by $\mathbf{p}_1, \ldots, \mathbf{p}_n$, then these equations can be expressed in matrix form as

$$\mathbf{A}\mathbf{p}_k = d_k \mathbf{p}_k \quad k = 1, \ldots, n \tag{8.8}$$

For a given k this is a homogeneous system of equations for the components of $\mathbf{p}_k$, and it has a non-trivial solution, yielding a non-zero vector $\mathbf{p}_k$ satisfying (8.8) precisely when the matrix $\mathbf{A} - d_k\mathbf{I}$ is singular. So to investigate the possibility of transforming **A** to diagonal form we need to study the equation

$$|x\mathbf{I} - \mathbf{A}| = 0 \tag{8.9}$$

This is called the **characteristic equation** of **A** and $|x\mathbf{I} - \mathbf{A}|$ is the **characteristic polynomial**. For an $n \times n$ matrix $\mathbf{A} = (a_{ij})$ it is

$$\varphi(x) = \begin{vmatrix} x - a_{11} & -a_{12} & -a_{13} & \cdots \\ -a_{21} & x - a_{22} & -a_{23} & \cdots \\ -a_{31} & -a_{32} & x - a_{33} & \cdots \\ \cdots & \cdots & \cdots & \cdots \end{vmatrix}$$

This makes it clear that $\varphi(x)$ is a polynomial of degree n with highest coefficient 1, which can be expanded as

$$\varphi(x) = x^n + c_1 x^{n-1} + \cdots + c_n \tag{8.10}$$

Here $c_1 = -(a_{11} + a_{22} + \cdots + a_{nn})$; the expression $-c_1$, the sum of the diagonal entries of $\mathbf{A}$, is called the **trace** of $\mathbf{A}$, $\mathrm{tr}(\mathbf{A})$. The constant term of $\varphi(x)$, obtained by putting $x = 0$, is $c_n = (-1)^n|\mathbf{A}|$. For example, for $n = 2$ we have

$$\begin{vmatrix} x - a_{11} & -a_{12} \\ -a_{21} & x - a_{22} \end{vmatrix} = x^2 - (a_{11} + a_{22})x + (a_{11}a_{22} - a_{12}a_{21})$$
$$= x^2 - \mathrm{tr}(\mathbf{A})x + \det(\mathbf{A})$$

The roots of (8.9) are called the **eigenvalues** of $\mathbf{A}$; older names are 'latent roots' and 'characteristic values', but the hybrid 'eigenvalue', using *Eigenwert*, the German for 'characteristic value', has found general acceptance. If α is an eigenvalue of $\mathbf{A}$, then there is a non-zero column vector $\mathbf{u}$ such that $\mathbf{Au} = \alpha\mathbf{u}$; such a vector $\mathbf{u}$ is called an **eigenvector** of $\mathbf{A}$ belonging to the eigenvalue α. The set of all eigenvalues of $\mathbf{A}$ is also called the **spectrum** of $\mathbf{A}$.

Equation (8.8) shows that a matrix similar to a diagonal matrix has n linearly independent eigenvectors; conversely, the existence of n linearly independent eigenvectors entails the equations (8.8) and hence (8.7). This result may be summed up as follows:

Theorem 8.2
A square matrix $\mathbf{A}$ *is similar to a diagonal matrix if and only if there is a basis for the whole space, formed by eigenvectors for* $\mathbf{A}$, *and the matrix with linearly independent eigenvectors as columns provides the matrix of transformation, while the eigenvalues are the entries of the diagonal matrix.*

It is important to note that the characteristic polynomial is unchanged by similarity transformation:

Theorem 8.3
Similar matrices have the same characteristic polynomial, hence the same trace and determinant, and the same eigenvalues.

Proof
If $\mathbf{A}' = \mathbf{P}^{-1}\mathbf{AP}$, then $\mathbf{P}^{-1}(x\mathbf{I} - \mathbf{A})\mathbf{P} = x\mathbf{I} - \mathbf{P}^{-1}\mathbf{AP} = x\mathbf{I} - \mathbf{A}'$, hence $|x\mathbf{I} - \mathbf{A}'| = |\mathbf{P}|^{-1}|x\mathbf{I} - \mathbf{A}|\cdot|\mathbf{P}| = |x\mathbf{I} - \mathbf{A}|$. The rest is clear.

As an example let us find the eigenvalues of

$$\mathbf{A} = \begin{pmatrix} 4 & -2 \\ 3 & -1 \end{pmatrix}$$

The characteristic equation is $x^2 - 3x + 2 = 0$; its roots are $x = 1, 2$. For each root we can find an eigenvector. For $x = 1$:

$$(\mathbf{A} - \mathbf{I})\mathbf{u} = \begin{pmatrix} 3 & -2 \\ 3 & -2 \end{pmatrix}\begin{pmatrix} u_1 \\ u_2 \end{pmatrix} = \mathbf{0}$$

A solution is $\mathbf{u} = (2, 3)^{\mathrm{T}}$. And for $x = 2$:

$$(\mathbf{A} - 2\mathbf{I})\mathbf{u} = \begin{pmatrix} 2 & -2 \\ 3 & -3 \end{pmatrix}\begin{pmatrix} u_1 \\ u_2 \end{pmatrix} = \mathbf{0}$$

A solution is $\mathbf{u} = (1, 1)^{\mathrm{T}}$. The two columns of eigenvectors found are linearly independent, so we can use the matrix with the eigenvectors as columns to transform $\mathbf{A}$ to diagonal form: $\mathbf{P} = \begin{pmatrix} 2 & 1 \\ 3 & 1 \end{pmatrix}$ has inverse $\mathbf{P}^{-1} = \begin{pmatrix} -1 & 1 \\ 3 & -2 \end{pmatrix}$, and so we have

$$\begin{pmatrix} -1 & 1 \\ 3 & -2 \end{pmatrix}\begin{pmatrix} 4 & -2 \\ 3 & -1 \end{pmatrix}\begin{pmatrix} 2 & 1 \\ 3 & 1 \end{pmatrix} = \begin{pmatrix} -1 & 1 \\ 3 & -2 \end{pmatrix}\begin{pmatrix} 2 & 2 \\ 3 & 2 \end{pmatrix} = \begin{pmatrix} 1 & 0 \\ 0 & 2 \end{pmatrix}$$

The linear independence of the eigenvectors in this example was no accident, but is a general property, expressed in the following result:

Theorem 8.4
Let $\mathbf{A}$ *be a square matrix. Then any set of eigenvectors belonging to distinct eigenvalues of* $\mathbf{A}$ *is linearly independent. If* $\mathbf{A}$ *is of order* n *with* n *distinct eigenvalues, then it is similar to a diagonal matrix.*

Proof
Let $\mathbf{u}_1, \ldots, \mathbf{u}_m$ be eigenvectors for which the corresponding eigenvalues $\alpha_1, \ldots, \alpha_m$ are distinct. If the $\mathbf{u}_i$ are linearly dependent, then some $\mathbf{u}_r$ is linearly dependent on $\mathbf{u}_1, \ldots, \mathbf{u}_{r-1}$:

$$\mathbf{u}_r = \lambda_1 \mathbf{u}_1 + \cdots + \lambda_{r-1}\mathbf{u}_{r-1} \tag{8.11}$$

By choosing the least value of r we can ensure that $\mathbf{u}_1, \ldots, \mathbf{u}_{r-1}$ are linearly independent. If we multiply this equation by $\mathbf{A}$ on the left and bear in mind that $\mathbf{A}\mathbf{u}_i = \alpha_i \mathbf{u}_i$, we obtain

$$\alpha_r \mathbf{u}_r = \sum_i \lambda_i \alpha_i \mathbf{u}_i \tag{8.12}$$

Hence by subtracting (8.12) from (8.11) multiplied by α_r we find

$$\sum_i \lambda_i(\alpha_r - \alpha_i)\mathbf{u}_i = \mathbf{0} \tag{8.13}$$

The λ_i do not all vanish, because $\mathbf{u}_r \neq \mathbf{0}$ and $\alpha_r - \alpha_i \neq 0$ for $i = 1, \ldots, r-1$, by hypothesis, so (8.13) is a non-trivial linear dependence relation between $\mathbf{u}_1, \ldots, \mathbf{u}_{r-1}$, contradicting the fact that they were chosen to be linearly independent. This contradiction shows that $\mathbf{u}_1,\ldots,\mathbf{u}_m$ are linearly independent.

If $\mathbf{A}$ has n distinct eigenvalues, we obtain an eigenvector for each, and these n eigenvectors are linearly independent, by what has just been proved. Let $\mathbf{P}$ be the matrix with these vectors as columns, then $\mathbf{P}$ is regular; denoting the eigenvalues by α_i and the corresponding eigenvector by $\mathbf{u}_i$, we have $\mathbf{A}\mathbf{u}_i = \mathbf{u}_i\alpha_i$ $(i = 1,\ldots,n)$ and so

$$\mathbf{AP} = \mathbf{PD}$$

where $\mathbf{D} = \text{diag}(\alpha_1,\ldots,\alpha_n)$. Hence $\mathbf{P}^{-1}\mathbf{AP} = \mathbf{D}$, and $\mathbf{A}$ has been transformed to diagonal form, so the proof is complete.

8.3 SIMILARITY REDUCTION TO TRIANGULAR FORM

We have already seen that not every matrix is similar to a diagonal matrix. As another example consider the matrix

$$\mathbf{J} = \begin{pmatrix} 0 & -1 \\ 1 & 0 \end{pmatrix}$$

Its characteristic equation is $x^2 + 1 = 0$, which has no real roots, so we cannot transform $\mathbf{J}$ to diagonal form over the real field $\mathbf{R}$. However, if we allow complex numbers, we have the eigenvalues $\pm\mathrm{i}$, with eigenvectors obtained by solving

$$\begin{pmatrix} -\mathrm{i} & 1 \\ -1 & -\mathrm{i} \end{pmatrix}\begin{pmatrix} u_1 \\ u_2 \end{pmatrix} = 0$$

$$\begin{pmatrix} \mathrm{i} & 1 \\ -1 & \mathrm{i} \end{pmatrix}\begin{pmatrix} v_1 \\ v_2 \end{pmatrix} = 0$$

A solution is $\mathbf{u} = (1, \mathrm{i})^{\mathrm{T}}$, $\mathbf{v} = (\mathrm{i}, 1)^{\mathrm{T}}$ and we have

$$\frac{1}{2}\begin{pmatrix} 1 & -\mathrm{i} \\ -\mathrm{i} & 1 \end{pmatrix}\begin{pmatrix} 0 & -1 \\ 1 & 0 \end{pmatrix}\begin{pmatrix} 1 & \mathrm{i} \\ \mathrm{i} & 1 \end{pmatrix} = \frac{1}{2}\begin{pmatrix} 1 & -\mathrm{i} \\ -\mathrm{i} & 1 \end{pmatrix}\begin{pmatrix} -\mathrm{i} & -1 \\ 1 & \mathrm{i} \end{pmatrix} = \begin{pmatrix} -\mathrm{i} & 0 \\ 0 & \mathrm{i} \end{pmatrix}$$

Once we admit complex numbers, every equation of positive degree in one unknown has a root. This is known as the **fundamental theorem of algebra**; we shall assume it here without proof. It is usually proved in complex analysis by means of Liouville's theorem, but there are also purely algebraic proofs – cf., for example, Cohn (1991, section 8.3).

As a consequence of the fundamental theorem of algebra every polynomial of degree n can be written as a product of linear factors. Thus for any $a_0, a_1,\ldots,a_n \in \mathbf{C}$, where $a_0 \neq 0$, there exist $\alpha_1,\ldots,\alpha_n \in \mathbf{C}$ such that

$$a_0x^n + a_1x^{n-1} + \cdots + a_n = a_0(x - \alpha_1)\cdots(x - \alpha_n). \tag{8.14}$$

In abbreviated notation this may be written

$$\sum_0^n a_{n-i}x^i = a_0 \prod_{i=1}^n (x - \alpha_i)$$

where $\prod$ (capital Greek pi) stands for 'product', just as $\sum$ stands for 'sum'. To prove (8.14), let us write the left-hand side as $a_0 f(x)$, so that f is a polynomial with highest term x^n. We note for $k = 1, 2, \ldots$ the identity

$$x^k - \alpha^k = (x - \alpha)(x^{k-1} + x^{k-2}\alpha + \cdots + x\alpha^{k-2} + \alpha^{k-1}) \qquad (8.15)$$

which is easily verified. Hence

$$\begin{aligned} f(x) - f(\alpha) &= \sum_i a_{n-i}(x^i - \alpha^i) \\ &= (x - \alpha)\sum_i a_{n-i} \sum_j x^j \alpha^{i-j-1} \end{aligned}$$

Thus we have, for any α,

$$f(x) - f(\alpha) = (x - \alpha)g(x) \qquad (8.16)$$

where g is a polynomial of degree $n - 1$. This is often called the **remainder formula** or the **remainder theorem**. Now by our assumption, the equation $f = 0$ has a root α_1: $f(\alpha_1) = 0$. Hence by (8.16),

$$f(x) = (x - \alpha_1)g(x)$$

for some polynomial g of degree $n - 1$. By induction on n we can write $g(x) = \prod_2^n (x - \alpha_i)$ and so $f(x) = \prod_1^n (x - \alpha_i)$. Now (8.14) follows by multiplying both sides by a_0. Thus an equation of degree n has n roots, not necessarily distinct, but counted with their proper multiplicities, as in (8.14). We also see that an equation of degree n cannot have more than n roots, for if f vanishes for $x = \alpha_i$ $(i = 1, \ldots, n)$ and $x = \beta$, where $\alpha_1, \ldots, \alpha_n, \beta$ are all distinct, than we have a factorization (8.14) and on putting $x = \beta$ in (8.14), we obtain

$$(\beta - \alpha_1)\cdots(\beta - \alpha_n) = 0$$

hence β must equal one of the α_i.

With the help of these results we can show that every square matrix is similar to a triangular matrix. By a **triangular** matrix we understand a square matrix in which all entries below the main diagonal are zero (upper triangular) or all entries above the main diagonal are zero (lower triangular). We remark that the eigenvalues of a triangular matrix are just its diagonal elements, for in this case $|x\mathbf{I} - \mathbf{A}| = (x - a_{11})\cdots(x - a_{nn})$, by expanding the triangular determinant (see section 5.11).

Theorem 8.5

Every square matrix $\mathbf{A}$ *over the complex numbers is similar to an upper triangular matrix, and also to a lower triangular matrix, with the eigenvalues of* $\mathbf{A}$ *on the main diagonal.*

Proof

Given an $n \times n$ matrix $\mathbf{A}$ over the complex numbers, we can choose an eigenvalue α_1 of $\mathbf{A}$ and an eigenvector $\mathbf{u}_1$ of $\mathbf{A}$ belonging to α_1. We complete $\mathbf{u}_1$ to a basis $\mathbf{u}_1, \ldots, \mathbf{u}_n$ for the space of n-vectors; then the matrix $\mathbf{U}$ with these vectors as columns is regular, with first column $\mathbf{u}_1$.

Consider the matrix $\mathbf{B} = \mathbf{U}^{-1}\mathbf{A}\mathbf{U}$; we have $\mathbf{AU} = \mathbf{UB}$ and comparing the first columns we find that $\mathbf{Au}_1 = \sum \mathbf{u}_i \mathbf{b}_{i1}$ (where $\mathbf{B} = (b_{ij})$). But $\mathbf{Au}_1 = \mathbf{u}_1 \alpha_1$, so the first column of $\mathbf{B}$ must be $(\alpha_1, 0, \ldots, 0)^{\mathrm{T}}$. Thus we have

$$\mathbf{U}^{-1}\mathbf{A}\mathbf{U} = \mathbf{B} = \begin{pmatrix} \alpha_1 & \mathbf{b} \\ 0 & \\ \vdots & \mathbf{B}_1 \\ 0 & \end{pmatrix}$$

where $\mathbf{b}$ is a row of length $n-1$ and $\mathbf{B}_1$ is a square matrix of order $n-1$. By induction on n there exists an invertible matrix $\mathbf{U}_1$ such that $\mathbf{U}_1^{-1}\mathbf{B}_1\mathbf{U}_1$ is upper triangular; hence

$$\begin{pmatrix} 1 & \mathbf{0} \\ \mathbf{0} & \mathbf{U}_1 \end{pmatrix}^{-1} \mathbf{U}^{-1}\mathbf{A}\mathbf{U} \begin{pmatrix} 1 & \mathbf{0} \\ \mathbf{0} & \mathbf{U}_1 \end{pmatrix} = \begin{pmatrix} \alpha_1 & \mathbf{b}\mathbf{U}_1 \\ 0 & \\ \vdots & \mathbf{C}_1 \\ 0 & \end{pmatrix}$$

where $\mathbf{C}_1$ is upper triangular. Thus $\mathbf{A}$ is similar to an upper triangular matrix $\mathbf{C}$, say. The diagonal elements of $\mathbf{C}$ are its eigenvalues and they are also the eigenvalues of $\mathbf{A}$, by Theorem 8.3, since $\mathbf{C}$ is similar to $\mathbf{A}$.

To complete the proof, we note that any upper triangular matrix is similar to a lower triangular matrix. We simply write down the rows in reverse order; this is a product of elementary row operations, corresponding to left multiplication by

$$\mathbf{T} = \begin{pmatrix} 0 & 0 & \cdots & 0 & 1 \\ 0 & 0 & \cdots & 1 & 0 \\ \cdots & & \cdots & \cdots & \\ 1 & 0 & \cdots & 0 & 0 \end{pmatrix} = (t_{ij})$$

where $t_{ij} = \delta_{j,n+1-i}$. Next we write the columns in reverse order, corresponding to right multiplication by $\mathbf{T}$. The result is a lower triangular matrix and is $\mathbf{TCT}$, which is similar to $\mathbf{C}$, because $\mathbf{T}^{-1} = \mathbf{T}$.

The remainder formula also allows us to evaluate the Vandermonde determinant (encountered in Exercise 5.6) in a methodical way. Taking the case of order 3 as an example, we have to calculate

$$V(x, y, z) = \begin{vmatrix} 1 & 1 & 1 \\ x & y & z \\ x^2 & y^2 & z^2 \end{vmatrix}$$

Clearly each term in the expansion of V has degree 3 in x, y, z, so V is a polynomial of total degree 3; it vanishes if $x = y$ or $x = z$ or $y = z$. By the remainder formula it follows that V is divisible by $y - x$; similarly it is divisible by $z - x$ and $z - y$, and so

$$V(x, y, z) = (y - x)(z - x)(z - y) \cdot c;$$

A comparison of degrees shows that c is a constant, and by comparing coefficients of yz^2 we see that $c = 1$. The same argument shows that the general Vandermonde determinant $V(x_1, \ldots, x_n)$ with (i,j)th entry x_i^{j-1} has the value

$$V(x_1, \ldots, x_n) = \prod_{i<j} (x_j - x_i)$$

8.4 THE CAYLEY–HAMILTON THEOREM

Any square matrix satisfies a polynomial equation. For an $n \times n$ matrix $\mathbf{A}$ may be regarded as a vector in an n^2-dimensional vector space, so its first $n^2 + 1$ powers $\mathbf{I}, \mathbf{A}, \mathbf{A}^2, \ldots, \mathbf{A}^{n^2}$ are linearly dependent. This yields an equation of degree at most n^2, but it is of interest that we can always find an equation of degree n, as the **Cayley–Hamilton theorem** states:

Theorem 8.6

Every square matrix satisfies its characteristic equation. Thus if $\varphi(x) = |x\mathbf{I} - \mathbf{A}|$, *then* $\varphi(\mathbf{A}) = \mathbf{0}$.

Proof

It is important to understand the assertion correctly. The statement is that if $|x\mathbf{I} - \mathbf{A}| = x^n + c_1 x^{n-1} + \cdots + c_n$, then

$$\mathbf{A}^n + c_1 \mathbf{A}^{n-1} + \cdots + c_n \mathbf{I} = \mathbf{0}$$

A naive 'proof', which replaces x by $\mathbf{A}$, giving $|\mathbf{A} - \mathbf{A}| = 0$, is not valid.

For a valid proof we go back to the remainder formula (8.16); here we can replace α by a square matrix $\mathbf{A}$, since the formula (8.15), on which it is based, still holds with $\mathbf{A}$ in place of α. Thus we have, for any polynomial f,

$$f(x)\mathbf{I} - f(\mathbf{A}) = (x\mathbf{I} - \mathbf{A})g(x) \tag{8.17}$$

for some polynomial g with matrix coefficients. Now we have by (5.15) from section 5.12,

$$(x\mathbf{I} - \mathbf{A}) \cdot \operatorname{adj}(x\mathbf{I} - \mathbf{A}) = |x\mathbf{I} - \mathbf{A}| \cdot \mathbf{I} = \varphi(x)\mathbf{I}$$

where $\varphi(x) = |x\mathbf{I} - \mathbf{A}|$ is the characteristic polynomial of $\mathbf{A}$ and $\operatorname{adj}(x\mathbf{I} - \mathbf{A})$ is a matrix whose entries are polynomials in x, and which can also be written as a polynomial in x with matrix coefficients. By (8.17) we have $\varphi(x)\mathbf{I} - \varphi(\mathbf{A}) = (x\mathbf{I} - \mathbf{A})\psi(x)$, for some polynomial ψ. Hence

$$\begin{aligned}\varphi(\mathbf{A}) &= \varphi(x)\mathbf{I} - (x\mathbf{I} - \mathbf{A})\psi(x)\\ &= (x\mathbf{I} - \mathbf{A})[\operatorname{adj}(x\mathbf{I} - \mathbf{A}) - \psi(x)\mathbf{I}]\end{aligned}$$

If the matrix in square brackets is non-zero, then the highest power of x occurring is of the form $x^r\mathbf{C}$ for some non-zero matrix $\mathbf{C}$. Then the right-hand side has the highest term $x^{r+1}\mathbf{C}$, which is not cancelled by any other term, whereas the left-hand side is independent of x. This contradiction shows that the right-hand side is zero, so $\varphi(\mathbf{A}) = 0$, as we wished to show.

The Cayley–Hamilton theorem can often be used to shorten calculations. As an illustration consider the matrix

$$\mathbf{A} = \begin{pmatrix} 2 & 3 \\ 3 & 5 \end{pmatrix}$$

Suppose we wish to find $\mathbf{A}^{-1}$. The characteristic equation is

$$x^2 - 7x + 1 = 0$$

Hence $\mathbf{A}^2 = 7\mathbf{A} - \mathbf{I}$, and $\mathbf{A} - 7\mathbf{I} + \mathbf{A}^{-1} = \mathbf{0}$, so

$$\mathbf{A}^{-1} = 7\mathbf{I} - \mathbf{A} = \begin{pmatrix} 5 & -3 \\ -3 & 2 \end{pmatrix}$$

For a 2×2 matrix it is in any case easy to find the inverse (see section 5.12), but for higher powers this is a shorter route. For example, suppose we want to find $\mathbf{A}^5$; we have $\mathbf{A}^4 = (7\mathbf{A} - \mathbf{I})^2 = 49\mathbf{A}^2 - 14\mathbf{A} + \mathbf{I} = 49(7\mathbf{A} - \mathbf{I}) - 14\mathbf{A} + \mathbf{I} = 329\mathbf{A} - 48\mathbf{I}$, $\mathbf{A}^5 = 329(7\mathbf{A} - \mathbf{I}) - 48\mathbf{A} = 2255\mathbf{A} - 329\mathbf{I}$, hence

$$\mathbf{A}^5 = 2255\begin{pmatrix} 2 & 3 \\ 3 & 5 \end{pmatrix} - 329\begin{pmatrix} 1 & 0 \\ 0 & 1 \end{pmatrix} = \begin{pmatrix} 4181 & 6765 \\ 6765 & 10\,946 \end{pmatrix}$$

8.5 THE REDUCTION OF REAL QUADRATIC FORMS

We shall now take up another apparently unrelated topic, which, however, will turn out to lead to an important special case of the similarity reduction.

By a **quadratic form** in the variables $x_1, \ldots, x_n$ we understand a polynomial in the x_i which is homogeneous of degree 2. Such a form can always be written as $\sum a_{ij}x_ix_j$ or, in matrix terms,

$$f = \mathbf{x}^{\mathrm{T}}\mathbf{A}\mathbf{x} \tag{8.18}$$

where $\mathbf{x} = (x_1, \ldots, x_n)^{\mathrm{T}}$ and $\mathbf{A}$ is a symmetric matrix, i.e. $\mathbf{A}^{\mathrm{T}} = \mathbf{A}$. For if we are given the form $\mathbf{x}^{\mathrm{T}}\mathbf{B}\mathbf{x}$, with any square matrix $\mathbf{B}$, then

$$\mathbf{x}^{\mathrm{T}}\mathbf{B}\mathbf{x} = (\mathbf{x}^{\mathrm{T}}\mathbf{B}\mathbf{x})^{\mathrm{T}} = \mathbf{x}^{\mathrm{T}}\mathbf{B}^{\mathrm{T}}\mathbf{x}$$

hence

$$\mathbf{x}^{\mathrm{T}}\mathbf{B}\mathbf{x} = \mathbf{x}^{\mathrm{T}}\mathbf{A}\mathbf{x}$$

where $\mathbf{A} = \frac{1}{2}(\mathbf{B} + \mathbf{B}^{\mathrm{T}})$, and $\mathbf{A}$ is a symmetric matrix. For example, the form $x_1^2 + 5x_2^2 - 2x_3^2 + 4x_1x_2 - x_1x_3 + 6x_2x_3$ corresponds to the symmetric matrix

$$\mathbf{A} = \begin{pmatrix} 1 & 2 & -1/2 \\ 2 & 5 & 3 \\ -1/2 & 3 & -2 \end{pmatrix}$$

where it has to be remembered that a_{ij} is *one-half* the coefficient of x_ix_j, because on expansion the term x_ix_j in (8.18) reads $a_{ij}x_ix_j + a_{ji}x_jx_i = 2a_{ij}x_ix_j$. It follows that if we have

$$\mathbf{x}^{\mathrm{T}}\mathbf{A}\mathbf{x} = \mathbf{x}^{\mathrm{T}}\mathbf{A}'\mathbf{x},$$

for all $\mathbf{x}$, where $\mathbf{A}$ and $\mathbf{A}'$ are symmetric matrices, then we can equate coefficients of x_ix_j and so obtain $\mathbf{A} = \mathbf{A}'$; without the condition of symmetry this conclusion no longer holds.

Quadratic forms are used in geometry to describe conics and quadrics, but they also occur very widely in mechanics, statistics, economics and elsewhere, so a reduction theory will have many uses. Our aim will be to find a transformation of coordinates for which the matrix of the form becomes fairly simple, for example, diagonal.

If we change to new coordinates $\mathbf{x}'$, related to the old ones by

$$\mathbf{x} = \mathbf{P}\mathbf{x}'$$

with a regular matrix $\mathbf{P}$, then our form becomes $f = \mathbf{x}'^{\mathrm{T}}\mathbf{A}'\mathbf{x}'$. To find $\mathbf{A}'$ we have

$$\mathbf{x}'^{\mathrm{T}}\mathbf{A}'\mathbf{x}' = \mathbf{x}^{\mathrm{T}}\mathbf{A}\mathbf{x} = (\mathbf{P}\mathbf{x}')^{\mathrm{T}}\mathbf{A}(\mathbf{P}\mathbf{x}') = \mathbf{x}'^{\mathrm{T}}\mathbf{P}^{\mathrm{T}}\mathbf{A}\mathbf{P}\mathbf{x}'$$

Since $\mathbf{A}'$ and $\mathbf{P}^{\mathrm{T}}\mathbf{A}\mathbf{P}$ are both symmetric, they must be equal:

$$\mathbf{A}' = \mathbf{P}^{\mathrm{T}}\mathbf{A}\mathbf{P} \tag{8.19}$$

Two symmetric matrices $\mathbf{A}$ and $\mathbf{A}'$, related as in (8.19), are said to be **congruent** and the corresponding transformation with matrix $\mathbf{P}$ is called a **congruence transformation**. Our first objective will be to show that every symmetric matrix is congruent to a diagonal matrix; we shall find it convenient to operate directly with the corresponding quadratic form f, which as we have seen comes to the same.

If f is the zero form, there is nothing to prove, so we may assume that not all coefficients vanish. If a diagonal coefficient is non-zero, we may renumber the variables so that $a_{11} \neq 0$. Now introduce new variables

$$\begin{aligned} x_1' &= \sum a_{1i}x_i \\ x_2' &= x_2 \\ &\vdots \\ x_n' &= x_n \end{aligned}$$

We have

$$x_1'^2 = \left(\sum a_{1i}x_i\right)^2 = a_{11}(a_{11}x_1^2 + 2a_{12}x_1x_2 + \cdots + 2a_{1n}x_1x_n) + g(x_2,\ldots,x_n)$$

where g is a quadratic form in $x_2,\ldots,x_n$. We see that the terms involving x_1 are the same in f and $a_{11}^{-1}x'^2$, so we can write

$$\begin{aligned} f &= a_{11}^{-1}x_1'^2 + h(x_2,\ldots,x_n) \\ &= a_{11}^{-1}x_1'^2 + h(x_2',\ldots,x_n') \end{aligned}$$

we remark that this step is just the completion of the square, familiar from the solution of quadratic equations.

Since h is a quadratic form in $n-1$ variables, we can apply induction on n to transform h to diagonal form, and this will accomplish the reduction for f. There still remains the case where all diagonal coefficients vanish; in that case we choose a non-zero coefficient, a_{12} say, and write

$$\begin{aligned} x_1' &= x_1 + x_2 \\ x_2' &= x_1 - x_2 \\ x_3' &= x_3 \\ &\vdots \\ x_n' &= x_n \end{aligned}$$

Now we have $2a_{12}x_1x_2 = \frac{1}{2}a_{12}(x_1'^2 - x_2'^2)$, so in terms of the new variables there is a non-zero diagonal coefficient and we can continue the reduction as before.

The end result is a quadratic form in diagonal form

$$f = a_1y_1^2 + \cdots + a_ny_n^2$$

Here we may assume the variables numbered so that $a_i > 0$ for $i = 1,\ldots,s$, $a_i < 0$ for $i = s+1,\ldots,r$ and $a_i = 0$ for $i = r+1,\ldots,n$. If we now put

$$z_i = \begin{cases} a_i^{1/2}y_i, & i = 1,\ldots,s \\ (-a_i)^{1/2}y_i, & i = s+1,\ldots,r \\ y_i, & i = r+1,\ldots,n \end{cases}$$

then f becomes

$$f = z_1^2 + \cdots + z_s^2 - z_{s+1}^2 - \cdots - z_r^2 \tag{8.20}$$

From (8.19), where $\mathbf{P}$ is regular, we see that $\mathbf{A}$ and $\mathbf{A}'$ have the same rank; in (8.20) the matrix is $\mathbf{I}_s \oplus -\mathbf{I}_{r-s} \oplus \mathbf{0}$, which has rank r. Thus the rank of f in (8.20) is r. Since a change of coordinates leaves the rank unchanged, it is independent of the mode of reduction and is an invariant, called the **rank** of the quadratic form. If $r = n$, the form is also called **non-degenerate** or sometimes **non-singular**.

The number $2s-r=s-(r-s)$, representing the excess of the number of positive over the number of negative terms in (8.20), is called the **signature**. Sometimes the pair of numbers $(s, r-s)$ is described as the signature; this is more usual in physics, for instance the relativity metric

$$\mathrm{d}s^2=\mathrm{d}x_0^2-\mathrm{d}x_1^2-\mathrm{d}x_2^2-\mathrm{d}x_3^2$$

has signature $(1,3)$.

Like the rank, the signature is independent of the mode of reduction. For if we also had

$$f=z_1'^2+\cdots+z_t'^2-z_{t+1}'^2-\cdots-z_r'^2 \tag{8.21}$$

where $t<s$, say, consider the $r-(s-t)$ equations

$$z_1'=\cdots=z_t'=0, \quad z_{s+1}=\cdots=z_r=0 \tag{8.22}$$

These are linear equations in $z_1,\ldots,z_r$ and their number is $r-(s-t)<r$, hence they have a non-trivial solution $z(0)$. By (8.22) and (8.20) we have

$$f(z(0))=z_1(0)^2+\cdots+z_s(0)^2>0$$

with strict inequality because $z_1(0),\ldots,z_s(0)$ are not all zero, by hypothesis. But (8.22), combined with (8.21), shows that $f(z(0))\leqslant 0$, and this is a contradiction. Hence $t\geqslant s$ and by symmetry, $s\geqslant t$, therefore $s=t$, as asserted. We can sum up our conclusion as follows:

Theorem 8.7
A real quadratic form f in $x_1,\ldots,x_n$ can always be transformed to the form

$$f=z_1^2+\cdots+z_s^2-z_{s+1}^2-\cdots-z_r^2$$

where the rank r and the signature $2s-r$ are independent of the mode of reduction.

The invariance of the signature is also known as **Sylvester's law of inertia**. The case $s=r$ is particularly important and it can be described by saying that the form satisfies $f(\mathbf{x})\geqslant 0$ for all $\mathbf{x}$; such a form (and its matrix) is called **positive**. If $f(\mathbf{x})>0$ for all $\mathbf{x}\neq\mathbf{0}$, then f and its matrix are called **positive definite**; clearly this is the case precisely when $s=r=n$. From theorem 8.7 we see that every positive definite matrix is congruent to the unit matrix $\mathbf{I}$.

8.6 QUADRATIC FORMS ON A METRIC SPACE

Our last conclusion in section 8.5 throws an interesting light on the definition of the scalar product of vectors. As we saw in sections 6.4–6.5, this could be defined in terms of the length of a vector, which in turn was taken to be the positive square root of the quadratic form $\sum x_i^2$. We now see that in place of $\sum x_i^2$ we could take any positive definite quadratic form, since the latter has been shown to be congruent to $\sum x_i^2$. This suggests the following definition. A real vector space V will be called a **metric space** if it is equipped with a positive definite quadratic form, $f(\mathbf{x})$, called the square of the **length** of the vector $\mathbf{x}$. By Theorem 8.7 we can always choose the coordinates in a metric space in such a way that the length of $\mathbf{x}$ is given by the positive square root of $\sum x_i^2$,

A coordinate transformation with matrix $\mathbf{P}$ preserves lengths precisely when $\mathbf{x}'^{\mathrm{T}}\mathbf{x}' = \mathbf{x}^{\mathrm{T}}\mathbf{x}$. Substituting from $\mathbf{x} = \mathbf{P}\mathbf{x}'$, we find

$$\mathbf{x}^{\mathrm{T}}\mathbf{x} = \mathbf{x}'^{\mathrm{T}}\mathbf{x}' = (\mathbf{P}\mathbf{x}')^{\mathrm{T}}\mathbf{P}\mathbf{x}' = \mathbf{x}'^{\mathrm{T}}\mathbf{P}^{\mathrm{T}}\mathbf{P}\mathbf{x}';$$

since this holds for all $\mathbf{x}'$, we must have $\mathbf{P}^{\mathrm{T}}\mathbf{P} = \mathbf{I}$, that is, $\mathbf{P}$ is orthogonal. Thus we see that the length-preserving transformations are just the transformations preserving orthonormal bases (cf. section 7.2).

We now go on to examine the behaviour of a quadratic form in a metric space under orthogonal transformations. If our form is specified by the symmetric matrix $\mathbf{A}$, its matrix after an orthogonal transformation with matrix $\mathbf{P}$ will be $\mathbf{P}^{\mathrm{T}}\mathbf{A}\mathbf{P}$, but since $\mathbf{P}^{\mathrm{T}}\mathbf{P} = \mathbf{I}$, this can also be written as $\mathbf{P}^{-1}\mathbf{A}\mathbf{P}$. Thus we are dealing again with a similarity reduction; however, in the special case when $\mathbf{A}$ is real symmetric, we shall find that all the difficulties disappear and we can always find a diagonal form. We shall prove the following reduction theorem for real quadratic forms:

Theorem 8.8

Let $\mathbf{A}$ be any real symmetric matrix of order n. Then all the eigenvalues $\alpha_1, \ldots, \alpha_n$ of $\mathbf{A}$ are real and there exists a proper orthogonal matrix $\mathbf{P}$ such that $\mathbf{P}^{\mathrm{T}}\mathbf{A}\mathbf{P} = \mathrm{diag}(\alpha_1, \ldots, \alpha_n)$.

Proof

We begin by showing that all the eigenvalues of $\mathbf{A}$ are real. Let α be an eigenvalue of $\mathbf{A}$, possibly complex, and $\mathbf{u}$ a corresponding eigenvector. Then

$$\mathbf{A}\mathbf{u} = \alpha\mathbf{u} \tag{8.23}$$

Taking complex conjugates (indicated by a bar, as usual), we have $\mathbf{A}\bar{\mathbf{u}} = \bar{\alpha}\bar{\mathbf{u}}$ and transposing, we obtain

$$\bar{\mathbf{u}}^{\mathrm{T}}\mathbf{A} = \bar{\alpha}\bar{\mathbf{u}}^{\mathrm{T}} \tag{8.24}$$

where $\mathbf{A}$ is unchanged, because it is real symmetric. Now multiply (8.23) by $\bar{\mathbf{u}}^{\mathrm{T}}$ on the left, (8.24) by $\mathbf{u}$ on the right and form the difference:

$$0 = \bar{\mathbf{u}}^{\mathrm{T}}\mathbf{A}\mathbf{u} - \bar{\mathbf{u}}^{\mathrm{T}}\mathbf{A}\mathbf{u} = \alpha\bar{\mathbf{u}}^{\mathrm{T}}\mathbf{u} - \bar{\alpha}\bar{\mathbf{u}}^{\mathrm{T}}\mathbf{u} = (\alpha - \bar{\alpha})\bar{\mathbf{u}}^{\mathrm{T}}\mathbf{u}$$

If $\mathbf{u} = (u_1, \ldots, u_n)^{\mathrm{T}}$, then $\bar{\mathbf{u}}^{\mathrm{T}}\mathbf{u} = \sum |u_i|^2$ and this is positive, because $\mathbf{u} \neq \mathbf{0}$. Hence $\alpha - \bar{\alpha} = 0$, and this shows α to be real.

We now take an eigenvalue α_1 of $\mathbf{A}$ and an eigenvector $\mathbf{u}_1$ belonging to α_1. Since $\mathbf{A}$ and α_1 are real, $\mathbf{u}_1$ can also be chosen to be real, and since $\mathbf{u}_1$ is only determined up to a scalar factor, we can take it to be normalized, so that $\mathbf{u}_1^T\mathbf{u}_1 = 1$. We can complete $\mathbf{u}_1$ to an orthonormal basis of n-space, $\mathbf{u}_1, \ldots, \mathbf{u}_n$. These columns form a matrix $\mathbf{U}$ which is orthogonal, $\mathbf{U}^T\mathbf{U} = \mathbf{I}$, and, as in the proof of Theorem 8.5, we can form $\mathbf{U}^{-1}\mathbf{A}\mathbf{U}$ with the first column $(\alpha_1, \ldots, 0)^T$. Now $\mathbf{A}$ is symmetric and $\mathbf{U}^{-1} = \mathbf{U}^T$, hence $(\mathbf{U}^T\mathbf{A}\mathbf{U})^T = \mathbf{U}^T\mathbf{A}\mathbf{U}$, so $\mathbf{U}^T\mathbf{A}\mathbf{U}$ is symmetric and it follows that

$$\mathbf{U}^{-1}\mathbf{A}\mathbf{U} = \begin{pmatrix} \alpha_1 & 0 & \cdots & 0 \\ 0 & & & \\ \vdots & & \mathbf{B}_1 & \\ 0 & & & \end{pmatrix}$$

where $\mathbf{B}_1$ is a real symmetric matrix of order $n-1$. By induction we can transform $\mathbf{B}_1$ to diagonal form by an orthogonal matrix and now we can complete the proof as for Theorem 8.5. Moreover, by changing the sign of the last column of $\mathbf{U}$ if necessary, we can ensure that the matrix of transformation is proper orthogonal. This completes the proof of Theorem 8.8.

In order to carry out the reduction we note that eigenvectors belonging to different eigenvalues must be orthogonal. For let α, β be distinct eigenvalues of $\mathbf{A}$, with eigenvectors $\mathbf{u}, \mathbf{v}$, respectively. We have $\mathbf{A}\mathbf{u} = \alpha\mathbf{u}$, $\mathbf{A}\mathbf{v} = \beta\mathbf{v}$, hence $\mathbf{v}^T\mathbf{A}\mathbf{u} = \alpha\mathbf{v}^T\mathbf{u}, \mathbf{u}^T\mathbf{A}\mathbf{v} = \beta\mathbf{u}^T\mathbf{v}$; transposing the last equation, we find $\mathbf{v}^T\mathbf{A}\mathbf{u} = \beta\mathbf{v}^T\mathbf{u}$, hence $(\alpha - \beta)\mathbf{v}^T\mathbf{u} = 0$; since $\beta \neq \alpha$, we conclude that $\mathbf{v}^T\mathbf{u} = 0$, as claimed.

Suppose that α is an m-fold root of the characteristic equation of $\mathbf{A}$, that is, $|x\mathbf{I} - \mathbf{A}|$ is divisible by $(x - \alpha)^m$ but no higher power; we claim that $\alpha\mathbf{I} - \mathbf{A}$ has rank $n - m$, or equivalently, the kernel of the linear mapping with matrix $\alpha\mathbf{I} - \mathbf{A}$ has dimension m. Let us denote this kernel by V_0:

$$V_0 = \{\mathbf{x} \in \mathbf{R}^n \mid \mathbf{A}\mathbf{x} = \alpha\mathbf{x}\}$$

V_0 is often called the **eigenspace** for α. We take an orthonormal basis $\mathbf{v}_1, \ldots, \mathbf{v}_r$ of V_0 and complete it to an orthonormal basis $\mathbf{v}_1, \ldots, \mathbf{v}_n$ of the whole space. Referred to this basis, $\mathbf{A}$ is transformed into a similar matrix $\mathbf{A}'$. Since the new basis is again orthonormal, the matrix $\mathbf{P}$ of the transformation is orthogonal, hence $\mathbf{A}' = \mathbf{P}^T\mathbf{A}\mathbf{P}$ is again symmetric. Now the (i,j)th entry of $\mathbf{A}'$ is $\mathbf{v}_i^T\mathbf{A}\mathbf{v}_j$. If $j \leqslant r < i$, then $\mathbf{A}\mathbf{v}_j = \alpha\mathbf{v}_j$, hence $\mathbf{v}_i^T\mathbf{A}\mathbf{v}_j = \alpha\mathbf{v}_i^T\mathbf{v}_j = 0$; similarly for $i \leqslant r < j$, while for $i,j \leqslant r$, $\mathbf{v}_i^T\mathbf{A}\mathbf{v}_j = \alpha\mathbf{v}_i^T\mathbf{v}_j = \delta_{ij}\alpha$. Thus $\mathbf{A}'$ has the form

$$\begin{pmatrix} \alpha & & & \\ & \ddots & & \mathbf{0} \\ & & \alpha & \\ \mathbf{0} & & & \mathbf{A}_1 \end{pmatrix}$$

where $\mathbf{A}_1$ is of order $n - r$. Further, $\alpha\mathbf{I} - \mathbf{A}_1$ is non-singular, for otherwise we could enlarge V_0. This shows that $r = \dim V_0$ is the precise multiplicity of α as eigenvalue of $\mathbf{A}$, that is, $m = r$.

This makes it clear how to accomplish the transformation to diagonal form for **A**. We first find all eigenvalues for **A**. For each eigenvalue we find the eigenspace and take an orthonormal basis for this eigenspace. The bases for all the different eigenspaces taken together form an orthonormal basis of the whole space, and if we transform by the matrix with this basis as columns we reach a diagonal form.

In practice, it is often enough to know the eigenvalues, without having to compute the actual matrix of transformation to diagonal form. But even when this matrix is needed, we usually find the eigenvalues first, and then choose eigenvectors for each eigenvalue, taking care to choose the vectors for multiple values to be orthogonal. We illustrate the process by some examples.

First, let **A** be the following matrix:

$$\mathbf{A} = \begin{pmatrix} 4 & 0 & -2 \\ 0 & 2 & -2 \\ -2 & -2 & 3 \end{pmatrix}$$

The characteristic equation is

$$\begin{vmatrix} x-4 & 0 & 2 \\ 0 & x-2 & 2 \\ 2 & 2 & x-3 \end{vmatrix} = 0$$

which comes to $x(x^2 - 9x + 18) = 0$. Hence the eigenvalues are $0, 3, 6$, and the matrix **A** is similar to $\mathrm{diag}(0, 3, 6)$. To obtain the transforming matrix **P** we have to find the eigenvectors.

$x = 0$. The eigenvector **u** is obtained by solving

$$\begin{pmatrix} 4 & 0 & -2 \\ 0 & 2 & -2 \\ -2 & -2 & 3 \end{pmatrix} \begin{pmatrix} u_1 \\ u_2 \\ u_3 \end{pmatrix} = 0$$

A normalized solution is $\mathbf{u} = \frac{1}{3}(1, 2, 2)^{\mathrm{T}}$.

$x = 3$. We have to solve the system

$$\begin{pmatrix} 1 & 0 & -2 \\ 0 & -1 & -2 \\ -2 & -2 & 0 \end{pmatrix} \begin{pmatrix} v_1 \\ v_2 \\ v_3 \end{pmatrix} = 0$$

A normalized solution is $\mathbf{v} = \frac{1}{3}(2, -2, 1)^{\mathrm{T}}$.

$x = 6$. The system is

$$\begin{pmatrix} -2 & 0 & -2 \\ 0 & -4 & -2 \\ -2 & -2 & -3 \end{pmatrix} \begin{pmatrix} w_1 \\ w_2 \\ w_3 \end{pmatrix} = 0.$$

A normalized solution is $\mathbf{w} = \frac{1}{3}(2, 1, -2)^{\mathrm{T}}$.

It follows that the matrix with columns $\mathbf{u}, \mathbf{v}, \mathbf{w}$ is orthogonal; in fact, we see that it is proper orthogonal (for example, by checking that $\mathbf{u} \wedge \mathbf{v} = \mathbf{w}$). Hence the transforming matrix is

$$\mathbf{P} = \frac{1}{3}\begin{pmatrix} 1 & 2 & 2 \\ 2 & -2 & 1 \\ 2 & 1 & -2 \end{pmatrix}$$

Next take the matrix $\mathbf{A}$ to be the following:

$$\mathbf{A} = \begin{pmatrix} 2 & 1 & -1 \\ 1 & 2 & -1 \\ -1 & -1 & 2 \end{pmatrix}$$

The characteristic equation is $(x-2)^3 - 3(x-2) - 2 = 0$. Since the equation $y^3 - 3y - 2 = 0$ has roots $-1, -1, 2$, the eigenvalues are $1, 1, 4$, and our matrix is similar to $\operatorname{diag}(1, 1, 4)$.

To find the transforming matrix, we determine the eigenvectors. For $x = 4$:

$$\begin{pmatrix} 2 & -1 & 1 \\ -1 & 2 & 1 \\ 1 & 1 & 2 \end{pmatrix}\begin{pmatrix} u_1 \\ u_2 \\ u_3 \end{pmatrix} = 0$$

This has the normalized solution $\mathbf{u} = \dfrac{1}{\sqrt{3}}(1, 1, -1)^{\mathrm{T}}$. And for $x = 1$:

$$\begin{pmatrix} -1 & -1 & 1 \\ -1 & -1 & 1 \\ 1 & 1 & -1 \end{pmatrix}\begin{pmatrix} v_1 \\ v_2 \\ v_3 \end{pmatrix} = 0$$

This system has rank 1 (as expected); in fact any vector orthogonal to $\mathbf{u}$ is a solution. We may take, for example, $\mathbf{v} = (1/\sqrt{2})(1, -1, 0)^{\mathrm{T}}$ and $\mathbf{w} = \mathbf{u} \wedge \mathbf{v} = -(1/\sqrt{6})(1, 1, 2)^{\mathrm{T}}$. The transforming matrix is

$$\mathbf{P} = \begin{pmatrix} 1/\sqrt{3} & 1/\sqrt{2} & -1/\sqrt{6} \\ 1/\sqrt{3} & -1/\sqrt{2} & -1/\sqrt{6} \\ -1/\sqrt{3} & 0 & -2/\sqrt{6} \end{pmatrix}$$

It is natural to ask whether any of this generalizes to the complex case; at first sight this seems to break down because all is based on the positive definiteness of $\sum x_i^2$. However, as Charles Hermite (1822–1901) observed, if we take instead the form

$$\sum x_i \bar{x}_i$$

an analogous theory can be developed for such forms. These forms and their associated matrices are called Hermitian; thus a matrix is **Hermitian** if its transpose equals its complex conjugate. The place of orthogonal transformations is taken by **unitary** matrices, i.e. matrices $\mathbf{P}$ satisfying $\bar{\mathbf{P}}^{\mathrm{T}}\mathbf{P} = \mathbf{I}$. The resulting theory is quite similar to the real case, but it will not be treated here (see, for example, Cohn, 1982, Chapter 8).

8.7 THE JORDAN NORMAL FORM

We now return to the general case of a matrix which need not be symmetric, to describe a normal form under similarity. We have already seen (in Theorem 8.4) that $\mathbf{A}$ is similar to a diagonal matrix whenever the characteristic equation of $\mathbf{A}$ has distinct roots. The conclusion holds more generally when there is an equation with distinct roots satisfied by $\mathbf{A}$. Let us first consider the special case of a matrix $\mathbf{A}$ such that

$$\mathbf{A}^2 = \mathbf{A}. \tag{8.25}$$

Such a matrix is said to be **idempotent**. It satisfies the quadratic equation $x^2 - x = 0$ with distinct roots 0, 1, and we wish to transform $\mathbf{A}$ to diagonal form. Consider the homogeneous linear system

$$(\mathbf{A} - \mathbf{I})\mathbf{x} = \mathbf{0}$$

and let $\mathbf{u}_1, \ldots, \mathbf{u}_r$ be a basis of its solution space. Since $(\mathbf{A} - \mathbf{I})\mathbf{A} = \mathbf{0}$, we can include among the $\mathbf{u}_i$ a basis for the column space of $\mathbf{A}$. Next we take the system

$$\mathbf{A}\mathbf{x} = \mathbf{0}$$

and let $\mathbf{u}_{r+1}, \ldots, \mathbf{u}_{r+s}$ be a basis for its solution space. Here we can include a basis for the column space of $\mathbf{I} - \mathbf{A}$, since $\mathbf{A}(\mathbf{I} - \mathbf{A}) = \mathbf{0}$. Now we have

$$\mathbf{I} = \mathbf{A} + (\mathbf{I} - \mathbf{A})$$

hence all the columns of $\mathbf{I}$ are linearly dependent on $\mathbf{u}_1, \ldots, \mathbf{u}_{r+s}$, so this set includes a basis of the whole space; in fact it must be a basis, since there can be no relations between eigenvectors belonging to different eigenvalues, by Theorem 8.4. If $\mathbf{U}$ is the matrix formed from these columns, then $\mathbf{U}$ is regular and

$$\mathbf{A}\mathbf{U} = \mathbf{U}\begin{pmatrix} \mathbf{I}_r & \mathbf{0} \\ \mathbf{0} & \mathbf{0} \end{pmatrix}$$

This accomplishes the similarity reduction of the idempotent matrix $\mathbf{A}$ to the diagonal form $\mathbf{I}_r \oplus \mathbf{0}$.

To deal with more general equations, let us pause for a moment to consider the factorization of polynomials. As we have seen, over $\mathbf{C}$ every polynomial f can be written as a product of linear factors, but these factors need not be distinct. In general, we have

$$f(x) = a_0 \prod_{i=1}^{r} (x - \alpha_i)^{m_i}$$

where $\alpha_1,\ldots,\alpha_r$ are distinct complex numbers, but the m_i may be greater than 1. We shall be concerned with polynomials without repeated factors, so all the m_i are 1, and, taking $a_0=1$ for simplicity, we can write our polynomial as

$$f(x)=(x-\alpha_1)\cdots(x-\alpha_r) \tag{8.26}$$

where the α_i are distinct. We shall also assume that $r>1$, since the trivial case $r=1$ will not be needed. For each $i=1,\ldots,r$ we define a polynomial $f_i(x)$ of degree $r-1$ as

$$f_i(x)=(x-\alpha_1)\cdots(x-\alpha_{i-1})(x-\alpha_{i+1})\cdots(x-\alpha_r)$$

Thus f_i satisfies the relation (which can also be used to define it):

$$(x-\alpha_i)f_i(x)=f(x) \tag{8.27}$$

We claim that the following relation holds identically in x:

$$\sum_i \frac{f_i(x)}{f_i(\alpha_i)}=1 \tag{8.28}$$

On the left of (8.28) we have a polynomial in x, of degree at most $r-1$. If we put $x=\alpha_1$, then the first term becomes $f_1(\alpha_1)/f_1(\alpha_1)=1$, while for $i>1$ we have $f_i(\alpha_1)/f_i(\alpha_i)=0$, because $f_i(\alpha_j)=0$ for $i\neq j$. Therefore the difference between the two sides in (8.28) vanishes for $x=\alpha_1$, and similarly for $x=\alpha_2,\ldots,\alpha_r$. So the difference is a polynomial of degree at most $r-1$ vanishing at r distinct values, hence it must vanish identically (as we saw in section 8.3), and this establishes the identity (8.28).

We can now achieve our aim, of diagonalizing a matrix satisfying an equation without repeated factors.

Theorem 8.9

Let $\mathbf{A}$ *be a square matrix satisfying an equation* $f=0$*, where the polynomial* f *can be written as a product of distinct linear factors. Then* $\mathbf{A}$ *is similar to a diagonal matrix.*

Proof

Let us factorize the polynomial f satisfied by $\mathbf{A}$ as in (8.26) and define $f_i(x)$ as in (8.27). Replacing x by $\mathbf{A}$ in (8.27), we obtain

$$(\mathbf{A}-\alpha_i\mathbf{I})f_i(\mathbf{A})=f(\mathbf{A})=\mathbf{0}$$

Thus on writing $\mathbf{A}_i=f_i(\mathbf{A})/f_i(\alpha_i)$, we have

$$(\mathbf{A}-\alpha_i\mathbf{I})\mathbf{A}_i=\mathbf{0}$$

It follows that $\mathbf{A}_i\mathbf{A}=\mathbf{A}_i\alpha_i$, hence $\mathbf{A}_i\varphi(\mathbf{A})=\mathbf{A}_i\varphi(\alpha_i)$ for any polynomial φ, and so

$$\mathbf{A}_i(\mathbf{A}_i - \mathbf{I}) = \mathbf{A}_i \frac{(f_i(\mathbf{A}) - f_i(\alpha_i))}{f_i(\alpha_i)} = \mathbf{0}$$

thus each $\mathbf{A}_i$ is idempotent. When $i \neq j$, $f_i(x)f_j(x)$ is divisible by $f(x)$, hence $\mathbf{A}_i\mathbf{A}_j = \mathbf{0}$ for $i \neq j$. Further, by (8.28), $\sum \mathbf{A}_i = \mathbf{I}$. This shows that the columns of the different $\mathbf{A}_i$ span the whole space, because the columns of $\mathbf{I}$ are linearly dependent on them. Therefore, we can select a basis for the whole space from the columns of the different $\mathbf{A}_i$. If $\mathbf{u}$ is a member of this basis, forming a column of $\mathbf{A}_i$, say, then $\mathbf{Au} = \alpha_i\mathbf{u}$, because $(\mathbf{A} - \alpha_i\mathbf{I})\mathbf{A}_i = \mathbf{0}$. Hence if $\mathbf{U}$ is the matrix formed from these columns, then $\mathbf{U}$ is regular and $\mathbf{AU} = \mathbf{UD}$, where $\mathbf{D}$ is diagonal, with the α_i along the main diagonal (possibly with repetitions). This shows $\mathbf{A}$ to be similar to a diagonal matrix and it completes the proof.

There still remains the general case of a matrix not satisfying the hypotheses of Theorem 8.9. Some preparation is necessary (and the rest of this section may be omitted without loss of continuity). We begin with a result on matrix equations.

Lemma 1

Given a matrix with block decomposition

$$\begin{pmatrix} \mathbf{A} & \mathbf{B} \\ \mathbf{0} & \mathbf{C} \end{pmatrix} \tag{8.29}$$

where $\mathbf{A}$, $\mathbf{C}$ *are square matrices of orders r, s, respectively, and* $\mathbf{B}$ *is* $r \times s$, *if the equation*

$$\mathbf{B} = \mathbf{AX} - \mathbf{XC} \tag{8.30}$$

has a solution $\mathbf{X}$, *an* $r \times s$ *matrix, then the matrix (8.29) is similar to* $\mathbf{A} \oplus \mathbf{C}$. *Such a solution always exists if there is a polynomial f such that* $f(\mathbf{A}) = \mathbf{0}$, *while* $f(\mathbf{C})$ *is regular.*

Proof

Suppose that $\mathbf{X} = \mathbf{H}$ is a solution of (8.30). Then $\begin{pmatrix} \mathbf{I} & \mathbf{H} \\ \mathbf{0} & \mathbf{I} \end{pmatrix}$ is regular with inverse $\begin{pmatrix} \mathbf{I} & -\mathbf{H} \\ \mathbf{0} & \mathbf{I} \end{pmatrix}$ and

$$\begin{pmatrix} \mathbf{I} & \mathbf{H} \\ \mathbf{0} & \mathbf{I} \end{pmatrix}\begin{pmatrix} \mathbf{A} & \mathbf{B} \\ \mathbf{0} & \mathbf{C} \end{pmatrix}\begin{pmatrix} \mathbf{I} & -\mathbf{H} \\ \mathbf{0} & \mathbf{I} \end{pmatrix} = \begin{pmatrix} \mathbf{I} & \mathbf{H} \\ \mathbf{0} & \mathbf{I} \end{pmatrix}\begin{pmatrix} \mathbf{A} & -\mathbf{AH} + \mathbf{B} \\ \mathbf{0} & \mathbf{C} \end{pmatrix} = \begin{pmatrix} \mathbf{A} & \mathbf{0} \\ \mathbf{0} & \mathbf{C} \end{pmatrix}$$

which proves the first part. It remains to solve (8.30). By the remainder formula we have

$$f(x) - f(y) = (x - y)\varphi(x, y)$$

for some polynomial $\varphi(x, y)$. Moreover, if $\varphi(x, y) = \sum x^i y^j c_{ij}$, then

$$f(x) - f(y) = (x - y) \sum x^i y^j c_{ij} = \sum x^i y^j (c_{i-1,j} - c_{i,j-1}) \tag{8.31}$$

Consider the matrix

$$\mathbf{M} = \sum_{i,j} \mathbf{A}^i \mathbf{B} \mathbf{C}^j c_{ij}$$

We have

$$\mathbf{AM} - \mathbf{MC} = \sum_{i,j} \mathbf{A}^i \mathbf{B} \mathbf{C}^j (c_{i-1,j} - c_{i,j-1}) = f(\mathbf{A})\mathbf{B} - \mathbf{B}f(\mathbf{C})$$

where we have used (8.31). By hypothesis $f(\mathbf{A}) = \mathbf{0}$ and $f(\mathbf{C})$ is invertible; further, $\mathbf{C}f(\mathbf{C})^{-1} = f(\mathbf{C})^{-1}\mathbf{C}$, so $\mathbf{X} = -\mathbf{M}f(\mathbf{C})^{-1}$ is a solution of (8.30), and the proof is complete.

With the help of this lemma we can establish a significant improvement of Theorem 8.5 (on the reduction to triangular form), but first another definition is needed. A matrix is said to be **nilpotent** if a power of it equals zero. There are non-zero nilpotent matrices, for example, $\begin{pmatrix} 0 & 1 \\ 0 & 0 \end{pmatrix}$; more generally, any triangular $n \times n$ matrix $\mathbf{C}$ with zeros on the main diagonal satisfies $\mathbf{C}^n = \mathbf{0}$. This is easily checked directly, but it also follows because all its eigenvalues are 0, so its characteristic polynomial must be x^n and the assertion follows from the Cayley–Hamilton theorem (Theorem 8.6). More generally, a triangular matrix $\mathbf{C}$ with all diagonal elements α satifies the equation $(\mathbf{C} - \alpha\mathbf{I})^n = \mathbf{0}$; for $\mathbf{C} - \alpha\mathbf{I}$ is triangular with zeros on the main diagonal, so we are in the previous case. We shall find that most of the remaining difficulty lies in dealing with nilpotent matrices; our next result accomplishes the reduction to that case.

Theorem 8.10

Let $\mathbf{A}$ *be a square matrix over* $\mathbf{C}$. *Then* $\mathbf{A}$ *is similar to a matrix of the form* $\mathbf{A}_1 \oplus \cdots \oplus \mathbf{A}_r$, *where each* $\mathbf{A}_i$ *has a single eigenvalue.*

Proof

By Theorem 8.5 we may assume that $\mathbf{A}$ is already in upper triangular form. Let the first r diagonal elements of $\mathbf{A}$ be α and all the others different from α. This is permissible, since the order of the eigenvalues in Theorem 8.5 was at our disposal. Now $\mathbf{A}$ has the form

$$\mathbf{A} = \begin{pmatrix} \mathbf{A}_1 & \mathbf{B} \\ 0 & \mathbf{C} \end{pmatrix}$$

where $\mathbf{A}_1$ is triangular with all elements on the main diagonal equal to α, and $\mathbf{C}$ is triangular with all diagonal elements different from α. If $\mathbf{A}_1$ is of order r, then the polynomial $f=(x-\alpha)^r$ has the property that $f(\mathbf{A}_1)=\mathbf{0}$, while $f(\mathbf{C})$ is triangular with non-zero elements on the main diagonal, hence invertible. Therefore we can use Lemma 1 to reduce $\mathbf{B}$ to $\mathbf{0}$. Now $\mathbf{C}$ is triangular of lower order than $\mathbf{A}$ and an induction on the order of $\mathbf{A}$ yields the result.

This result reduces the problem to the case of a matrix with a single eigenvalue, and by considering $\mathbf{A}-\alpha\mathbf{I}$ we can concentrate on the case of a nilpotent matrix. Here we need a lemma on the normal form for such a matrix. For any number α we denote by $\mathbf{J}_r(\alpha)$ the matrix of order r with all entries on the main diagonal equal to α, those on the diagonal above the main diagonal equal to 1 and the rest zero. This matrix $\mathbf{J}_r(\alpha)$ is called the **Jordan block** for the eigenvalue α. Thus, for example,

$$\mathbf{J}_4(\alpha)=\begin{pmatrix}\alpha & 1 & 0 & 0\\ 0 & \alpha & 1 & 0\\ 0 & 0 & \alpha & 1\\ 0 & 0 & 0 & \alpha\end{pmatrix}$$

Lemma 2

Let $\mathbf{N}$ be a nilpotent matrix. Then $\mathbf{N}$ is similar to a diagonal sum of Jordan blocks for the eigenvalue 0: $\mathbf{T}^{-1}\mathbf{N}\mathbf{T}=\mathbf{P}=(p_{ij})$, *where*

$$p_{ij}=\gamma_i\delta_{i,j-1}$$

and $\gamma_i=0$ *or* 1.

Proof

Suppose that $\mathbf{N}^r=\mathbf{0}$, $\mathbf{N}^{r-1}\neq\mathbf{0}$; then there is a vector $\mathbf{u}$ such that $\mathbf{N}^{r-1}\mathbf{u}\neq\mathbf{0}$. We claim that $\mathbf{N}^{r-1}\mathbf{u},\mathbf{N}^{r-2}\mathbf{u},\ldots,\mathbf{N}\mathbf{u},\mathbf{u}$ are linearly independent. For if not, let $\mathbf{N}^{i-1}\mathbf{u}$ be a vector in this sequence linearly dependent on earlier ones:

$$\mathbf{N}^{i-1}\mathbf{u}=\lambda_1\mathbf{N}^{r-1}\mathbf{u}+\cdots+\lambda_{r-i}\mathbf{N}^{i}\mathbf{u}.$$

Multiplying by $\mathbf{N}^{r-i}$, we obtain $\mathbf{N}^{r-1}\mathbf{u}=\mathbf{0}$, which is a contradiction. Thus the $\mathbf{N}^{r-i}\mathbf{u}$ are linearly independent and we can find a basis $\mathbf{u}_1,\ldots,\mathbf{u}_n$ of the whole space such that $\mathbf{u}_i=\mathbf{N}^{r-i}\mathbf{u}$ $(i=1,\ldots,r)$. Referred to this basis, $\mathbf{N}$ takes the form

$$\mathbf{N}'=\begin{pmatrix}\mathbf{A} & \mathbf{B}\\ \mathbf{0} & \mathbf{C}\end{pmatrix} \tag{8.32}$$

where $\mathbf{A}=\mathbf{J}_r(0)$, and by an induction on n we may suppose that $\mathbf{C}=(c_{ij})$, where $c_{j-1,j}=\gamma_j=0$ or 1 and $c_{ij}=0$ if $i\neq j-1$. Here $\mathbf{C}$ is a square matrix of order $s=n-r$ and we may suppose that $r>1$, since otherwise $\mathbf{N}=\mathbf{0}$ and there is nothing to prove.

We shall use Lemma 1 to find a similarity transformation which will reduce **B** to **0**. Thus we must solve the matrix equation

$$\mathbf{B} = \mathbf{AX} - \mathbf{XC} \tag{8.33}$$

Here we may assume that the last row of **B** is zero, for on writing $\mathbf{B} = (b_{ij})$, consider b_{rj}. If $\gamma_j = 1$, we can make a similarity transformation by $\mathbf{I} + b_{rj}\mathbf{E}_{r,r+j-1}$. This amounts to subtracting b_{rj} times the $(j-1)$th row of **C** from the last row of **B** and adding b_{rj} times the last column of **A** to the $(j-1)$th column of **B**. The first part reduces b_{rj} to zero, while the latter operation does not affect the last row of **B** because the last row of **A** is zero. If $\gamma_j = 0$, then the $(j-1)$th row of **C** is zero and so $\mathbf{N}'\mathbf{e}_{r+j-1} = b_{1j}\mathbf{e}_1 + \cdots + b_{rj}\mathbf{e}_r$, hence

$$\mathbf{0} = \mathbf{N}'^{r}\mathbf{e}_{r+j-1} = \sum_i b_{ij}\mathbf{N}'^{r-1}\mathbf{e}_i = b_{rj}\mathbf{e}_1$$

and so $b_{rj} = 0$. Thus the last row of **B** is now zero and (8.33) takes the form

$$\begin{aligned} b_{ij} &= x_{i+1,j} - x_{i,j-1}\gamma_j, \quad && i = 1,\ldots,r-1; \quad j = 2,\ldots,s \\ b_{i1} &= x_{i+1,1}, && i = 1,\ldots,r-1 \end{aligned}$$

where we have put $\mathbf{X} = (x_{ij})$. These equations can be solved for x_{ij} by taking $i = 1,\ldots,r-1$ in succession, with arbitrary values for x_{1j}. With these values (8.33) holds, and using Lemma 1 we can reduce **B** to zero in (8.32); thus Lemma 2 is proved.

With the help of this lemma we can complete our reduction. Given any square matrix **A** over the complex numbers, let $\alpha_1,\ldots,\alpha_r$ be its distinct eigenvalues. By Theorem 8.10, **A** is similar to a diagonal sum

$$\mathbf{A}_1 \oplus \cdots \oplus \mathbf{A}_r \tag{8.34}$$

where $\mathbf{A}_i$, the component corresponding to α_i, is a matrix with the unique eigenvalue α_i; hence $\mathbf{A}_i - \alpha_i\mathbf{I}$ is nilpotent. The order of $\mathbf{A}_i$ is the multiplicity of α_i as root of the characteristic equation of **A**; it is also called the **algebraic multiplicity** of α_i. The dimension of $\ker(\mathbf{A}_i - \alpha_i\mathbf{I})$ is called the **geometric multiplicity**. It is clear that

$$\text{geometric multiplicity} \leqslant \text{algebraic multiplicity} \tag{8.35}$$

with equality precisely when the corresponding matrix $\mathbf{A}_i$ is scalar: $\mathbf{A}_i = \alpha_i\mathbf{I}$. In any case $\mathbf{A}_i - \alpha_i\mathbf{I}$ is nilpotent and, by Lemma 2, $\mathbf{A}_i - \alpha_i\mathbf{I}$ is similar to a diagonal sum of Jordan blocks $\mathbf{J}_t(0)$, hence $\mathbf{A}_i$ is similar to $\mathbf{A}'_i$, where

$$\mathbf{A}'_i = \mathbf{J}_{t_{i1}}(\alpha_i) \oplus \cdots \oplus \mathbf{J}_{t_{i\lambda_i}}(\alpha_i) \tag{8.36}$$

Here we may take the components ordered by size, so that $t_{i1} \geqslant t_{i2} \geqslant \cdots \geqslant t_{i\lambda_i}$. It can be shown (see, for example, Cohn, 1982, section 11.4) that the similarity class of **A** is completely described by its eigenvalues and the orders t_{ij}. These numbers are usually written in the form

$$[(t_{11}t_{12}\cdots t_{1\lambda_1})(t_{21}\cdots t_{2\lambda_2})\cdots(t_{r1}\cdots t_{r\lambda_r})]$$

where each parenthesis refers to an eigenvalue α_i. This expression is called the **Segre characteristic** of $\mathbf{A}$ (after Corrado Segre, 1863–1924). We remark that the geometric multiplicity of α_i is λ_i, while the algebraic multiplicity is $\sum_\nu t_{i\nu}$. Our findings may be summed up as follows:

Theorem 8.11
Every square matrix $\mathbf{A}$ *over the complex numbers is similar to a diagonal sum of Jordan blocks* $\mathbf{J}_t(\alpha)$, *where* α *ranges over the eigenvalues of* $\mathbf{A}$, *and* t *assumes the numbers indicated by the Segre characteristic.*

The form for $\mathbf{A}$ described here is called the **Jordan normal form**.

For example, the Segre characteristic of a matrix similar to a diagonal matrix is of the form $[(1^{m_1})(1^{m_2})\cdots(1^{m_r})]$.

A 5×5 matrix with characteristic polynomial x^5 is similar to exactly one of the following: (i) $\mathbf{J}_5$; (ii) $\mathbf{J}_4\oplus\mathbf{0}$; (iii) $\mathbf{J}_3\oplus\mathbf{J}_2$; (iv) $\mathbf{J}_3\oplus\mathbf{0}_2$; (v) $\mathbf{J}_2\oplus\mathbf{J}_2\oplus\mathbf{0}$; (vi) $\mathbf{J}_2\oplus\mathbf{0}_3$; (vii) $\mathbf{0}_5$. Here we have put $\mathbf{J}_r=\mathbf{J}_r(0)$ and $\mathbf{0}_r$ is the $r\times r$ zero matrix. The corresponding Segre characteristics are (i) $[(5)]$; (ii) $[(4\ \ 1)]$; (iii) $[(3\ \ 2)]$; (iv) $[(3\ \ 1^2)]$; (v) $[(2^2\ \ 1)]$; (vi) $[(2\ \ 1^3)]$; (vii) $[(1^5)]$. The minimal polynomials (that is, polynomials of least degree satisfied by the matrix) are x^5 for (i); x^4 for (ii); x^3 for (iii), (iv); x^2 for (v), (vi); and x for (vii).

The Jordan normal form was first described by Camille Jordan (1838–1922) in his *Traité des substitutions et équations algébriques*, which appeared in 1870.

EXERCISES

8.1. Show that if $\mathbf{P}^{-1}\mathbf{AP}=\mathbf{A}$ for all regular matrices $\mathbf{P}$, then $\mathbf{A}=\alpha\mathbf{I}$.

8.2. Give a direct proof that $\begin{pmatrix}0&1\\0&0\end{pmatrix}$ is not similar to a diagonal matrix.

8.3. Show that the symmetric matrix $\begin{pmatrix}1&\mathrm{i}\\\mathrm{i}&1\end{pmatrix}$ is similar to a diagonal matrix.

8.4. Transform the following matrices to diagonal form, where possible, and find the matrix of transformation:

$$\begin{pmatrix}-3&1\\2&-2\end{pmatrix},\begin{pmatrix}3&-2\\2&-1\end{pmatrix},\begin{pmatrix}11&17&5\\-4&-6&-2\\-6&-10&-2\end{pmatrix},\begin{pmatrix}-2&-8&-2\\2&6&1\\2&4&3\end{pmatrix}$$

8.5. For $\mathbf{A}=\begin{pmatrix}4&-3\\3&-2\end{pmatrix}$, show that $\mathbf{A}^n=n\mathbf{A}-(n-1)\mathbf{I}$, for all integers n.

8.6. Let $\alpha_1,\ldots,\alpha_n$ be distinct numbers and define $f=\prod_1^n(x-\alpha_i)=(x-\alpha_i)f_i$ as in section 8.7. Show that for any $b_1,\ldots,b_n$, $\sum b_if_i(x)/f_i(\alpha_i)$ is a polynomial of degree less than n which assumes the value b_i when $x=\alpha_i$ $(i=1,\ldots,n)$. (This is known as the **Lagrange interpolation formula**, cf. section 9.5.)

8.7. Show that: (i) every square matrix is similar to itself; (ii) if $\mathbf{P}$ is similar to $\mathbf{Q}$, then $\mathbf{Q}$ is similar to $\mathbf{P}$; (iii) if $\mathbf{P}$ is similar to $\mathbf{Q}$ and $\mathbf{Q}$ is similar to $\mathbf{R}$, then $\mathbf{P}$ is similar to $\mathbf{R}$.

(These properties are usually expressed by saying that the relation of similarity between square matrices is **reflexive, symmetric** and **transitive**. Any relation with these properties is called an **equivalence relation**. For example, equality '$=$' is an equivalence and this exercise shows that similarity is an equivalence; this follows from Theorem 8.1, but is also easily proved directly.)

8.8. Let V be the vector space of polynomials of degree at most n in x. Show that $f(x) \mapsto x^n f(1/x)$, for any $f \in V$, defines a linear mapping $\mathbf{T}$ whose square is the identity. Find the eigenvalues of $\mathbf{T}$ and the corresponding eigenspaces.

8.9. If $\mathbf{A}, \mathbf{B}$ are $n \times n$ matrices such that $\mathbf{x}^{\mathrm{T}}\mathbf{A}\mathbf{x} = \mathbf{x}^{\mathrm{T}}\mathbf{B}\mathbf{x}$ for all $\mathbf{x}$, show that $\mathbf{A}^{\mathrm{T}} + \mathbf{A} = \mathbf{B}^{\mathrm{T}} + \mathbf{B}$, but that $\mathbf{A}$ and $\mathbf{B}$ need not be equal. (The matrix $\mathbf{C} = \mathbf{A} - \mathbf{B}$ is **skew symmetric**, that is, it satisfies $\mathbf{C}^{\mathrm{T}} = -\mathbf{C}$.)

8.10. For any square matrices $\mathbf{A}, \mathbf{B}, \mathbf{C}$ of the same order show that $\mathrm{tr}(\mathbf{ABC}) = \mathrm{tr}(\mathbf{BCA}) = \mathrm{tr}(\mathbf{CAB})$, and give an example where $\mathrm{tr}(\mathbf{ABC}) \neq \mathrm{tr}(\mathbf{BAC})$.

8.11. Let $\mathbf{A}$ be an $n \times n$ matrix and suppose that there is a vector $\mathbf{v}$ such that $\mathbf{v}, \mathbf{A}\mathbf{v}, \ldots, \mathbf{A}^{n-1}\mathbf{v}$ are linearly independent. Find the matrix similar to $\mathbf{A}$ referred to these vectors as a basis and show that the characteristic polynomial of $\mathbf{A}$ is $f(x) = x^n - a_n x^{n-1} - \cdots - a_1$, where the coefficients are found from the relation $\mathbf{A}^n\mathbf{v} = a_1\mathbf{v} + a_2\mathbf{A}\mathbf{v} + \cdots + a_n\mathbf{A}^{n-1}\mathbf{v}$. (The matrix found here is called the **companion matrix** of f; this exercise shows how to form a matrix with a given polynomial as characteristic polynomial.)

8.12. Let f be a given polynomial of degree n with zeros $\alpha_1, \ldots, \alpha_n$ (not necessarily distinct). By evaluating $\mathbf{C}^r$, where $\mathbf{C}$ is the companion matrix of f (cf. Exercise 8.11), find the value of $\sum \alpha_i^r$ for low values of r. (*Hint*: Show that $\mathbf{C}^r$ has eigenvalues $\alpha_1^r, \ldots, \alpha_n^r$ and recall that $\mathrm{tr}(\mathbf{C})$ is the sum of the eigenvalues of $\mathbf{C}$.)

8.13. Let $\mathbf{N} = (n_{ij})$, where $n_{i,i+1} = 1$ and the rest are zero. Find a matrix $\mathbf{T}$ which is invertible such that $\mathbf{T}^{-1}\mathbf{N}\mathbf{T} = \mathbf{N}^{\mathrm{T}}$.

8.14. Show (by using Exercise 8.13 and the Jordan normal form) that every square matrix is similar to its transpose.

8.15. Construct up to similarity all matrices of order 6 with characteristic polynomial $(x-1)^2(x-2)^4$.

8.16. Construct up to similarity all matrices of order 6 with minimal polynomial $(x-1)^2(x+1)^2$. (The **minimal polynomial** of a matrix $\mathbf{A}$ is the polynomial of least degree satisfied by $\mathbf{A}$.)

8.17. Construct up to similarity all matrices of order 6 with minimal polynomial $(x-1)^3$ and characteristic polynomial $(x-1)^6$.

8.18. Show that the equation of degree n with distinct roots $\alpha_1, \ldots, \alpha_n$ can be written $V(x, \alpha_1, \ldots, \alpha_n) = 0$.

9

Applications I. Algebra and geometry

In this chapter and the next we give some illustrations of how the methods developed in Chapters 1–8 can be applied. In most cases we merely sketch the outlines, to whet the reader's appetite, with a reference to more thorough treatments.

9.1 QUADRICS IN SPACE

A **quadric surface** or simply **quadric** is a surface in 3-space given by an equation of the second degree in the coordinates. It will be convenient to express all vectors in terms of coordinates referred to a right-handed rectangular coordinate system in space. Instead of speaking of 'the vector whose components are a_1, a_2, a_3' we shall simply speak of 'the vector (a_1, a_2, a_3)', leaving the coordinate system to be understood. As before, we shall abbreviate the coordinates by a single letter; for example, to say that $\mathbf{u}$ is a unit vector is to say that its components u_1, u_2, u_3 satisfy $u_1^2 + u_2^2 + u_3^2 = 1$, or, in matrix notation, $\mathbf{u}^{\mathrm{T}}\mathbf{u} = 1$, where $\mathbf{u}^{\mathrm{T}} = (u_1, u_2, u_3)$ and $\mathbf{u}$ is the corresponding column.

The most general equation of the second degree (in 3-space) is

$$a_{11}x_1^2 + a_{22}x_2^2 + a_{33}x_3^2 + 2a_{12}x_1x_2 + 2a_{13}x_1x_3 + 2a_{23}x_2x_3 \\ + 2a_1x_1 + 2a_2x_2 + 2a_3x_3 + \alpha = 0$$

where $a_{11}, \ldots, a_3, \alpha$ are real numbers. In matrix form this equation can be written as

$$\mathbf{x}^{\mathrm{T}}\mathbf{A}\mathbf{x} + 2\mathbf{a}^{\mathrm{T}}\mathbf{x} + \alpha = 0 \tag{9.1}$$

where $\mathbf{A} = (a_{ij})$ is a symmetric matrix and $\mathbf{a}^{\mathrm{T}} = (a_1, a_2, a_3)$.

A complete description of quadrics would take us too far afield, so we shall limit ourselves to the case where the matrix $\mathbf{A}$ is regular. In that case the quadric (9.1) is called **central**, because such a quadric has a centre of symmetry. This centre may be determined as follows.

If we make a parallel translation of our coordinate system by a vector **b**, then the point with position vector **x** is described by a vector **y**, related to **x** by the equation

$$\mathbf{x} = \mathbf{y} + \mathbf{b}$$

Expressed in the new system, equation (9.1) becomes

$$\mathbf{y}^{\mathrm{T}}\mathbf{A}\mathbf{y} + \mathbf{y}^{\mathrm{T}}\mathbf{A}\mathbf{b} + \mathbf{b}^{\mathrm{T}}\mathbf{A}\mathbf{y} + 2\mathbf{a}^{\mathrm{T}}\mathbf{y} + \beta = 0$$

where $\beta = \mathbf{b}^{\mathrm{T}}\mathbf{A}\mathbf{b} + 2\mathbf{a}^{\mathrm{T}}\mathbf{b} + \alpha$. By the symmetry of **A** this may be written

$$\mathbf{y}^{\mathrm{T}}\mathbf{A}\mathbf{y} + 2\mathbf{y}^{\mathrm{T}}(\mathbf{A}\mathbf{b} + \mathbf{a}) + \beta = 0 \qquad (9.2)$$

Now the regularity of **A** is precisely the condition for a unique vector **b** to exist, satisfying $\mathbf{A}\mathbf{b} + \mathbf{a} = \mathbf{0}$. If we choose **b** in this way, equation (9.2) reduces to

$$\mathbf{y}^{\mathrm{T}}\mathbf{A}\mathbf{y} + \beta = 0 \qquad (9.3)$$

This equation contains only terms of degree 2 or 0 in the components of **y**, so if $\mathbf{y} = \mathbf{c}$ is a point on the quadric, then $\mathbf{y} = -\mathbf{c}$ also lies on the surface. This is expressed by saying that the surface is **symmetric** about the origin of coordinates as **centre**. The regularity of **A** ensures that there exists exactly one centre. Central quadrics include, for example, spheres, but exclude such quadrics as a cylinder (with a line of symmetry) and a paraboloid (with no points of symmetry, at least in general).

In (9.2) we have seen the effect of a translation of coordinates on the equation of our quadric, and (9.3) shows how for a central quadric the equation may be simplified by choosing the origin of coordinates at the centre. We now want to consider the effect of a rotation of the coordinate system on the equation of the quadric. We shall assume for simplicity that the quadric is central and is referred to the centre as origin, so that its equation is

$$\mathbf{x}^{\mathrm{T}}\mathbf{A}\mathbf{x} + \alpha = 0 \qquad (9.4)$$

In a rotation of coordinates the new coordinates **y** are related to the old ones **x** by an equation

$$\mathbf{x} = \mathbf{P}\mathbf{y} \qquad (9.5)$$

where **P** is a proper orthogonal matrix (cf. section 7.3, which also showed that every such matrix defines a rotation). If the equation in the new coordinates is

$$\mathbf{y}^{\mathrm{T}}\mathbf{B}\mathbf{y} + \alpha = 0$$

then, as (9.5) shows, we have

$$\mathbf{B} = \mathbf{P}^{\mathrm{T}}\mathbf{A}\mathbf{P}$$

Thus we have a congruence transformation with an orthogonal matrix. By

Theorem 8.8 we can choose **P** so that **B** is diagonal, with the eigenvalues of **A** as diagonal entries. We state the result as follows:

Theorem 9.1
Let Q be a quadric, given by the equation

$$\mathbf{x}^{\mathrm{T}}\mathbf{A}\mathbf{x} + \alpha = 0 \tag{9.4}$$

in a right-handed rectangular coordinate system. Then in a suitable coordinate system (again right-handed rectangular) Q has the equation

$$\alpha_1 y_1^2 + \alpha_2 y_2^2 + \alpha_3 y_3^2 + \alpha = 0 \tag{9.6}$$

where $\alpha_1, \alpha_2, \alpha_3$ are real and are the eigenvalues of **A**. *In particular, every central quadric can be written in the form (9.4) and hence in the form (9.6).*

The axes of the new coordinate system are called the **principal axes** of the quadric, and the transformation is often called a transformation to principal axes.

By way of illustration let us consider the examples of section 8.6 in the present context. The first quadric considered was:

$$4x_1^2 + 2x_2^2 + 3x_3^2 - 4x_1x_3 - 4x_2x_3 - 3 = 0$$

This has as symmetric matrix

$$\mathbf{A} = \begin{pmatrix} 4 & 0 & -2 \\ 0 & 2 & -2 \\ -2 & -2 & 3 \end{pmatrix}$$

(where we must remember to halve the off-diagonal entries). The eigenvalues of **A** were found to be $0, 3, 6$, and its equation, referred to principal axes, is $3y_2^2 + 6y_3^2 - 3 = 0$, or

$$y_2^2 + 2y_3^2 - 1 = 0.$$

Each plane parallel to the 23-plane in the y-system cuts the surface in the same ellipse. It is therefore known as an **elliptic cylinder**. We note that it is actually a non-central quadric (it has a whole line of symmetry), even though its equation can be expressed in the form (9.4).

The other quadric considered in section 8.6 was:

$$x_1^2 + x_2^2 + x_3^2 + x_1x_2 - x_1x_3 - x_2x_3 - 1 = 0$$

If we multiply the equation by 2 (to avoid fractions) we obtain the symmetric matrix

$$\mathbf{A} = \begin{pmatrix} 2 & 1 & -1 \\ 1 & 2 & -1 \\ -1 & -1 & 2 \end{pmatrix}$$

As we saw in section 8.6, the eigenvalues are 1, 1, 4; the equation of our quadric, referred to principal axes, is

$$y_1^2 + y_2^2 + 4y_3^2 - 2 = 0 \tag{9.7}$$

The sections parallel to the 12-plane are circles (or empty), other sections are ellipses; thus we have an **ellipsoid of revolution**. When we constructed the orthogonal matrix of transformation, we found the eigenspace for $x = 1$ to be 2-dimensional, and we could take any pair of orthogonal vectors (forming a right-handed system with the third vector) in this plane. This corresponds to the fact that a rotation about the 3-axis in the y-system leaves the principal axes form (9.7) intact.

To find the tangent plane at a point of a quadric we consider, more generally, the points of intersection with a given line, say

$$\mathbf{x} = \mathbf{p} + \lambda\mathbf{u} \tag{9.8}$$

For a fixed λ, the point $\mathbf{p} + \lambda\mathbf{u}$ lies on the quadric (9.1) if and only if

$$(\mathbf{p} + \lambda\mathbf{u})^{\mathrm{T}}\mathbf{A}(\mathbf{p} + \lambda\mathbf{u}) + 2\mathbf{a}^{\mathrm{T}}(\mathbf{p} + \lambda\mathbf{u}) + \alpha = 0$$

Rearranging this equation in powers of λ, we obtain

$$\lambda^2\mathbf{u}^{\mathrm{T}}\mathbf{A}\mathbf{u} + 2\lambda\mathbf{u}^{\mathrm{T}}(\mathbf{A}\mathbf{p} + \mathbf{a}) + \mathbf{p}^{\mathrm{T}}\mathbf{A}\mathbf{p} + 2\mathbf{a}^{\mathrm{T}}\mathbf{p} + \alpha = 0 \tag{9.9}$$

The two roots of this equation, real or complex, give the two intersections of the line (9.8) with the quadric (9.1). In particular, (9.9) has 0 as a root precisely when $\mathbf{p}$ lies on the quadric. Suppose that $\mathbf{p}$ is a point on the quadric and $\mathbf{q}$ is any point other than $\mathbf{p}$ on the tangent plane at $\mathbf{p}$. Then the line joining $\mathbf{p}$ and $\mathbf{q}$ has coincident intersections with the quadric. Putting $\mathbf{u} = \mathbf{q} - \mathbf{p}$ in (9.8), we can express this by saying that both roots of (9.9) are zero, that is, $\mathbf{p}$ lies on the quadric and

$$(\mathbf{q} - \mathbf{p})^{\mathrm{T}}(\mathbf{A}\mathbf{p} + \mathbf{a}) = 0$$

As $\mathbf{q}$ varies, we obtain the tangent plane of the quadric at $\mathbf{p}$, whose equation is therefore

$$(\mathbf{x} - \mathbf{p})^{\mathrm{T}}(\mathbf{A}\mathbf{p} + \mathbf{a}) = 0 \tag{9.10}$$

The normal to this plane at $\mathbf{p}$ is called the **normal to the quadric at p**. Its direction is given by the vector $\mathbf{A}\mathbf{p} + \mathbf{a}$, hence the equation of the normal is

$$\mathbf{x} = \mathbf{p} + \mu(\mathbf{A}\mathbf{p} + \mathbf{a})$$

From this equation and (9.10) we see that the tangent plane and normal are defined at any point $\mathbf{p}$ of the quadric, unless $\mathbf{A}\mathbf{p} + \mathbf{a} = \mathbf{0}$. As we saw, this equation is satisfied only by the centre of the quadric. Thus in a central quadric, every point has a tangent plane and a normal, except the centre itself. As we shall see later, the centre does not usually lie on the quadric. An example where it does is the cone, whose centre is its vertex.

To find the tangent plane of the ellipsoid considered above at the point $\mathbf{p}=(0,1,1)$ we have, by (9.10), $\mathbf{x}^{\mathrm{T}}\mathbf{A}\mathbf{p}=\mathbf{p}^{\mathrm{T}}\mathbf{A}\mathbf{p}=2$, which simplifies to

$$x_2+x_3=2$$

9.2 THE CLASSIFICATION OF CENTRAL QUADRICS

We have seen that every central quadric may, in a suitable coordinate system, be expressed in the form

$$\lambda_1x_1^2+\lambda_2x_2^2+\lambda_3x_3^2+c=0 \tag{9.11}$$

With the help of this form it is easy to classify the different central quadrics.

Suppose first that $c\neq 0$; this means that the origin does not lie on the surface. On dividing (9.11) by $-c$ and taking the constant term to the right-hand side we may write the quadric as

$$\alpha_1x_1^2+\alpha_2x_2^2+\alpha_3x_3^2=1 \tag{9.12}$$

The different cases will be distinguished by the signs of the αs.

(i) $\alpha_1,\alpha_2,\alpha_3$ are all positive. Put $a_i=1/\sqrt{\alpha_i}\,(i=1,2,3)$, so that (9.12) becomes

$$x_1^2/a_1^2+x_2^2/a_2^2+x_3^2/a_3^2=1$$

The intercepts on the 1-axis are $\pm a_1$, and similarly for the other axes; further, the sections by the coordinate planes are ellipses, and the surface represented is called an **ellipsoid** (Fig. 9.1) with semi-axes a_1,a_2,a_3. If two eigenvalues are equal, say $\alpha_1=\alpha_2$, then we have an **ellipsoid of revolution**, **prolate** (elongated) or **oblate** (flattened) according as $\alpha_3<\alpha_1$ or $\alpha_3>\alpha_1$. For example, the Earth has the shape of an oblate ellipsoid of revolution. If all three eigenvalues are equal, we have a **sphere**. Thus a sphere of centre $\mathbf{a}$ and radius r has the equation $|\mathbf{x}-\mathbf{a}|=r$. Generally an equation

$$|\mathbf{x}|^2+2\mathbf{a}^{\mathrm{T}}\mathbf{x}+c=0$$

represents a sphere provided that $|\mathbf{a}|^2>c$, and its centre is then $-\mathbf{a}$, while its radius is $(|\mathbf{a}|^2-c)^{1/2}$.

(ii) One eigenvalue is negative, say $\alpha_3<0<\alpha_1,\alpha_2$. The quadric meets the 1- and 2-axes, but not the 3-axis. It consists of a single piece which is cut by planes parallel to the 12-plane in an ellipse and by planes parallel to the

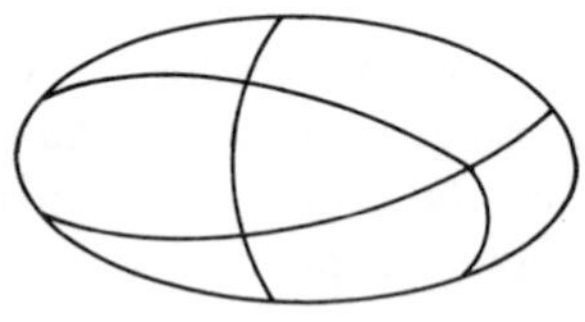

Fig. 9.1

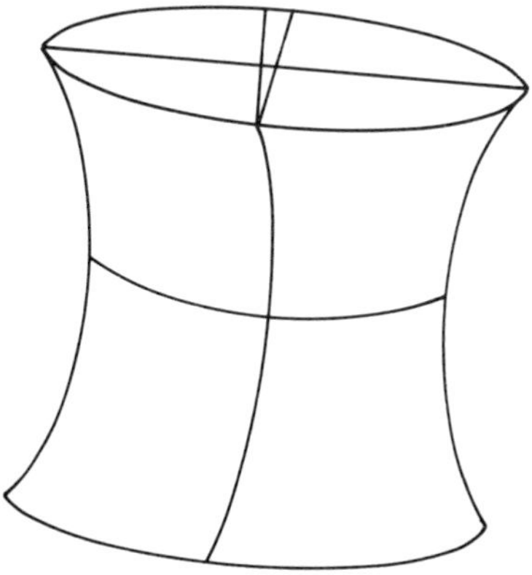

Fig. 9.2

13-plane or the 23-plane in a hyperbola. The quadric is called a **hyperboloid of one sheet** (Fig. 9.2). Although it is a curved surface, there are straight lines passing through every point on it. It may be constructed by taking two equal circular bands, joining them by a number of straight line staves so as to form a cylinder and then giving the circles a slight twist. The result is a hyperboloid of revolution (rather like a certain type of napkin ring).

(iii) Two eigenvalues are negative, say $\alpha_1 > 0 > \alpha_2, \alpha_3$. This is a **hyperboloid of two sheets** (Fig. 9.3). It consists of two separate pieces and meets the 1-axis, but not the 2- or 3-axes. A plane parallel to the 12-plane or the 13-plane cuts the surface in a hyperbola, while a plane parallel to the 23-plane cuts it in an ellipse or not at all.

No other cases are possible, since in a central quadric $\alpha_1, \alpha_2, \alpha_3 \neq 0$ and in the remaining case where $\alpha_1, \alpha_2, \alpha_3$ are all negative, the equation (9.12) is not satisfied by any point (with real coordinates), and so does not define a real surface.

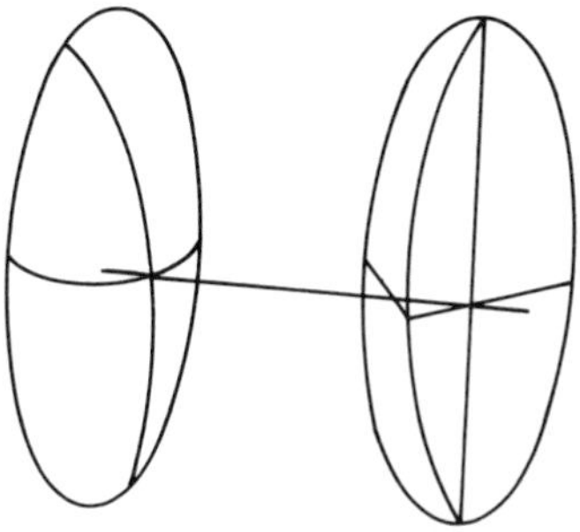

Fig. 9.3

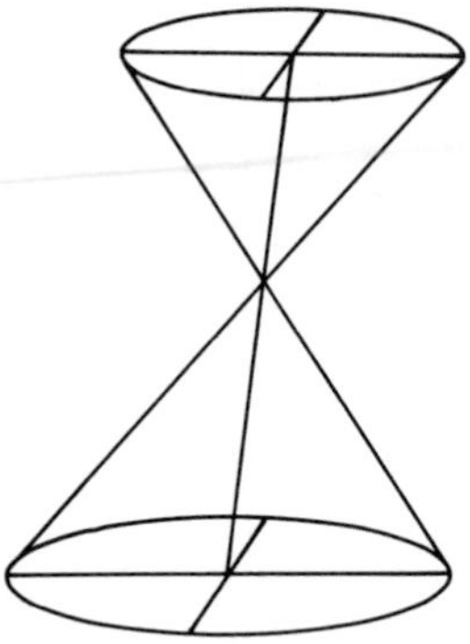

Fig. 9.4

If in (9.11) $c = 0$, then either $\lambda_1, \lambda_2, \lambda_3$ are all of the same sign, in which case the origin is the only point with real coordinates on the surface, or two are of one sign and one of the other. By permuting the coordinate axes and if necessary multiplying the equation by -1, we may suppose that $\lambda_1, \lambda_2 > 0$ and $\lambda_3 < 0$. The resulting quadric is a **cone** whose vertex ($=$ centre) is at the origin (Fig. 9.4).

9.3 POSITIVITY CRITERIA

It is useful to have a means of recognizing the type of a quadric from its equation without having to reduce it to principal axes. The problem is essentially one of determining the sign of the eigenvalues of a symmetric matrix. Before answering this question we give a criterion for all eigenvalues to be positive.

We recall that all the eigenvalues of the symmetric matrix $\mathbf{A}$ are positive precisely when $\mathbf{A}$ is positive definite. Let $\mathbf{A}$ be given by

$$\mathbf{A} = \begin{pmatrix} a_{11} & a_{12} & a_{13} \\ a_{21} & a_{22} & a_{23} \\ a_{31} & a_{32} & a_{33} \end{pmatrix} \tag{9.13}$$

where $a_{ij} = a_{ji}$. Then the numbers

$$d_0 = 1, \quad d_1 = a_{11}, \quad d_2 = \begin{vmatrix} a_{11} & a_{12} \\ a_{21} & a_{22} \end{vmatrix}, \quad d_3 = |\mathbf{A}|$$

are called the **principal minors** of the matrix $\mathbf{A}$. We note that each d_i (apart from the first and last) is the minor of the entry in the bottom right-hand corner of d_{i+1}. Further, d_3, being equal to $|\mathbf{A}|$, is just the product of all the eigenvalues of $\mathbf{A}$. Now we have the following criterion for positive definiteness:

Theorem 9.2
A real symmetric matrix of order 3 is positive definite if and only if all its principal minors are positive.

Of course, d_0 may be ignored here, since it is always 1. We remark that the theorem in fact holds for matrices of any order, with the appropriate definition of higher principal minors (cf., for example, Cohn, 1982, section 8.3).

Proof
Let the matrix $\mathbf{A}$ be given by (9.13) and denote by α_{ij} the minor of a_{ij} in $\mathbf{A}$, so that $d_1 = a_{11}, d_2 = \alpha_{33}$. Suppose now that $\mathbf{A}$ is positive definite. Then all the eigenvalues of $\mathbf{A}$ are positive and so $d_3 = |\mathbf{A}| > 0$. Next we have

$$a_{11}x_1^2 + a_{22}x_2^2 + 2a_{12}x_1x_2 = \mathbf{x}^{\mathrm{T}}\mathbf{A}\mathbf{x} \tag{9.14}$$

where $\mathbf{x} = (x_1, x_2, 0)^{\mathrm{T}}$. The right-side of (9.14) is positive for all x_1, x_2 not both zero, and this shows that the 2×2 matrix

$$\begin{pmatrix} a_{11} & a_{12} \\ a_{21} & a_{22} \end{pmatrix}$$

is positive definite. Hence its determinant d_2 is positive, and similarly, $d_1 = a_{11}$ must be positive, because $a_{11}x_1^2 > 0$ for all $x_1 \neq 0$. This proves the necessity of the condition.

Conversely, assume that d_1, d_2, d_3 are all positive. We have

$$\begin{aligned}\mathbf{x}^{\mathrm{T}}\mathbf{A}\mathbf{x} &= a_{11}x_1^2 + a_{22}x_2^2 + a_{33}x_3^2 + 2a_{12}x_1x_2 + 2a_{13}x_1x_3 + 2a_{23}x_2x_3 \\ &= \frac{1}{a_{11}}(a_{11}x_1 + a_{12}x_2 + a_{13}x_3)^2 + \frac{1}{a_{11}}(\alpha_{33}x_2^2 + \alpha_{22}x_3^2 + 2\alpha_{23}x_2x_3)\end{aligned}$$

Hence

$$\mathbf{x}^{\mathrm{T}}\mathbf{A}\mathbf{x} = \frac{1}{a_{11}}(\Sigma_i a_{1i}x_i)^2 + \frac{1}{a_{11}\alpha_{33}}(\alpha_{33}x_2 + \alpha_{23}x_3)^2 + \frac{\beta}{\alpha_{33}}x_3^2 \tag{9.15}$$

where

$$\beta = (\alpha_{33}\alpha_{22} - \alpha_{23}^2)/a_{11} \tag{9.16}$$

The expression on the right of (9.15) may be written $\mathbf{x}^{\mathrm{T}}\mathbf{S}^{\mathrm{T}}\mathbf{B}\mathbf{S}\mathbf{x}$, where

$$\mathbf{B} = \begin{pmatrix} 1/a_{11} & 0 & 0 \\ 0 & 1/a_{11}\alpha_{33} & 0 \\ 0 & 0 & \beta/\alpha_{33} \end{pmatrix} \quad \mathbf{S} = \begin{pmatrix} a_{11} & a_{12} & a_{13} \\ 0 & \alpha_{33} & \alpha_{23} \\ 0 & 0 & 1 \end{pmatrix}$$

Here $\mathbf{S}$ may not be orthogonal, so the new coordinate system will in general be oblique, but this does not affect the argument. We have $\mathbf{A} = \mathbf{S}^{\mathrm{T}}\mathbf{B}\mathbf{S}$, and taking determinants in this equation, we see that

$$\beta = |\mathbf{A}| \tag{9.17}$$

Putting $\mathbf{y} = \mathbf{Sx}$, we may rewrite (9.15) as

$$\mathbf{x}^{\mathrm{T}}\mathbf{Ax} = \mathbf{y}^{\mathrm{T}}\mathbf{By} = \frac{1}{d_1}y_1^2 + \frac{1}{d_1 d_2}y_2^2 + \frac{d_3}{d_2}y_3^2 \tag{9.18}$$

Since d_1, d_2, d_3 are all positive, the right-hand side of (9.18) is positive for all $y \neq 0$, therefore $\mathbf{x}^{\mathrm{T}}\mathbf{Ax} > 0$ for all $\mathbf{x} \neq \mathbf{0}$, and this shows $\mathbf{A}$ to be positive definite, so that the proof is complete.

The proof for matrices of higher order is similar, using a generalization of (9.15), which is due to Carl Gustav Jacob Jacobi (1804–1851).

As an example we take the form

$$9x_1^2 + 13x_2^2 + 7x_3^2 + 12x_1x_2 - 6x_1x_3 + 2x_2x_3$$

Its principal minors are $d_1 = 9$, $d_2 = 81$, $d_3 = 405$, hence the form is positive definite. Secondly, consider the form

$$9x_1^2 + 13x_2^2 + 7x_3^2 + 24x_1x_2 + 6x_1x_3 + 2x_2x_3$$

Here $d_1 = 9, d_2 = -27, d_3 = -243$, therefore this form is not positive definite; in fact we get a negative value for $\mathbf{x} = (1, -1, 0)^{\mathrm{T}}$.

With the help of the above criterion we can obtain an expression for the number of negative eigenvalues of a symmetric matrix. We shall observe the convention that multiple eigenvalues are counted as often as they occur.

We shall need the fact that in a real symmetric matrix, when a principal minor is zero, then adjacent ones are of opposite sign. This is easily verified in the two cases needed here: If $d_1 = 0$, then

$$d_2 = \begin{vmatrix} 0 & a \\ a & b \end{vmatrix} = -a^2$$

If $d_2 = 0$, our matrix has the form

$$\begin{pmatrix} a & \lambda a & b \\ \lambda a & \lambda^2 a & c \\ b & c & d \end{pmatrix}$$

and its determinant is easily seen to be $-a(c - \lambda b)^2$. For the general case (not needed here) we refer to Cohn (1991, Chapter 2, Exercise 16); see also Exercise 9.10 below.

Theorem 9.3

Let $\mathbf{A}$ be a regular (real) symmetric 3×3 matrix and $d_0 = 1, d_1, d_2, d_3$ its principal minors. Then the number of negative eigenvalues of $\mathbf{A}$ is equal to the number of changes in sign in the sequence d_0, d_1, d_2, d_3. Here any zero occurring between two non-zero numbers is omitted; if two successive zeros occur between two numbers of the same sign, the zeros are counted with opposite sign; if the zeros occur between two numbers of opposite signs they are omitted.

We remark that not more than two consecutive zeros can occur, since d_0 and d_3 are both non-zero. To give an example, let us write v for the number of sign changes as defined in the theorem. If the minors are $1, 3, 0, -4$, then $v = 1$; if they are $1, 0, 0, 2$, then $v = 2$, while for $1, 0, 0, -7$ we have $v = 1$.

To prove the theorem, suppose first that $d_3 > 0$. Then the eigenvalues $\lambda_1, \lambda_2, \lambda_3$ satisfy $\lambda_1\lambda_2\lambda_3 > 0$, so the number of negative eigenvalues is even. If there are no negative eigenvalues, then d_1 and d_2 are positive, by Theorem 9.2, and the result follows because there are now no changes of sign. If there are negative eigenvalues, then d_1 and d_2 cannot both be positive, again by Theorem 9.2, and the number of sign changes is then 2, whatever the actual values of d_1 and d_2.

Next let $d_3 < 0$. Then $-\mathbf{A}$ has the principal minors

$$d_0, -d_1, d_2, -d_3 \tag{9.19}$$

and the eigenvalues

$$-\lambda_1, -\lambda_2, -\lambda_3 \tag{9.20}$$

Since $-d_3 > 0$, the first part of the proof applies to $-\mathbf{A}$. Either the minors (9.19) are all positive, and then the eigenvalues (9.20) are all positive, or there are two changes of sign in (9.19) and two negative values in (9.20). Accordingly, either d_0, d_1, d_2, d_3 are alternately positive and negative and $\lambda_1, \lambda_2, \lambda_3$ are all negative, or at least one of the inequalities $d_1 \geqslant 0$, $d_2 \leqslant 0$ holds and $\mathbf{A}$ has a negative eigenvalue. Clearly there are three sign changes in the former case and one in the latter. Since d_3 cannot vanish (by the regularity of $\mathbf{A}$), all the possibilities are exhausted and the theorem is established.

This theorem can again be extended to matrices of higher order, if we assume in addition that $\mathbf{A}$ has not more than two successive vanishing principal minors. It was first proved by Georg Frobenius (1849–1917).

9.4 SIMULTANEOUS REDUCTION OF TWO QUADRATIC FORMS

The reduction to diagonal form of a quadratic form on a metric space may be described in terms of its matrix as follows. We are given a real symmetric matrix $\mathbf{A}$ and require a regular matrix $\mathbf{P}$ such that $\mathbf{P}^T\mathbf{P} = \mathbf{I}$ and $\mathbf{P}^T\mathbf{A}\mathbf{P}$ is a diagonal matrix. Consider the following slightly more general problem. Given two real symmetric matrices $\mathbf{A}, \mathbf{B}$, find a regular matrix $\mathbf{P}$ such that $\mathbf{P}^T\mathbf{A}\mathbf{P}$ and $\mathbf{P}^T\mathbf{B}\mathbf{P}$ are both diagonal; of course, we cannot expect $\mathbf{P}^T\mathbf{A}\mathbf{P}$ to be $\mathbf{I}$, unless $\mathbf{A}$ is positive definite. But even in this form the general problem cannot always be solved. For example, if $\mathbf{A} = \begin{pmatrix} 1 & 0 \\ 0 & -1 \end{pmatrix}$, $\mathbf{B} = \begin{pmatrix} 0 & 1 \\ 1 & 0 \end{pmatrix}$, suppose that $\mathbf{P} = \begin{pmatrix} p & q \\ r & s \end{pmatrix}$ is a regular matrix such that $\mathbf{P}^T\mathbf{A}\mathbf{P}$ and $\mathbf{P}^T\mathbf{B}\mathbf{P}$ are both diagonal.

We have

$$\mathbf{P}^{\mathrm{T}}\mathbf{A}\mathbf{P} = \begin{pmatrix} p & r \\ q & s \end{pmatrix}\begin{pmatrix} p & q \\ -r & -s \end{pmatrix} = \begin{pmatrix} p^2 - r^2 & pq - rs \\ pq - rs & q^2 - s^2 \end{pmatrix}$$

$$\mathbf{P}^{\mathrm{T}}\mathbf{B}\mathbf{P} = \begin{pmatrix} p & r \\ q & s \end{pmatrix}\begin{pmatrix} r & s \\ p & q \end{pmatrix} = \begin{pmatrix} 2pr & ps + qr \\ ps + qr & 2qs \end{pmatrix}$$

So we require that $pq - rs = ps + qr = 0$. If $q = 0$ or $r = 0$, then $ps = 0$ and so $\mathbf{P}$ would not be regular. Hence $q, r \neq 0$ and our equations can be written

$$p/r - s/q = 0, \quad (p/r)(s/q) = -1$$

Clearly there can be no solution in real numbers, since it would entail that $(p/r)^2 + 1 = 0$.

Thus we need to restrict $\mathbf{A}, \mathbf{B}$ further; a natural restriction is to require one of them, say $\mathbf{A}$, to be positive definite and to be transformed to $\mathbf{I}$. In this form the problem is easily solved, for we can now make a reduction to the case treated in Theorem 8.8.

Theorem 9.4

Let $\mathbf{A}, \mathbf{B}$ be two real symmetric matrices of the same order, where $\mathbf{A}$ is positive definite. Then there exists a regular matrix $\mathbf{P}$ such that

$$\mathbf{P}^{\mathrm{T}}\mathbf{A}\mathbf{P} = \mathbf{I} \quad \textit{and} \quad \mathbf{P}^{\mathrm{T}}\mathbf{B}\mathbf{P} \textit{ is diagonal} \tag{9.21}$$

where the diagonal elements of $\mathbf{P}^{\mathrm{T}}\mathbf{B}\mathbf{P}$ are the roots of the equation

$$|x\mathbf{A} - \mathbf{B}| = 0. \tag{9.22}$$

Equation (9.22) (as well as the special case $\mathbf{A} = \mathbf{I}$) is often called the **secular equation**, because it arises in the calculation of secular perturbations of planetary orbits. We shall meet some illustrations from mechanics in Chapter 10.

To prove the assertion, let $\mathbf{P}_1$ be a regular matrix such that $\mathbf{D} = \mathbf{P}_1^{\mathrm{T}}\mathbf{A}\mathbf{P}_1$ is diagonal. Such a matrix $\mathbf{P}_1$ exists by Theorem 8.7; as we saw, it can even be taken orthogonal, but this is immaterial. Since $\mathbf{A}$ is positive definite, the diagonal entries of $\mathbf{D}$ are positive, so there is a real matrix $\mathbf{P}_2$ such that $\mathbf{P}_2^{-2} = \mathbf{D}$; we simply take $\mathbf{P}_2$ diagonal with entries $d_i^{-1/2}$ along the main diagonal, where d_i is the corresponding diagonal entry of $\mathbf{D}$. Now $(\mathbf{P}_1\mathbf{P}_2)^{\mathrm{T}}\mathbf{A}\mathbf{P}_1\mathbf{P}_2 = \mathbf{P}_2^{\mathrm{T}}\mathbf{D}\mathbf{P}_2 = \mathbf{I}$ and $(\mathbf{P}_1\mathbf{P}_2)^{\mathrm{T}}\mathbf{B}\mathbf{P}_1\mathbf{P}_2 = \mathbf{B}'$ say, where $\mathbf{B}'$ is again real symmetric. Now we apply Theorem 8.8 to find an orthogonal matrix $\mathbf{P}_3$ such that $\mathbf{P}_3^{\mathrm{T}}\mathbf{B}'\mathbf{P}_3$ is diagonal. Since $\mathbf{P}_3^{\mathrm{T}}\mathbf{I}\mathbf{P}_3 = \mathbf{I}$, we find that (9.21) holds with the matrix $\mathbf{P} = \mathbf{P}_1\mathbf{P}_2\mathbf{P}_3$. Finally, the diagonal entries of $\mathbf{P}^{\mathrm{T}}\mathbf{B}\mathbf{P}$ are the roots of $|x\mathbf{I} - \mathbf{P}^{\mathrm{T}}\mathbf{B}\mathbf{P}| = 0$. But we have

$$|x\mathbf{I} - \mathbf{P}^{\mathrm{T}}\mathbf{B}\mathbf{P}| = |x\mathbf{P}^{\mathrm{T}}\mathbf{A}\mathbf{P} - \mathbf{P}^{\mathrm{T}}\mathbf{B}\mathbf{P}| = |\mathbf{P}^{\mathrm{T}}|\cdot|x\mathbf{A} - \mathbf{B}|\cdot|\mathbf{P}|$$

hence the required elements are the roots of $|x\mathbf{A} - \mathbf{B}| = 0$, and the proof is complete.

In making the reduction we need not go through the separate steps of the proof, but can proceed as before. We first solve the secular equation (9.22) and then, for each root λ, find the complete solution of the system

$$(\lambda\mathbf{A}-\mathbf{B})\mathbf{u}=\mathbf{0} \tag{9.23}$$

If the solution space is r-dimensional, then in the case $r>1$ we choose a basis $\mathbf{u}_1,\ldots,\mathbf{u}_r$ such that $\mathbf{u}_i^{\mathrm{T}}\mathbf{A}\mathbf{u}_j=\delta_{ij}$. Since $\mathbf{B}\mathbf{u}_i=\lambda\mathbf{A}\mathbf{u}_i$, we then have $\mathbf{u}_i^{\mathrm{T}}\mathbf{B}\mathbf{u}_j=\lambda\mathbf{u}_i^{\mathrm{T}}\mathbf{A}\mathbf{u}_j=\lambda\delta_{ij}$. Further, if $\mathbf{u},\mathbf{v}$ are solution vectors corresponding to different roots, say $\mathbf{B}\mathbf{u}=\lambda\mathbf{A}\mathbf{u}$, $\mathbf{B}\mathbf{v}=\mu\mathbf{A}\mathbf{v}$ and, of course, $\mathbf{u},\mathbf{v}\neq\mathbf{0}$, then $\mathbf{v}^{\mathrm{T}}\mathbf{B}\mathbf{u}=\lambda\mathbf{v}^{\mathrm{T}}\mathbf{A}\mathbf{u}$, $\mathbf{u}^{\mathrm{T}}\mathbf{B}\mathbf{v}=\mu\mathbf{u}^{\mathrm{T}}\mathbf{A}\mathbf{v}$, hence

$$0=\mathbf{u}^{\mathrm{T}}\mathbf{B}\mathbf{v}-\mathbf{v}^{\mathrm{T}}\mathbf{B}\mathbf{u}=(\lambda-\mu)\mathbf{u}^{\mathrm{T}}\mathbf{A}\mathbf{v}$$

and so $\mathbf{u}^{\mathrm{T}}\mathbf{A}\mathbf{v}=0$ whenever $\lambda\neq\mu$. By choosing a basis in this way for the solution space of (9.23) for each root λ of (9.22) we obtain a basis of the whole space, which can then be used for the columns of the transforming matrix. Sometimes we shall not trouble to normalize the basis; thus $\mathbf{u}_i^{\mathrm{T}}\mathbf{A}\mathbf{u}_i$ may differ from 1; of course, we must still ensure that $\mathbf{u}_i^{\mathrm{T}}\mathbf{A}\mathbf{u}_j=0$ for $i\neq j$. Then $\mathbf{A},\mathbf{B}$ will both be transformed to diagonal form, but not necessarily to $\mathbf{I}$.

As an example, let us take

$$\mathbf{A}=\begin{pmatrix}14&-5&-10\\-5&2&3\\-10&3&9\end{pmatrix}\quad \mathbf{B}=\begin{pmatrix}7&-1&-8\\-1&0&1\\-8&1&11\end{pmatrix}$$

The secular polynomial is

$$\begin{vmatrix}14x-7&-5x+1&-10x+8\\-5x+1&2x&3x-1\\-10x+8&3x-1&9x-11\end{vmatrix}$$

$$=\begin{vmatrix}-x-4&x+1&-x+5\\-5x+1&2x&3x-1\\6&-x-1&3x-9\end{vmatrix}=\begin{vmatrix}-x+2&0&2x-4\\-5x+1&2x&3x-1\\6&-x-1&3x-9\end{vmatrix}$$

$$=\begin{vmatrix}-x+2&0&0\\-5x+1&2x&-7x+1\\6&-x-1&3x+3\end{vmatrix}$$

$$=(-x+2)(6x^2+6x-7x^2-7x+x+1)=(x-2)(x^2-1).$$

Hence the roots of the secular equation are $1,-1,2$. We next find the corresponding vectors. For $x=1$:

$$\begin{pmatrix}7&-4&-2\\-4&2&2\\-2&2&-2\end{pmatrix}\begin{pmatrix}u_1\\u_2\\u_3\end{pmatrix}=0$$

Solving this system, we find $\mathbf{u} = \alpha(2,3,1)^T$, where α has to be chosen such that $\mathbf{u}^T\mathbf{A}\mathbf{u} = 1$. In fact we find that $\alpha^2 = 1$, so we may take $\alpha = 1$.

For $x = -1$:

$$\begin{pmatrix} -21 & 6 & 18 \\ 6 & -2 & -4 \\ 18 & -4 & -20 \end{pmatrix}\begin{pmatrix} v_1 \\ v_2 \\ v_3 \end{pmatrix} = 0$$

The solution is $\mathbf{v} = (2,4,1)^T$, where the constant factor can again be taken to be 1.

For $x = 2$:

$$\begin{pmatrix} 21 & -9 & -12 \\ -9 & 4 & 5 \\ -12 & 5 & 7 \end{pmatrix}\begin{pmatrix} w_1 \\ w_2 \\ w_3 \end{pmatrix} = 0$$

The solution is $\mathbf{w} = (1,1,1)^T$. Hence the transforming matrix is

$$\mathbf{P} = \begin{pmatrix} 2 & 2 & 1 \\ 3 & 4 & 1 \\ 1 & 1 & 1 \end{pmatrix}$$

and the diagonal form is $\mathbf{P}^T\mathbf{A}\mathbf{P} = \mathbf{I}$, $\mathbf{P}^T\mathbf{B}\mathbf{P} = \mathrm{diag}(1,-1,2)$.

The simultaneous reduction of two quadratic forms when neither is positive definite, cannot always be performed (as we have seen), and criteria for reducibility have been developed in the theory of elementary divisors; in particular, Karl Weierstrass (1815–97) showed in 1858 that the reduction could always be made when one of the forms was positive definite (Theorem 9.4); this case is of importance in mechanics (section 10.3).

9.5 THE POLAR FORM

For complex numbers there is a well-known expression in 'polar form'. Any non-zero complex number c can be uniquely written in the form

$$c = re^{i\alpha}$$

where r is a positive real number and α is an angle between 0 and 2π radians. For matrices there is a corresponding representation involving Hermitian and unitary matrices. Since we have restricted ourselves here to real matrices, we shall describe the analogue for real matrices. In order to obtain this form we shall need to extract the square root of a positive matrix. We begin by describing the Lagrange interpolation formula, of which a special case occurred in Section 8.7 (see also Exercise 8.6).

Given r distinct numbers $\alpha_1, \ldots, \alpha_r$, let us define f as the polynomial with these numbers as zeros:

$$f(x) = (x - \alpha_1)\cdots(x - \alpha_r)$$

and define f_i for $i = 1, \ldots, r$ by the equation

$$(x - \alpha_i) f_i(x) = f(x)$$

Thus f_i is a polynomial of degree $r - 1$ such that $f_i(\alpha_j) = 0$ for $j \neq i$. Now with any numbers $b_1, \ldots, b_r$ form

$$F(x) = \sum b_i \frac{f_i(x)}{f_i(\alpha_i)}. \tag{9.24}$$

This is a polynomial of degree at most $r - 1$ and for $x = \alpha_1$ it reduces to $F(\alpha_1) = b_1 f_1(\alpha_1)/f_1(\alpha_1) = b_1$, because $f_i(\alpha_1) = 0$ for $i > 1$. Similarly $F(\alpha_i) = b_i$ for $i = 2, \ldots, r$, so F is a polynomial of degree at most $r - 1$, assuming the value b_i for $x = \alpha_i$ $(i = 1, \ldots, r)$, and it is the only such polynomial, for if there were two, their difference would be a polynomial of degree less than r and vanishing at $x = \alpha_1, \ldots, \alpha_r$, so this difference must be zero (cf. section 8.3). This construction (9.24) is called the **Lagrange interpolation formula** (after Joseph-Louis Lagrange, 1736–1813). With the help of this formula we can extract the square root of a positive matrix and actually write it as a polynomial:

Theorem 9.5
Let $\mathbf{A}$ *be any real symmetric positive matrix. Then there exists exactly one real symmetric positive matrix* $\mathbf{B}$ *such that* $\mathbf{B}^2 = \mathbf{A}$. *Further,* $\mathbf{B}$ *can be expressed as a polynomial in* $\mathbf{A}$ *and* $\mathbf{B}$ *is positive definite whenever* $\mathbf{A}$ *is.*

Proof
Let $\alpha_1, \ldots, \alpha_r$ be the distinct eigenvalues of $\mathbf{A}$, put $\beta_i = \alpha_i^{1/2}$ and let F be the polynomial of degree at most $r - 1$ such that $F(\alpha_i) = \beta_i$, formed as in (9.24). We claim that $\mathbf{A} = \mathbf{B}^2$, where $\mathbf{B} = F(\mathbf{A})$. To prove the claim, we take an orthonormal basis of eigenvectors of $\mathbf{A}$, which exists by Theorem 8.8. If $\mathbf{u}$ is any eigenvector belonging to the eigenvalue α_1, then $\mathbf{Au} = \alpha_1 \mathbf{u}$; by induction on k we find $\mathbf{A}^k\mathbf{u} = \alpha_1^k \mathbf{u}$, and hence by linearity, $f(\mathbf{A})\mathbf{u} = f(\alpha_1)\mathbf{u}$ for any polynomial f. In particular, we have

$$\mathbf{Bu} = F(\mathbf{A})\mathbf{u} = F(\alpha_1)\mathbf{u} = \beta_1 \mathbf{u}$$

hence

$$\mathbf{B}^2\mathbf{u} = \beta_1^2 \mathbf{u} = \alpha_1 \mathbf{u} = \mathbf{Au}.$$

This holds for all members of our orthonormal basis, so we conclude that $\mathbf{A} = \mathbf{B}^2$, as claimed. From its construction as a polynomial in $\mathbf{A}, \mathbf{B}$ is real symmetric positive, and if $\mathbf{A}$ is positive definite, then all the α_i are positive, hence so are all the β_i and so $\mathbf{B}$ is then positive definite. It only remains to prove the uniqueness of $\mathbf{B}$.

Let $\mathbf{C}$ be another square root of $\mathbf{A}$, which is real symmetric positive and choose a basis of eigenvectors of $\mathbf{C}$. Since $\mathbf{B} = F(\mathbf{A}) = F(\mathbf{C}^2)$, they are also eigenvectors of $\mathbf{B}$ and if $\mathbf{Cu} = \gamma_1 \mathbf{u}$ say, then $\mathbf{Au} = \mathbf{C}^2\mathbf{u} = \gamma_1^2 \mathbf{u}$, hence γ_1^2 is an eigenvalue of **A**, say $\gamma_1^2 = \alpha_1 = \beta_1^2$. Since β_1, γ_1 are both positive (or 0) it follows that $\beta_1 = \gamma_1$. Now $\mathbf{Bu} = F(\mathbf{A})\mathbf{u} = F(\alpha_1)\mathbf{u} = \beta_1\mathbf{u} = \mathbf{Cu}$; this again holds for all members of the orthonormal basis, hence $\mathbf{C} = \mathbf{B}$, as we wished to show.

We can now describe the polar form for matrices:

Theorem 9.6
Let **A** *be a real invertible matrix. Then there exists a positive definite matrix* **P** *and an orthogonal matrix* **U** *such that*

$$\mathbf{A} = \mathbf{UP}.$$

Moreover, **P** *and* **U** *are uniquely determined by* **A**.

Proof
The matrix $\mathbf{A}^T\mathbf{A}$ is positive definite, for it is clearly symmetric and $\mathbf{x}^T\mathbf{A}^T\mathbf{A}\mathbf{x} = (\mathbf{Ax})^T\mathbf{Ax} \geqslant 0$; if equality holds here, then $\mathbf{Ax} = \mathbf{0}$, hence $\mathbf{x} = \mathbf{0}$. It follows by Theorem 9.5 that there is a unique positive definite matrix **P** such that $\mathbf{A}^T\mathbf{A} = \mathbf{P}^2$. Put $\mathbf{U} = \mathbf{AP}^{-1}$; then

$$\mathbf{U}^T\mathbf{U} = (\mathbf{AP}^{-1})^T\mathbf{AP}^{-1} = \mathbf{P}^{-1}\mathbf{A}^T\mathbf{AP}^{-1} = \mathbf{P}^{-1}\mathbf{P}^2\mathbf{P}^{-1} = \mathbf{I}$$

hence **U** is orthogonal and we have $\mathbf{A} = \mathbf{UP}$. If we also had $\mathbf{A} = \mathbf{U}_1\mathbf{P}_1$ where $\mathbf{P}_1$ is positive definite, and $\mathbf{U}_1$ is orthogonal, then $\mathbf{A}^T\mathbf{A} = \mathbf{P}_1\mathbf{U}_1^T\mathbf{U}_1\mathbf{P}_1 = \mathbf{P}_1^2$; by the uniqueness of Theorem 9.5, $\mathbf{P}_1 = \mathbf{P}$ and so also $\mathbf{U}_1 = \mathbf{U}$, and the proof is complete.

There is a second type of polar form that is sometimes of interest, in which a regular matrix is expressed as a product of an orthogonal matrix by a triangular matrix.

Theorem 9.7
Let **A** *be a real invertible matrix. Then there exists an upper triangular matrix* **T** *with positive diagonal entries and an orthogonal matrix* **U** *such that*

$$\mathbf{A} = \mathbf{UT}$$

where both **T** *and* **U** *are uniquely determined by* **A**.

Proof
We again take the positive definite matrix $\mathbf{A}^T\mathbf{A}$ and transform it to **I** by an upper triangular matrix. In fact the reduction as in section 8.5 yields an upper triangular matrix, with positive diagonal entries. Here we have to bear in mind that the (1, 1)th entry of $\mathbf{A}^T\mathbf{A}$ is positive, by positive definiteness, so a renumbering of variables is not necessary. At the first stage we have a reduction to diagonal form by an upper triangular matrix, and this is followed by a reduction to **I** by a diagonal matrix with positive diagonal entries, so

we have an upper triangular matrix **T** with positive diagonal entries such that

$$\mathbf{A}^T\mathbf{A} = \mathbf{T}^T\mathbf{T}$$

If we also have $\mathbf{A}^T\mathbf{A} = \mathbf{T}_1^T\mathbf{T}_1$ for another such matrix $\mathbf{T}_1$, then $\mathbf{S} = \mathbf{T}_1\mathbf{T}^{-1}$ is upper triangular with positive diagonal entries and satisfies

$$\mathbf{S}^T\mathbf{S} = (\mathbf{T}_1\mathbf{T}^{-1})^T \cdot \mathbf{T}_1\mathbf{T}^{-1} = (\mathbf{T}^{-1})^T\mathbf{T}_1^T\mathbf{T}_1\mathbf{T}^{-1} = (\mathbf{T}^{-1})^T\mathbf{T}^T\mathbf{T}\mathbf{T}^{-1} = \mathbf{I}$$

hence $\mathbf{S}^{-1} = \mathbf{S}^T$. It follows that **S** must be diagonal, with positive diagonal entries s_i such that $s_i^2 = 1$; since s_i is positive, $s_i = 1$ and so $\mathbf{S} = \mathbf{I}$, i.e. $\mathbf{T}_1 = \mathbf{T}$, so **T** is unique.

Now put $\mathbf{U} = \mathbf{A}\mathbf{T}^{-1}$; then **U** is easily seen to be orthogonal and we have

$$\mathbf{A} = \mathbf{U}\mathbf{T}$$

as claimed. Here **T** is unique, hence so is **U**, and the proof is complete.

9.6 LINEAR PROGRAMMING

In Chapters 2 and 4 we treated systems of linear equations, but frequently one needs to solve a system of linear inequalities. A systematic way of obtaining such solutions was developed during the Second World War by George B. Dantzig (1914–) and has since found widespread application. The problem is best illustrated by an example.

A machine shop produces two types of engine cowling, denoted by A and B. Their manufacture is in three stages – pressing, milling and grinding – which take the following times in minutes:

	A	B
Pressing	15	5
Milling	10	30
Grinding	10	20

The profit on each item of A is £4 and of B £5. In an 8-hour day, how many of each type should be produced to maximize the profit? If the number of items of each type is x for A and y for B, then we are trying to maximize $4x + 5y$, subject to the inequalities

$$\begin{aligned} 15x + 5y &\leqslant 480 \quad & x \geqslant 0 \\ 10x + 30y &\leqslant 480 \quad & y \geqslant 0 \\ 10x + 20y &\leqslant 480 \end{aligned}$$

Of course, we have to include the restriction that x and y cannot be negative. If we interpret x and y as coordinates in the plane, each inequality restricts the point (x, y) to lie in the region on one side of the straight line obtained

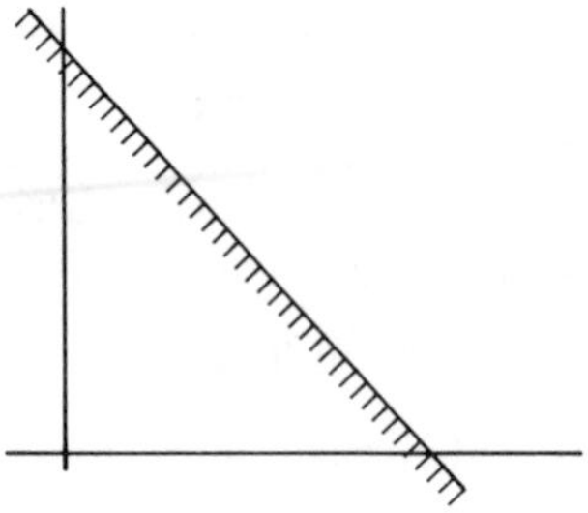

Fig. 9.5

by replacing $\leqslant$ by $=$ in the inequality; this is called a **half-plane**. For example, the first inequality is equivalent to $3x + y \leqslant 96$, and it defines the half-plane shown in Fig. 9.5. In the same way each of the other inequalities defines a half-plane and the solution must lie in the region which is the intersection of all these half-planes. This region is called the set of **feasible solutions**, or the **feasible set**. In our case the set is illustrated in Fig. 9.6, lying within the positive quadrant (because $x \geqslant 0$, $y \geqslant 0$). Further, as an intersection of half-planes the feasible set has the property that for any two points P, Q in the set it contains all points of the line segment PQ; this property is expressed by saying that the feasible set is **convex** (Fig. 9.7).

We notice that in our example the third inequality is no restriction; it follows from the second and the condition $y \geqslant 0$. To solve the problem we have to find a point in the feasible set to maximize $4x + 5y$. To achieve this we take the lines $4x + 5y = c$ for varying c and find the largest c for which the line meets the feasible set. These lines form a family of parallel lines and we have to find the line touching the feasible set, so as to have the set on the same side as the origin. We see that the maximum is attained at a corner. The corners are found by replacing the inequalities by equations and solving them two at a time. Thus, omitting the third inequality (which we saw follows

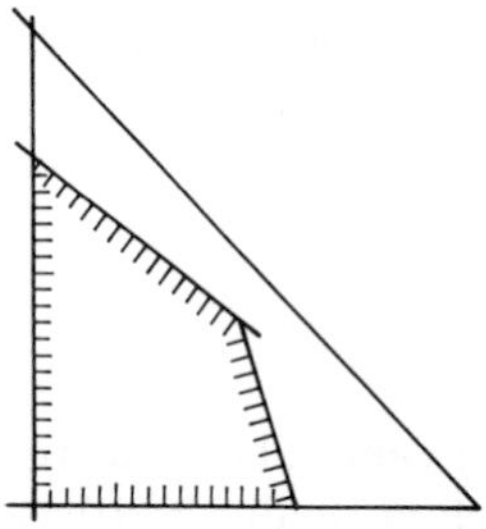

Fig. 9.6

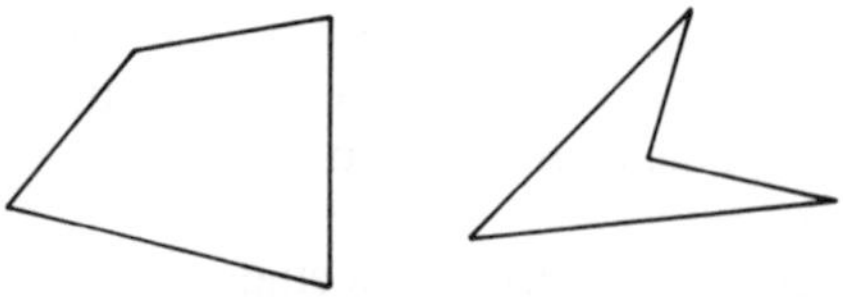

Fig. 9.7

from the others), we are left with the equations

$$3x + y = 96 \quad x = 0$$
$$x + 3y = 48 \quad y = 0$$

The corner points are (32, 0), (0, 16), (30,6). The value of $4x + 5y$ at these points is 128, 80, 150, respectively. Of these the last is largest, so we should make 30 items A and 6 items B, at a profit of £150. If the actual prices vary, say A costs £α and B costs £β, then the 'cost' line $\alpha x + \beta y$ has gradient $-\alpha/\beta$, and the equations defining the feasible set show that when $\alpha/\beta \geqslant 3$ we should only manufacture A, while for $\alpha/\beta \leqslant 1/3$ we should concentrate on B.

The general linear programming problem can be stated as follows. We have n variables $x_1, \ldots, x_n$ and our task is to *maximize* a function $f = \sum c_i x_i$, the **objective function**, subject to a number of **constraints**, $g_i = \sum a_{ij} x_j \leqslant b_i$, $x_j \geqslant 0$. In matrix language we have to maximize the objective function $\mathbf{cx}$, subject to the constraints $\mathbf{Ax} \leqslant \mathbf{b}, \mathbf{x} \geqslant \mathbf{0}$, where the vector relation $\mathbf{x} \geqslant \mathbf{0}$ means that the inequality holds for all components: $x_j \geqslant 0$ $(j = 1, \ldots, n)$.

There are several variants of this problem which can be reduced to the same form. Thus if it is required to *minimize* a function f, this can be accomplished by maximizing $-f$. If a constraint has the form of an equality $g = b$, it can be replaced by the inequalities $g \leqslant b$, $-g \leqslant -b$. Alternatively, the constraints can be taken in the form of equations $g_i = b_i$; if an inequality $g \leqslant b$ is given, we introduce a new variable y, called a **slack variable**, and replace $g \leqslant b$ by $g + y = b$, $y \geqslant 0$. Finally, if a variable x is not restricted to be $\geqslant 0$, we can introduce new variables x', x'', where $x = x' - x''$ and $x' \geqslant 0$, $x'' \geqslant 0$.

Thus for our basic problem we are given an $m \times n$ matrix $\mathbf{A}$, a column vector $\mathbf{b} \in {}^m\mathbf{R}$ and a row vector $\mathbf{c} \in \mathbf{R}^n$, and we have to find $\mathbf{x} \in {}^n\mathbf{R}$ to maximize $\mathbf{cx}$, subject to

$$\mathbf{Ax} = \mathbf{b} \tag{9.25}$$

$$\mathbf{x} \geqslant \mathbf{0} \tag{9.26}$$

The solutions of (9.25) and (9.26) are just the feasible solutions, and we have to find the feasible solution $\mathbf{x}$ which maximizes $\mathbf{cx}$. To solve (9.25) we need to express $\mathbf{b}$ as a linear combination of the columns of $\mathbf{A}$, and such a solution will exist precisely when $\text{rk}(\mathbf{A}) = \text{rk}(\mathbf{A}, \mathbf{b}) = r$, say. When this condition holds, we can express $\mathbf{b}$ as a linear combination of any r linearly independent columns of $\mathbf{A}$. Such a solution, also satisfying (9.26), is called a **basic feasible solution** (BFS). It corresponds to a solution vector $\mathbf{x} \geqslant 0$ in which $n - r$ components are zero, and it represents a corner point of the polyhedron of solutions. So in principle our problem is solved by computing $\mathbf{cx}$ for all corner points (BFSs) in turn, but the work can be shortened by the **simplex algorithm**, which we shall now describe.

We again consider (9.25) and (9.26); moreover, we may assume that $\mathbf{b} \geqslant \mathbf{0}$, for if $b_j < 0$ for some j, we need only multiply the jth equation of (9.25) by -1 to change the sign of b_j. We next reduce the system to echelon form. By renumbering the variables we can permute the columns and so reduce the matrix $\mathbf{A}$ to the form

$$\begin{pmatrix} \mathbf{I} & \mathbf{C} \\ \mathbf{0} & \mathbf{0} \end{pmatrix}. \tag{9.27}$$

Extra care is needed in this reduction, because we may multiply rows only by positive constants. This will lead to problems if a pivot is negative; if this happens, we can exchange it against another pivot in the same row, which is positive. Such a change can always be carried out if there is a feasible solution, for not all entries in a row can be negative, because $\mathbf{b} \geqslant \mathbf{0}$.

To take an example, let

$$\mathbf{A} = \begin{pmatrix} 3 & 4 & 10 \\ 1 & 2 & 3 \\ 4 & 6 & 13 \end{pmatrix}$$

This has the reduction

$$\begin{matrix} 3 & 4 & 10 \\ 1 & 2 & 3 \\ 4 & 6 & 13 \end{matrix} \longrightarrow \begin{matrix} 1 & 2 & 3 \\ 0 & -2 & 1 \\ 0 & 0 & 0 \end{matrix} \longrightarrow \begin{matrix} 1 & 3 & 2 \\ 0 & 1 & -2 \\ 0 & 0 & 0 \end{matrix} \longrightarrow \begin{matrix} 1 & 0 & 8 \\ 0 & 1 & -2 \\ 0 & 0 & 0 \end{matrix}$$

At the second stage we encounter a negative pivot in the (2, 2)th position, so we interchange the second and third columns and then continue the reduction.

Since our system is assumed to be consistent, any zero row in (9.27) corresponds to an equation $0 = 0$ and so can be omitted. The rank is now equal to the number of rows, thus if the system is $m \times n$, then it will be of rank m. When $m = n$, there is a unique solution of (9.25), which provides the required maximum if it also satisfies (9.26). There remains the case $m < n$; here we can write our system as

$$(\mathbf{I} \;\; \mathbf{C})\begin{pmatrix} \mathbf{x}' \\ \mathbf{x}'' \end{pmatrix} = \mathbf{b},$$

where $\mathbf{x}' = (x_1, \ldots, x_m)^{\mathrm{T}}$, $\mathbf{x}'' = (x_{m+1}, \ldots, x_n)^{\mathrm{T}}$. A BFS is obtained by taking $\mathbf{x}'' = \mathbf{0}$, for this expresses $\mathbf{b}$ as a linear combination of the columns of $\mathbf{I}$: $\mathbf{b} = \mathbf{Ib}$, hence $\mathbf{x}' = \mathbf{b}$ in this case.

Our task now is to find other BFSs. To find them, let us take a pair of suffixes r, s such that $1 \leqslant r \leqslant m < s$ and $c_{rs} \neq 0$ and carry out the reduction with c_{rs} as pivot. This reduction transforms row i (Ri) to R$i - (c_{is}/c_{rs}) \cdot$ Rr and so b_i becomes $b_i - (c_{is}/c_{rs}) \cdot b_r$. This will again be a feasible solution provided that

$$b_i - (c_{is}/c_{rs}) \cdot b_r \geqslant 0 \quad (i = 1, \ldots, m). \tag{9.28}$$

For $c_{is} = 0$ this reduces to $b_i \geqslant 0$, which is true by hypothesis. When $c_{is} \neq 0$, we require

$$b_i/c_{is} \geqslant b_r/c_{rs} \quad \text{if } c_{is} > 0 \tag{9.29}$$

$$b_i/c_{is} \leqslant b_r/c_{rs} \quad \text{if } c_{is} < 0 \tag{9.30}$$

By (9.29), b_r/c_{rs} has to be least among all the b_i/c_{is} with $c_{is} > 0$, while (9.30) requires that $b_r/c_{rs} \geqslant b_i/c_{is}$ whenever $c_{is} < 0$. Since $b_i > 0$ in any case, this means that for a given s we must choose r such that

$$c_{rs} \neq 0 \text{ and } b_r/c_{rs} \text{ assumes its least non-negative value as } r \text{ varies} \tag{9.31}$$

The value to be maximized changes from $\sum_1^m c_i b_i$ to $\sum_1^m c_i b_i - \sum_1^m c_i c_{is} b_r/c_{rs}$. To obtain a larger value we require that $\sum_1^m c_i c_{is} \leqslant 0$, because $b_r/b_{rs} > 0$ by (9.31). So we first choose s such that $\sum_1^m c_i c_{is} \leqslant 0$ if possible, and then choose r to satisfy (9.31). We reach a maximum when no such s can be found. Geometrically we have a convex polyhedron with one vertex at the origin and lying entirely within the positive octant (given by (9.26)). Starting from the origin, we move to an adjacent vertex with the largest value for $\mathbf{cx}$, and proceed in this way until no further increase can be made.

As an example, let us maximize $x_1 + x_2 + x_3 + x_4$ subject to

$$\begin{aligned} x_1 + 2x_2 + 4x_3 + 7x_4 &= 17 \\ x_1 + x_2 - x_3 + 5x_4 &= 12 \end{aligned}$$

We write down the corresponding scheme and perform the reduction to echelon form:

$$\left.\begin{array}{cccc|c} 1 & 2 & 4 & 7 & 17 \\ 1 & 1 & -1 & 5 & 12 \end{array}\right. \to \left.\begin{array}{cccc|c} 1 & 2 & 4 & 7 & 17 \\ 0 & -1 & -5 & -2 & -5 \end{array}\right. \to \left.\begin{array}{cccc|c} 1 & 2 & 4 & 7 & 17 \\ 0 & 1 & 5 & 2 & 5 \end{array}\right.$$

$$\to \begin{array}{cccc|c} 1 & 0 & -6 & 3 & 7 \\ 0 & 1 & 5 & 2 & 5 \end{array}$$

At the second stage we have a negative row, which we can convert to a positive row by multiplying by -1. The BFS (7,5,0,0) yields the value $7 + 5 = 12$. We next look for a column s with $\sum c_i c_{is} < 0$. The third column gives $-6 + 5 = -1$, and the least positive value of b_r/c_{rs} is 5/5, for which $r = 2$, so we make a reduction with the (2,3)th entry as pivot, giving

$$\begin{array}{cccc|c} 1 & 6/5 & 0 & 27/5 & 13 \\ 0 & 1 & 5 & 2 & 5 \end{array}$$

The corresponding BFS is $(13, 0, 1, 0)$ and $13 + 1 = 14$. Now there is no s to satisfy $\sum c_i c_{is} < 0$, hence the required maximum is 14, attained for $x = (13, 0, 1, 0)$.

We remark that for our original problem:

$$\text{maximize } \mathbf{cx}, \text{ subject to } \mathbf{Ax} \leqslant \mathbf{b}, \mathbf{x} \geqslant \mathbf{0}$$

the reduction to echelon form is unnecessary. For if we introduce slack variables $\mathbf{y} = (y_1, \ldots, y_m)^{\mathrm{T}}$, the problem takes the form:

$$\text{maximize } \mathbf{cx}, \text{ subject to } (\mathbf{I}\ \ \mathbf{A})\begin{pmatrix} \mathbf{y} \\ \mathbf{x} \end{pmatrix} = \mathbf{b}, \mathbf{x} \geqslant \mathbf{0}, \quad \mathbf{y} \geqslant \mathbf{0}$$

with its matrix already in reduced echelon form. The BFS $\mathbf{x} = \mathbf{0}$ gives $\mathbf{y} = \mathbf{b}$, $\mathbf{cx} = \mathbf{0}$ and we can now apply the simplex algorithm as before to find the maximum. Some care is needed in applying the method; it sometimes happens that after a finite number of steps one returns to the original solution. Such a 'cycle' occurs only rarely and can often be avoided by a judicious choice of pivot.

Here we have only presented the barest outline of what is now an extensive subject. The interested reader is advised to consult some of the many expositions, such as Dantzig (1963), Luenberger (1973).

9.7 THE METHOD OF LEAST SQUARES

Suppose we are given observational data in the form of number pairs (i.e. 2-vectors) (a_i, b_i), $i = 1, 2, \ldots, n$. When plotted on a graph they are expected to lie in a straight line, but this is not in fact achieved, owing to observational errors (Fig. 9.8). The problem is to find a straight line which best fits these data. If the line to be found is taken in the form

$$y = mx + c \tag{9.32}$$

then we have to determine m and c to satisfy

$$\begin{aligned} b_1 &= ma_1 + c \\ &\cdots \\ b_n &= ma_n + c \end{aligned} \tag{9.33}$$

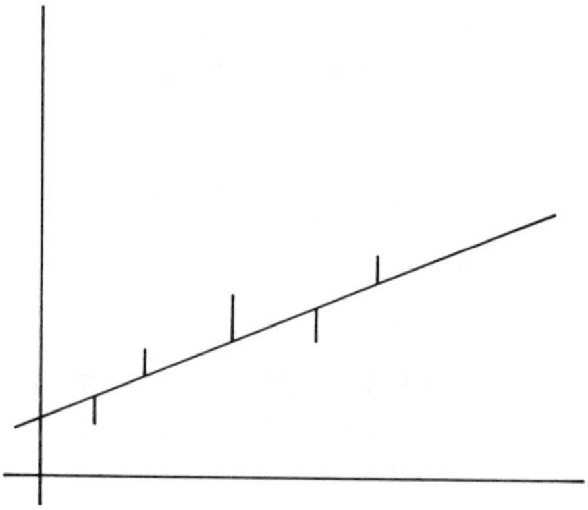

Fig. 9.8

These are n equations in the two unknowns m and c. Let us first consider the case $n = 2$. Then the system has the matrix

$$\mathbf{A} = \begin{pmatrix} a_1 & 1 \\ a_2 & 1 \end{pmatrix} \tag{9.34}$$

whose determinant is $a_1 - a_2$. Thus the equations have a unique solution, provided that $a_1 \neq a_2$, corresponding to the fact that through two distinct points in the graph there is a unique line of the form (9.32), provided that the line is not vertical (parallel to the 2-axis).

However, in general n will be large, the system (9.33) will have no exact solution and we merely try to find a 'closest fit', by finding a solution which will make the differences $b_i - (ma_i + c)$ as small as possible. One way of doing so is to minimize the sum of the squares

$$\varphi(m, c) = [b_1 - (ma_1 + c)]^2 + \cdots + [b_n - (ma_n + c)]^2$$

This is the **method of least squares**, first developed by Gauss and independently by Legendre (Adrien-Marie Legendre, 1752–1833). To formulate the problem in terms of matrices, let us write (9.33) in the form

$$\mathbf{b} = \mathbf{A}\mathbf{p} \tag{9.35}$$

where $\mathbf{b} = (b_1, \ldots, b_n)^{\mathrm{T}}$, $\mathbf{p} = (m, c)^{\mathrm{T}}$ and $\mathbf{A}$ is the $n \times 2$ matrix with ith row $(a_i, 1)$ (as in (9.34) for $n = 2$). Our aim will be to minimize

$$\varphi(m, c) = (\mathbf{b} - \mathbf{A}\mathbf{p})^{\mathrm{T}}(\mathbf{b} - \mathbf{A}\mathbf{p})$$

On expansion we have

$$\varphi(m, c) = \mathbf{b}^{\mathrm{T}}\mathbf{b} - \mathbf{p}^{\mathrm{T}}\mathbf{A}^{\mathrm{T}}\mathbf{b} - \mathbf{b}^{\mathrm{T}}\mathbf{A}\mathbf{p} + \mathbf{p}^{\mathrm{T}}\mathbf{A}^{\mathrm{T}}\mathbf{A}\mathbf{p} \tag{9.36}$$

Let us assume that the points (a_i, b_i) do not all lie in a single vertical line; this means that the a_i are not all equal, hence $\mathbf{A}$ has rank 2 and so $\mathbf{A}\mathbf{x} \neq \mathbf{0}$ for $\mathbf{x} \neq \mathbf{0}$. Since a sum of real square terms, not all zero, is positive, it follows that $\mathbf{x}^{\mathrm{T}}\mathbf{A}^{\mathrm{T}}\mathbf{A}\mathbf{x} = (\mathbf{A}\mathbf{x})^{\mathrm{T}}\mathbf{A}\mathbf{x} > 0$ for $\mathbf{x} \neq \mathbf{0}$, so $\mathbf{A}^{\mathrm{T}}\mathbf{A}$ is positive definite and in particular, it is regular. We can therefore determine a 2-vector $\mathbf{q}$ such that

$$\mathbf{A}^{\mathrm{T}}\mathbf{A}\mathbf{q} = \mathbf{A}^{\mathrm{T}}\mathbf{b} \tag{9.37}$$

We express the vector $\mathbf{p}$ to be found as $\mathbf{p} = \mathbf{q} + \mathbf{r}$; then it remains to determine the vector $\mathbf{r}$. If in (9.36) we replace $\mathbf{p}$ by $\mathbf{q} + \mathbf{r}$ and $\mathbf{A}^{\mathrm{T}}\mathbf{b}$ by $\mathbf{A}^{\mathrm{T}}\mathbf{A}\mathbf{q}$, we obtain

$$\begin{aligned}\varphi(m, c) &= \mathbf{b}^{\mathrm{T}}\mathbf{b} - (\mathbf{q} + \mathbf{r})^{\mathrm{T}}\mathbf{A}^{\mathrm{T}}\mathbf{A}\mathbf{q} - \mathbf{q}^{\mathrm{T}}\mathbf{A}^{\mathrm{T}}\mathbf{A}(\mathbf{q} + \mathbf{r}) + (\mathbf{q} + \mathbf{r})^{\mathrm{T}}\mathbf{A}^{\mathrm{T}}\mathbf{A}(\mathbf{q} + \mathbf{r}) \\ &= \mathbf{b}^{\mathrm{T}}\mathbf{b} - (\mathbf{q} + \mathbf{r})^{\mathrm{T}}\mathbf{A}^{\mathrm{T}}\mathbf{A}\mathbf{q} + \mathbf{r}^{\mathrm{T}}\mathbf{A}^{\mathrm{T}}\mathbf{A}(\mathbf{q} + \mathbf{r}) \\ &= \mathbf{b}^{\mathrm{T}}\mathbf{b} - \mathbf{q}^{\mathrm{T}}\mathbf{A}^{\mathrm{T}}\mathbf{A}\mathbf{q} + \mathbf{r}^{\mathrm{T}}\mathbf{A}^{\mathrm{T}}\mathbf{A}\mathbf{r}\end{aligned}$$

Since $\mathbf{r}^{\mathrm{T}}\mathbf{A}^{\mathrm{T}}\mathbf{A}\mathbf{r} > 0$ for $\mathbf{r} \neq \mathbf{0}$, this function attains its least value for $\mathbf{r} = \mathbf{0}$. Thus our problem is solved by taking $\mathbf{r} = \mathbf{0}$, and then

$$\mathbf{p} = \mathbf{q} = (\mathbf{A}^T\mathbf{A})^{-1}\mathbf{A}^T\mathbf{b}.$$

As an example consider the points (1, 1.2), (2, 2.9), (3, 4.8), (4, 7.2), (5, 9.1). We have

$$\mathbf{A}^T\mathbf{b} = \begin{pmatrix} 1 & 2 & 3 & 4 & 5 \\ 1 & 1 & 1 & 1 & 1 \end{pmatrix} \begin{pmatrix} 1.2 \\ 2.9 \\ 4.8 \\ 7.2 \\ 9.1 \end{pmatrix} = \begin{pmatrix} 95.7 \\ 25.2 \end{pmatrix}$$

$$\mathbf{A}^T\mathbf{A} = \begin{pmatrix} 55 & 15 \\ 15 & 5 \end{pmatrix} = 5\begin{pmatrix} 11 & 3 \\ 3 & 1 \end{pmatrix}$$

and $\det(\mathbf{A}^T\mathbf{A}) = 25 \times 2 = 50$, and so

$$\mathbf{A}(\mathbf{A}^T\mathbf{A})^{-1} = \frac{5}{50}\begin{pmatrix} 1 & -3 \\ -3 & 11 \end{pmatrix} = \begin{pmatrix} 0.1 & -0.3 \\ -0.3 & 1.1 \end{pmatrix}$$

Now the required approximation is

$$\mathbf{p} = \begin{pmatrix} 2.01 \\ -0.99 \end{pmatrix}$$

Thus the line of closest fit is given by

$$y = 2.01x - 0.99$$

EXERCISES

9.1. Transform to diagonal form, by an orthogonal transformation: (i) $3x_1^2 + 6x_2^2 - 2x_3^2 + 4x_1x_2 - 12x_1x_3 + 6x_2x_3$; (ii) $3x_1^2 + 3x_2^2 + 6x_3^2 + 2x_1x_2 - 4x_1x_3 - 4x_2x_3$; (iii) $x_1x_2 + x_1x_3 + x_2x_3$.

9.2. Show that the points of intersection of the quadric $\mathbf{x}^T\mathbf{A}\mathbf{x} + 2\mathbf{a}^T\mathbf{x} + \alpha = 0$ and the line $\mathbf{x} = \mathbf{p} + \lambda\mathbf{u}$ are obtained by solving the equation $(\mathbf{p} + \lambda\mathbf{u})^T \cdot \mathbf{A}(\mathbf{p} + \lambda\mathbf{u}) + 2\mathbf{a}^T(\mathbf{p} + \lambda\mathbf{u}) + \alpha = 0$ for λ. Deduce that this line is tangent to the quadric at the point $\mathbf{p}$ if $\mathbf{p}^T\mathbf{A}\mathbf{p} + 2\mathbf{a}^T\mathbf{p} + \alpha = 0$ and $\mathbf{u}^T(\mathbf{A}\mathbf{p} + \mathbf{a}) = 0$. Hence show that the equation of the tangent plane to the quadric at the point $\mathbf{p}$ is $(\mathbf{x} - \mathbf{p})^T(\mathbf{A}\mathbf{p} + \mathbf{a}) = 0$. Interpret the special case $\mathbf{A}\mathbf{p} + \mathbf{a} = \mathbf{0}$.

9.3. Given a central quadric, if a line through the centre meets the quadric in points P and P', show that the tangent planes at P and at P' are parallel.

9.4. For each of the following quadrics, find the tangent planes at the points where the 1-axis meets the quadric:

(i) $x_1^2 + x_2^2 + 2x_1x_3 + 2x_2x_3 = 1$;

(ii) $2x_1^2 - x_2^2 + 2x_3^2 + 2x_1x_2 - 4x_1x_3 + 2x_2x_3 = 18$;
(iii) $4x_1^2 + 5x_2^2 + 5x_3^2 + 6x_1x_2 + 6x_1x_3 + 4x_2x_3 = 1$.

9.5. Find the eigenvalues of the matrices corresponding to the quadrics in Exercise 9.4 and hence classify these quadrics.

9.6. Classify the following quadrics:
(i) $2x_1^2 + 17x_2^2 - 12x_1x_2 + 4x_1x_3 - 10x_2x_3 - 1 = 0$;
(ii) $x_1^2 + 13x_2^2 + 28x_3^2 + 4x_1x_2 + 10x_1x_3 + 26x_2x_3 - 67 = 0$,
(iii) $x_1^2 + 6x_2^2 - x_3^2 - 2x_1x_2 + 2x_1x_3 - 12x_2x_3 + 20x_2 - 6x_3 = 0$;
(iv) $x_1^2 + 9x_2^2 - 3x_3^2 - 8x_1x_2 - 2x_1x_3 + 8x_2x_3 - 2x_1 - 6x_2 + 2x_3 - 6 = 0$;
(v) $x_1^2 + 3x_2^2 + 5x_3^2 - 2x_1x_2 - 4x_1x_3 + 4x_2x_3 - 4x_1 + 8x_2 + 10x_3 + 8 = 0$.

9.7. Let **A** be a square matrix with eigenvalues α_i $(i = 1, \ldots, n)$. Show that the eigenvalues of $\mathbf{A} - c\mathbf{I}$, for any scalar c, are $\alpha_i - c$ $(i = 1, \ldots, n)$. Why cannot this argument be used to find the eigenvalues of $\mathbf{A} + c\mathbf{B}$, for general **A**, **B**?

9.8. Let **A** be a real square matrix such that $\mathbf{A}^{\mathrm{T}} = -\mathbf{A}$. Show that all eigenvalues of **A** are purely imaginary. Show that if α is an eigenvalue of an orthogonal matrix **A**, then α^{-1} is also an eigenvalue of **A**.

9.9. Let **A** be any 3×3 real symmetric matrix and for any α, not an eigenvalue of **A**, denote by $\delta(\alpha)$ the number of sign changes in the sequence of principal minors of $\mathbf{A} - \alpha\mathbf{I}$. Using Exercise 9.7 and Theorem 9.3, show that the number of eigenvalues of **A** that are less than α is $\delta(\alpha)$. Deduce that for any α and β which are not eigenvalues of **A** and such that $\alpha < \beta$, the number of eigenvalues of **A** between α and β is $\delta(\beta) - \delta(\alpha)$.

9.10. With the notation of section 9.3 show that the principal minors d_0, d_1, d_2, d_3 of a symmetric matrix $\mathbf{A} = (a_{ij})$ of order 3 satisfy the relations

$$d_0 d_3 = d_1 a_{22} - a_{12}^2$$

$$d_1 d_2 = d_2 a_{22} - \alpha_{23}^2$$

where α_{23} is the minor of a_{23}. Deduce that if $d_3 \neq 0$ and exactly one of d_1, d_2 vanishes, then its neighbours in the sequence of principal minors have opposite signs.

9.11. Let **A** be any regular symmetric matrix of order 3 and let $\mathbf{A} = \mathbf{P}^{\mathrm{T}}\mathbf{D}\mathbf{P}$, where **D** is a diagonal matrix and $\mathbf{P} = (\mathbf{u}, \mathbf{v}, \mathbf{w})$ is an orthogonal matrix. Show that the principal minors of **A** are $1, \mathbf{u}^{\mathrm{T}}\mathbf{D}\mathbf{u}$, $|\mathbf{D}|$. $\mathbf{w}^{\mathrm{T}}\mathbf{D}^{-1}\mathbf{w}, |\mathbf{A}|$.

9.12. Find the principal minors and hence the number of negative eigenvalues for the quadrics in Exercises 9.1, 9.4 and 9.6, and check your results.

9.13. Let $\mathbf{A}, \mathbf{B}$ be two real symmetric matrices of order n. Show that they can be simultaneously transformed to diagonal form, provided that $\mathbf{A} + c\mathbf{B}$ is positive definite, for some c. Is this sufficient condition also necessary?

9.14. Find the square root of the matrix

$$\mathbf{A} = \begin{pmatrix} 14 & -20 & 10 \\ -20 & 44 & -20 \\ 10 & -20 & 14 \end{pmatrix}$$

given that one eigenvalue is 64.

9.15. Express the matrix

$$\begin{pmatrix} 7 & 6 & 7 \\ 2 & 3 & -1 \\ -1 & 3 & 2 \end{pmatrix}$$

in the first polar form: orthogonal times positive definite.

9.16. Express the matrix

$$\begin{pmatrix} 2 & 4 & 11 \\ 1 & 5 & -2 \\ -2 & 2 & 1 \end{pmatrix}$$

in the second polar form: orthogonal times triangular.

9.17. Show that every regular (real) matrix $\mathbf{A}$ can be written as $\mathbf{A} = \mathbf{PU}$, where $\mathbf{P}$ is positive definite and $\mathbf{U}$ is orthogonal, and that $\mathbf{P}, \mathbf{U}$ are unique. Similarly for an upper triangular matrix $\mathbf{P}$.

9.18. Maximize the function $2x_1 + x_2 + x_3 + x_4$ subject to $3x_1 + 2x_2 - x_3 = 5$, $2x_1 + x_3 + 3x_4 = 19$, $5x_1 + x_2 - 2x_3 + 2x_4 = 13$, $x_i \geqslant 0$.

9.19. Use the method of least squares to find the straight line closest to the points (1.5, 5.4), (1.7, 6.3), (2.0, 6.8), (2.3, 8.1), (2.6, 8.8).

9.20. Describe a method of finding the parabola $y = ax^2 + bx + c$ of closest fit to a number of points, of which no four lie in the same parabola (observe that any three points, not on the same straight line, determine a unique parabola).

10

Applications II. Calculus, mechanics, economics

The methods of linear algebra have been applied in many different fields and here we can only give a few examples by way of illustration. Foremost is the calculus of functions of several variables: the derivative is a linear transformation and extreme values of a real function are described by a quadratic form. The Jordan normal form turns out to be of use in solving linear differential equations, and the simultaneous reduction of quadratic forms can be used to simplify mechanical problems; in fact much of the theory was first developed in this context. But the range of applications is much wider, as an illustration from economics shows.

The remaining sections deal with a method of obtaining the inverse matrix as the sum of a convergent series and an operator treatment of linear difference equations.

10.1 FUNCTIONAL DEPENDENCE AND JACOBIANS

The derivative of a function is essentially a linear mapping. This point of view is particularly useful in dealing with functions of several variables, but we shall first look at the simplest case, of a single variable. Here we have

$$y = f(x)$$

which can be interpreted geometrically as a curve in the xy-plane. We shall assume that f is **smooth**, that is, its derivative exists and is continuous. If $x = a$, $y = b$ is a point on this curve, then

$$y - b = \mathrm{D}f(a)(x - a) \tag{10.1}$$

is the equation of the tangent to the curve at the point (a, b), where $\mathrm{D}f(a) = (\mathrm{d}f/\mathrm{d}x)_{x=a}$. In fact, the derivative $\mathrm{D}f(a)$ can be defined as the unique number (assumed to exist), for which

$$f(x) = f(a) + (x - a)(\mathrm{D}f(a) + \eta),$$

where $\eta \to 0$ as $x \to a$.

Suppose now that f is a mapping from $\mathbf{R}^n$ (or more generally, an open subset of $\mathbf{R}^n$) to $\mathbf{R}^m$; in terms of coordinates f has m components, which are functions of n variables. We fix a point $\mathbf{x} = \mathbf{a}$ in $\mathbf{R}^n$ and write $f(\mathbf{a}) = \mathbf{b}$. Under suitable smoothness assumptions on f we can describe the tangent plane at this point by the same equation (10.1), where now $\mathbf{x} - \mathbf{a} \in \mathbf{R}^n$, $\mathbf{y} - \mathbf{b} \in \mathbf{R}^m$ and $\mathrm{D}f(\mathbf{a})$ is an $m \times n$ matrix. Of course, 'tangent plane' means here an n-dimensional linear subspace of an $(m+n)$-dimensional space. In detail, we have

$$\mathrm{D}f(\mathbf{a}) = (f_{ij}), \quad \text{where } f_{ij} = (\partial f_i/\partial x_j)_{\mathbf{x}=\mathbf{a}}. \tag{10.2}$$

This is called the **Jacobian matrix** of f; if its entries exist and are continuous in a neighbourhood of $\mathbf{a}$, f will be called **smooth** at $\mathbf{x} = \mathbf{a}$.

A key result in this context, the **inverse mapping theorem**, states that if f is a smooth mapping from an open subset D of $\mathbf{R}^n$ to $\mathbf{R}^n$ such that $f(\mathbf{a}) = \mathbf{b}$ for some $\mathbf{a} \in D$ and if the Jacobian matrix (10.2) is regular at $\mathbf{a}$, then there is a smooth mapping g, defined on a neighbourhood of $\mathbf{b}$, which is inverse to f, thus $f(g(\mathbf{y})) = \mathbf{y}$, $g(f(\mathbf{x})) = \mathbf{x}$.

For example, let $n = 2$ and let f be given by

$$y_1 = x_1 + x_2$$
$$y_2 = x_1 x_2$$

Then the Jacobian matrix is

$$\begin{pmatrix} 1 & 1 \\ x_2 & x_1 \end{pmatrix}$$

It is regular, provided that $x_2 \neq x_1$ and the inverse function is given by

$$x_1 = \tfrac{1}{2}(y_1 + \sqrt{y_1^2 - 4y_2})$$
$$x_2 = \tfrac{1}{2}(y_1 - \sqrt{y_1^2 - 4y_2})$$

Locally, near any point not on the line $x_1 = x_2$ the inverse is unique, but we cannot expect uniqueness globally, since the value $y = f(x)$ is unchanged when x_1 and x_2 are interchanged.

As another illustration, from complex function theory, consider functions $f(z)$ from $\mathbf{C}$ to $\mathbf{C}$, where $\mathbf{C}$ is regarded as a 2-dimensional vector space over $\mathbf{R}$. If $\mathrm{D}f$ is to reduce to multiplication by a complex number, then the Jacobian matrix of f must be of the form $\begin{pmatrix} a & -b \\ b & a \end{pmatrix}$, as we saw in section 7.5. Hence on writing our function as $f(z) = u + \mathrm{i}v$, $z = x + \mathrm{i}y$, where u, v, x, y are real, we must have

$$\partial u/\partial x = \partial v/\partial y$$
$$\partial u/\partial y = -\partial v/\partial x$$

These are just the well known Cauchy–Riemann equations which are necessary and sufficient for f to be a holomorphic function of z (cf. Ahlfors 1979).

10.2 EXTREMA, HESSIANS AND THE MORSE LEMMA

Let us consider a smooth function f from $\mathbf{R}^n$ to $\mathbf{R}$. A point $\mathbf{a} \in \mathbf{R}^n$ is said to be a **critical point** if $\mathrm{D}f(\mathbf{a}) = 0$, where $\mathrm{D}f(\mathbf{a}) = (\partial f/\partial x_i)_{\mathbf{x}=\mathbf{a}}$ as in section 10.1. A well-known result of calculus states that any maximum or minimum of a smooth function is at a critical point; thus it is of interest to have a means of finding these points. A smooth function f with smooth derivatives can at a critical point $\mathbf{x} = \mathbf{a}$ be expressed as

$$f(\mathbf{x}) = f(\mathbf{a}) + \sum (x_i - a_i)(x_j - a_j) g_{ij}(\mathbf{x}) \tag{10.3}$$

where $g_{ij}(\mathbf{x}) \to \partial^2 f/\partial x_i \partial x_j$ as $\mathbf{x} \to \mathbf{a}$. The critical point $\mathbf{a}$ is called **non-degenerate** if the quadratic form in (10.3) is non-degenerate, that is, of rank n at $\mathbf{x} = \mathbf{a}$. If we define the **Hessian matrix** of f (after Otto Hesse, 1811–74) as

$$\mathrm{H}f = (\partial^2 f/\partial x_i \partial x_j)$$

then it is clear that the form is non-degenerate precisely when the Hessian matrix is regular. When this is the case, we can, by Theorem 8.7, make a transformation of variables such that

$$f(\mathbf{u}) = f(\mathbf{a}) - u_1^2 - \cdots - u_s^2 + u_{s+1}^2 + \cdots + u_n^2 \tag{10.4}$$

where the negative terms have been put first. This statement – depending on a proof of (10.3) which is not difficult, but will not be given here (see, for example, Poston and Stewart 1978) – is known as the **Morse lemma** (after Marston Morse, 1892–1977), and a critical point, where f has the form (10.4) is called a **saddle-point**, more precisely a **Morse s-saddle**. Thus a Morse 0-saddle is a minimum, a Morse n-saddle is a maximum, while intermediate values give a maximum in some directions and a minimum in others (as illustrated for $n = 2$ by the shape of a saddle).

If the diagonalization process is applied to a function at a degenerate critical point, the resulting function has the form

$$f = f(a) - z_1^2 - \cdots - z_s^2 + z_{s+1}^2 + \cdots + z_r^2 + h(z_1, \ldots, z_n)$$

where h has zero Hessian matrix. This form is not very useful, because h still depends on all the z_i; it is better to adjust the functions g_{ij} in (10.3) so that for $i, j > 1$, g_{ij} does not depend on x_1. In this way it is possible to make the transformation as before so as to obtain the form

$$f = f(\mathbf{a}) - z_1^2 - \cdots - z_s^2 + z_{s+1}^2 + \cdots + z_r^2 + h(z_{r+1}, \ldots, z_n)$$

This is the content of the **splitting lemma**, whose proof may be found in Poston and Stewart (1978). This lemma is of particular use when $n - r$ (the 'corank') is small compared with n.

10.3 NORMAL MODES OF VIBRATION

The simultaneous reduction of two quadratic forms has an important application in mechanics, to finding the normal modes of vibration of a mechanical system. Let us first recall the simplest kind of mechanical system with one degree of freedom executing simple harmonic motion, say a simple pendulum. A mass m is fixed to a light string of length r and swings in a plane. When the string makes an angle θ with the vertical (Fig 10.1), the mass has potential energy $mgr(1 - \cos\theta)$ (where g is the acceleration due to gravity). Its speed is $v = r\dot{\theta}$, where $\dot{\theta} = \mathrm{d}\theta/\mathrm{d}t$, so its kinetic energy is $\frac{1}{2}mv^2 = \frac{1}{2}mr^2\dot{\theta}^2$. The energy equation, expressing the conservation of total energy, states that

$$mgr(1 - \cos\theta) + \tfrac{1}{2}mr^2\dot{\theta}^2 = \text{const.} \tag{10.5}$$

Differentiating with respect to t and dividing by mr^2, we obtain

$$\ddot{\theta}\dot{\theta} + \frac{g}{r}(\sin\theta)\dot{\theta} = 0$$

and if this is to hold for all $\dot{\theta}$, we must have

$$\ddot{\theta} + \frac{g}{r}\sin\theta = 0$$

For small-amplitude oscillations we may replace $\sin\theta$ by θ and so find

$$\ddot{\theta} + \frac{g}{r}\theta = 0 \tag{10.6}$$

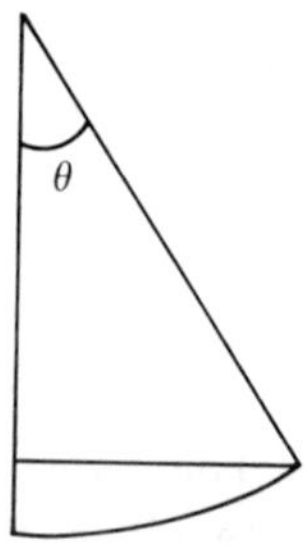

Fig. 10.1

Clearly $g/r > 0$ and on writing $g/r = \omega^2$, we obtain the solution

$$\theta = a\cos\omega t + b\sin\omega t,$$

where a and b are constants to be determined by the initial conditions. For example, if the pendulum starts from rest in the position $\theta = \theta_0$, then $\theta = \theta_0 \cos\omega t$. This is the simple harmonic motion with amplitude θ_0 associated with an equation (10.6), when $g/r > 0$.

In the general case we have a mechanical system with a number of variables $q_1, \ldots, q_n$, the **generalized coordinates**, and the energy equation again states that the total energy

$$E = T + V$$

is constant, where T is the kinetic and V the potential energy. The kinetic energy is a quadratic function in the velocities $\dot{q}_i$ and $T > 0$ whenever the velocities are not all zero; thus T is positive definite. The potential energy V is a function of the q_i, and to study small displacements from the equilibrium position we expand V in powers of the q_i:

$$V = \sum c_i q_i + \sum b_{ij} q_i q_j + \cdots$$

where the dots indicate higher terms and $V = 0$ for $q_i = 0$ $(i = 1, \ldots, n)$. For equilibrium the point $q_i = 0$ must be a critical point; $\partial V/\partial q_i = 0$ at $q_i = 0$. Hence $c_i = 0$; moreover, the equilibrium is stable if V has a minimum, i.e. V is positive definite: $V > 0$ for all small displacements from equilibrium. Thus, neglecting higher terms, we have

$$T = \sum a_{ij}\dot{q}_i\dot{q}_j$$
$$V = \sum b_{ij} q_i q_j$$

Here the matrices $\mathbf{A} = (a_{ij})$ and $\mathbf{B} = (b_{ij})$ are symmetric and $\mathbf{A}$ is positive definite; if the equilibrium is stable, then $\mathbf{B}$ is also positive definite. In that case the roots of the secular equation

$$|x\mathbf{A} - \mathbf{B}| = 0 \tag{10.7}$$

are all positive. If these roots are $\lambda_1, \ldots, \lambda_n$, then by a suitable change of coordinates $q_i = \sum s_{ij} q'_j$ (using Theorem 9.4) we can bring T and V to simultaneous diagonal form; dropping the dash from q'_i we have

$$T = \sum \dot{q}_i^2$$
$$V = \sum \lambda_i q_i^2$$

The energy equation now takes the form $\sum \dot{q}^2 + \sum \lambda_i q_i^2 = \text{const.}$, which on differentiating yields

$$2\sum \dot{q}_i\ddot{q}_i + 2\sum \lambda_i \dot{q}_i q_i = 0 \tag{10.8}$$

Since the q_i are independent variables, the coefficients of each $\dot{q}_i$ must vanish and we obtain

$$\ddot{q}_i + \lambda_i q_i = 0$$

Assuming stable equilibrium, we have $\lambda_i > 0$, and putting $\omega_i = \lambda_i^{1/2}$, we obtain the solution

$$q_i = a_i \cos \omega_i t + b_i \sin \omega_i t,$$

where a_i, b_i are constants determined by the initial conditions. We see, in particular, that the system can vibrate with a single q_i varying and all the others zero. These coordinates q_i in (10.8) are called **normal coordinates**, giving rise to **normal modes of vibration** of the system.

To illustrate the situation let us take a double pendulum. A mass m is suspended by a light rod of length a, to swing in a vertical plane; hinged to it is another light rod of length b with a mass m' attached to its end, to swing in the same vertical plane (Fig. 10.2). Find the oscillations about the equilibrium position.

Let θ, φ be the angles made by the rods with the vertical. The kinetic energy is given by

$$2T = m(a\dot{\theta})^2 + m'(a\dot{\theta} + b\dot{\varphi})^2 = Ma^2\dot{\theta}^2 + 2m'ab\dot{\theta}\dot{\varphi} + m'b^2\dot{\varphi}^2$$

where $M = m + m'$. For the potential energy we have

$$V = gma(1 - \cos\theta) + gm'[a(1 - \cos\theta) + b(1 - \cos\varphi)],$$

which on expanding and omitting higher terms, leads to

$$\begin{aligned} 2V &= mga\theta^2 + m'g(a\theta^2 + b\varphi^2) \\ &= Mga\theta^2 + m'gb\varphi^2 \end{aligned}$$

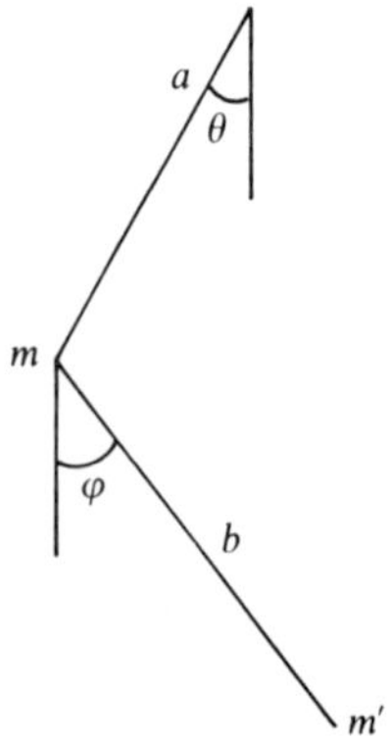

Fig. 10.2

Thus we have the matrices (absorbing the factor 2)

$$\mathbf{A}=\begin{pmatrix} Ma^2 & m'ab \\ m'ab & m'b^2 \end{pmatrix} \quad \mathbf{B}=\begin{pmatrix} Mag & 0 \\ 0 & m'bg \end{pmatrix}$$

$$x\mathbf{A}-\mathbf{B}=\begin{pmatrix} Ma(xa-g) & m'xab \\ m'xab & m'b(xb-g) \end{pmatrix}$$

The secular polynomial is $Mm'ab(xa-g)(xb-g)-m'^2a^2b^2x^2$, which simplifies to

$$m'ab[mabx^2-Mg(a+b)x+Mg^2]=0$$

Let us take the following numerical values: $m=2$, $m'=1$, $a=5$, $b=7.5$, $g=10$ (in cgs units). Then we obtain

$$ab[75x^2-375x+300]=75ab[x^2-5x+4]=0$$

The roots are 1, 4 and

$$\mathbf{A}=\begin{pmatrix} 75 & 37.5 \\ 37.5 & 56.25 \end{pmatrix} \quad \mathbf{B}=\begin{pmatrix} 150 & 0 \\ 0 & 75 \end{pmatrix}$$

Dividing by 75/4, we obtain

$$\mathbf{A}'=\begin{pmatrix} 4 & 2 \\ 2 & 3 \end{pmatrix} \quad \mathbf{B}'=\begin{pmatrix} 8 & 0 \\ 0 & 4 \end{pmatrix}$$

$\mathbf{A}'-\mathbf{B}'=\begin{pmatrix} -4 & 2 \\ 2 & -1 \end{pmatrix}$ has solution vector $\begin{pmatrix} 1 \\ 2 \end{pmatrix}$, $4\mathbf{A}'-\mathbf{B}'=\begin{pmatrix} 8 & 8 \\ 8 & 8 \end{pmatrix}$ has solution vector $\begin{pmatrix} -1 \\ 1 \end{pmatrix}$. If we use the transforming matrix $\mathbf{P}=\begin{pmatrix} 1 & -1 \\ 2 & 1 \end{pmatrix}$ (without normalizing the columns), we find

$$\mathbf{P}^{\mathrm{T}}\mathbf{A}'\mathbf{P}=\begin{pmatrix} 24 & 0 \\ 0 & 3 \end{pmatrix}$$

$$\mathbf{P}^{\mathrm{T}}\mathbf{B}'\mathbf{P}=\begin{pmatrix} 24 & 0 \\ 0 & 12 \end{pmatrix}$$

Taking normal coordinates to be q_1, q_2, we have

$$\theta=q_1-q_2$$
$$\varphi=2q_1+q_2$$
$$q_1=\tfrac{1}{3}(\theta+\varphi)$$
$$q_2=\tfrac{1}{3}(\varphi-2\theta)$$

If the system starts from rest, the equations in normal coordinates are

$$\ddot{q}_1 + q_1 = 0$$
$$\ddot{q}_2 + 4q_2 = 0$$
$$\dot{q}_1 = \dot{q}_2 = 0 \quad \text{when } t = 0.$$

This gives $q_1 = \lambda \cos t, q_2 = \mu \cos 2t$, hence

$$\theta = \lambda \cos t - \mu \cos 2t$$
$$\varphi = 2\lambda \cos t + \mu \cos 2t$$

For example, $\mu = 0$ gives a motion in which φ has twice the amplitude of θ, while for $\lambda = 0$, the pendulum swings twice as fast, with the two bobs going in opposite directions, but with the same amplitude. Moreover, the most general motion arises from a superposition of these two special cases.

10.4 LINEAR DIFFERENTIAL EQUATIONS WITH CONSTANT COEFFICIENTS AND ECONOMIC MODELS

In section 10.3 we applied the simultaneous reduction of two quadratic forms to the solution of systems of differential equations. Another type of system of differential equations requires the similarity reduction of a single matrix. Consider a system of linear differential equations with constant coefficients in one independent variable t:

$$\dot{x}_i = \sum a_{ij} x_j + b_i, \tag{10.9}$$

where $\dot{x}_i = \mathrm{d}x_i/\mathrm{d}t$ and the a_{ij} and b_i are constants. If derivatives of higher orders occur, the system can be reduced to the same form by introducing further variables. Thus if $\mathrm{d}^n y/\mathrm{d}t^n$ occurs, we introduce new variables $y_i = \mathrm{d}^i y/\mathrm{d}t^i$ $(i = 0, 1, \ldots, n-1)$; then $\mathrm{d}^n y/\mathrm{d}t^n = \dot{y}_{n-1}$ and this can be expressed in terms of $y_0, y_1, \ldots, y_{n-1}$ and the other variables. On writing $\mathbf{A} = (a_{ij})$, $\mathbf{x} = (x_1, \ldots, x_n)^{\mathrm{T}}$, $\mathbf{b} = (b_1, \ldots, b_n)^{\mathrm{T}}$, we can express the system (10.9) in matrix form:

$$\dot{\mathbf{x}} = \mathbf{A}\mathbf{x} + \mathbf{b} \tag{10.10}$$

For an equilibrium position we must have $\dot{\mathbf{x}} = \mathbf{0}$, so that $\mathbf{A}\mathbf{x} + \mathbf{b} = \mathbf{0}$. Such a position exists and is unique precisely when $\mathbf{A}$ is regular. Assuming this and replacing $\mathbf{x}$ by $\mathbf{y} = \mathbf{x} + \mathbf{A}^{-1}\mathbf{b}$, we can reduce our system to the form $\dot{\mathbf{y}} = \mathbf{A}\mathbf{y}$. Thus we assume from now on that $\mathbf{b} = \mathbf{0}$ in (10.10), so that $\mathbf{x} = \mathbf{0}$ is an equilibrium position:

$$\dot{\mathbf{x}} = \mathbf{A}\mathbf{x} \tag{10.11}$$

Consider first the 1-dimensional case: $\dot{x} = ax$, which has the solution $x = c\mathrm{e}^{at}$. More precisely, for complex a, say $a = \alpha + \beta\mathrm{i}$ (α, β real), the general solution is $\mathrm{e}^{\alpha t}(b \cos \beta t + c \sin \beta t)$. For stable equilibrium we must have $\alpha \leqslant 0$, and then we have vibrations, decaying exponentially for $\alpha < 0$, while for $\alpha = 0$ we have simple harmonic motion.

Let us now return to the general case (10.11). If $\mathbf{A}$ is diagonal, say $\mathbf{A} = \mathrm{diag}(a_1, \ldots, a_n)$, our system takes the form

$$\dot{x}_i = a_i x_i \quad (i = 1, \ldots, n),$$

with solution

$$x_i = c_i \mathrm{e}^{a_i t}$$

This makes it clear how (10.11) is to be solved in general. Suppose first that $\mathbf{A}$ is similar to a diagonal matrix $\mathbf{D}$, say $\mathbf{P}^{-1}\mathbf{AP} = \mathbf{D}$ for an invertible matrix $\mathbf{P} = (p_{ij})$. We transform the variables by $\mathbf{P}$: $\mathbf{x} = \mathbf{Py}$. Since $\mathbf{P}$, like $\mathbf{A}$, has constant entries, we have $\dot{\mathbf{P}} = \mathbf{0}$, so (10.11) becomes $\mathbf{P}\dot{\mathbf{y}} = \mathbf{APy}$, and on replacing $\mathbf{P}^{-1}\mathbf{AP}$ by $\mathbf{D}$, we obtain

$$\dot{\mathbf{y}} = \mathbf{Dy}$$

Here $\mathbf{D} = \mathrm{diag}(\lambda_1, \ldots, \lambda_n)$, where the λ_i are the eigenvalues of $\mathbf{A}$. Hence we have $y_i = c_i \mathrm{e}^{\lambda_i t}$ and we obtain for the solution of (10.11),

$$x_i = \sum p_{ij} y_j = \sum p_{ij} c_j \mathrm{e}^{\lambda_j t}$$

In general, we may not be able to transform $\mathbf{A}$ to diagonal form, but in any case we can reach a form $\mathbf{P}^{-1}\mathbf{AP} = \mathbf{B} = (b_{ij})$, where $\mathbf{B}$ is in Jordan normal form. Let us take a single Jordan block $\mathbf{B} = \mathbf{J}_r(\lambda)$ and solve $\dot{\mathbf{y}} = \mathbf{By}$. The equation for y_r is $\dot{y}_r = \lambda y_r$, hence $y_r = c_r \mathrm{e}^{\lambda t}$. We shall show by induction on $r - j$ that we have

$$y_j = f_j(t)\mathrm{e}^{\lambda t} \tag{10.12}$$

where f_j is a polynomial of degree $r - j$ in t, with constant coefficients, and $f_{j-1} = t f_j + c_{j-1}$. For $j = r$ this has just been proved. If (10.12) holds, we have the equation for y_{j-1}:

$$\dot{y}_{j-1} = \lambda y_{j-1} + y_j$$

Hence

$$\frac{\mathrm{d}}{\mathrm{d}t}[y_{j-1}\mathrm{e}^{-\lambda t}] = \dot{y}_{j-1}\mathrm{e}^{-\lambda t} - \lambda y_{j-1}\mathrm{e}^{-\lambda t} = y_j \mathrm{e}^{-\lambda t} = f_j(t)$$

where we have used (10.12) at the last step. It follows that $y_{j-1}\mathrm{e}^{-\lambda t} = f_{j-1}(t)$, with $\dot{f}_{j-1} = f_j$, which proves the induction step. In this way we obtain the complete solution of the system (10.11), since we can, by Theorem 8.11, transform the matrix $\mathbf{A}$ of the system to a diagonal sum of Jordan blocks.

Let us take a couple of examples. First:

$$\dot{x}_1 = 3x_1 - 5x_2$$
$$\dot{x}_2 = 2x_2$$

Here the matrix is already in triangular form, so the equations can be solved recursively. We have $x_2 = be^{2t}, \dot{x}_1 = 3x_1 - 5be^{2t}$, $d(x_1e^{-3t})/dt = -5be^{-t}$, hence $x_1e^{-3t} = 5be^{-t} + a$, so the complete solution is

$$x_1 = ae^{3t} + 5be^{2t}$$
$$x_2 = be^{2t}$$

For another example, let us solve:

$$\dot{x}_1 = x_1 - x_2$$
$$\dot{x}_2 = x_1 + 3x_2$$

subject to $x_1 = 6$, $x_2 = 4$ when $t = 0$. Here $\dot{\mathbf{x}} = \mathbf{Ax}$, where $\mathbf{A} = \begin{pmatrix} 1 & -1 \\ 1 & 3 \end{pmatrix}$; the characteristic polynomial is $(\lambda - 1)(\lambda - 3) + 1 = \lambda^2 - 4\lambda + 4 = (\lambda - 2)^2$, so the eigenvalues are 2, 2 and the eigenvector (unique up to a scalar factor) is $\begin{pmatrix} 1 \\ -1 \end{pmatrix}$, obtained by solving $(2\mathbf{I} - \mathbf{A})\mathbf{u} = 0$. We take any vector linearly independent of this one, say $\begin{pmatrix} 1 \\ 1 \end{pmatrix}$ and form $\mathbf{P} = \begin{pmatrix} 1 & 1 \\ -1 & 1 \end{pmatrix}$; then $\mathbf{P}^{-1}\mathbf{AP} = \begin{pmatrix} 2 & -2 \\ 0 & 2 \end{pmatrix}$, and so

$$x_1 = y_1 + y_2 \quad x_2 = -y_1 + y_2 \quad y_1 = \tfrac{1}{2}(x_1 - x_2) \quad y_2 = \tfrac{1}{2}(x_1 + x_2)$$
$$y_2 = be^{2t} \quad y_1 = ae^{2t} - bte^{2t} \quad x_1 = (a + b - 2bt)e^{2t} \quad x_2 = (b - a + 2bt)e^{2t}$$

For $t = 0$ we have $a + b = 6$, $b - a = 4$, so $b = 5$, $a = 1$ and hence

$$x_1 = (6 - 10t)e^{2t}$$
$$x_2 = (4 + 10t)e^{2t}$$

Sometimes the preceding theory allows one to make certain qualitative statements about the solutions. We illustrate this point by two examples from economics (for which I am indebted to N.J. Rau).

The demand x_1 for a certain product is a function of the price p; clearly x_1 falls with rising price. We shall assume a linear relation

$$x_1 = a_0 - a_1 p \tag{10.13}$$

where a_0, a_1 are positive constants. The supply x_2 is a function of the expected price q. It is generally an increasing function and we take it to have the form

$$x_2 = b_0 + b_1 q \tag{10.14}$$

where b_0, b_1 are again positive constants. If $a_0 \leqslant b_0$, then the whole demand can be supplied for $p = q = 0$ and the problem is trivially solved. We shall therefore assume that $b_0 < a_0$.

The expected price is related to the actual price by the equation

$$\dot{q} = \sigma(p - q) \tag{10.15}$$

where σ is a positive constant representing the speed of adjustment. It expresses the fact that when the actual price exceeds the expected price, the supplier moves price expectations upwards, while for a lower actual price the expected price is lowered.

The actual price changes by a rate depending on the excess of demand over supply:

$$\dot{p} = \theta(x_1 - x_2) \tag{10.16}$$

where θ is a positive constant. By eliminating x_1 and x_2 (that is, substituting from (10.13) and (10.14), we obtain the equation

$$\dot{p} = \theta(a_0 - b_0) + \theta(-a_1 p - b_1 q)$$

This equation and (10.15) can together be written in matrix form

$$\begin{pmatrix} \dot{p} \\ \dot{q} \end{pmatrix} = \mathbf{A}\begin{pmatrix} p \\ q \end{pmatrix} + \theta\begin{pmatrix} a_0 - b_0 \\ 0 \end{pmatrix}$$

where

$$\mathbf{A} = \begin{pmatrix} -\theta a_1 & -\theta b_1 \\ \sigma & -\sigma \end{pmatrix}$$

The unique equilibrium point $(\bar{p}, \bar{q})$ is given by

$$\bar{p} = \bar{q} = \frac{a_0 - b_0}{a_1 + b_1}$$

The eigenvalues of $\mathbf{A}$ are the roots in λ of the equation

$$(\lambda + \theta a_1)(\lambda + \sigma) + \sigma\theta b_1 = 0$$

They are either complex conjugate or real, so for an eigenvalue with negative real part (representing a solution tending to stable equilibrium) we require their sum to be negative. But this sum is the trace of $\mathbf{A}$, and since a_1, σ, θ are all positive, we always have

$$\theta a_1 + \sigma > 0$$

Thus there is always a stable equilibrium position and $p \to \bar{p}$, $q \to \bar{q}$ as $t \to \infty$. We remark that the condition that p, q tend to the same limit ultimately depends on (10.15).

As a second example we take a stylized model of the housing market. As our unit we take the 'standard house', with **stock price** p, which is its market value at a given time, while the **flow price** c is the cost per unit of time of living in a given unit of housing. An owner-occupier finds that the cost of being housed is equal to the mortgage repayments plus the cost of repairs minus its appreciation in value (if any). The latter depends on the change in price, while the other items depend directly on the cost. Thus we have

$$c = rp + \delta p - (\dot{p})^e, \tag{10.17}$$

where r is the interest rate (determining the mortgage), δ is the depreciation rate, and $(\dot{p})^e$ is the rate of capital gain.

As a first approximation let us take $e = 1$; then (10.17) may be written as

$$\dot{p} = (r + \delta)p - c \tag{10.18}$$

The total housing stock is a decreasing function of the cost c, and will be taken to be

$$H = a_0 - a_1 c \tag{10.19}$$

where a_0, a_1 are positive constants. From (10.18) and (10.19) we have

$$\dot{p} = \frac{H}{a_1} + (r + \delta)p - \frac{a_0}{a_1} \tag{10.20}$$

Let us consider how H changes with time. The flow of new houses on to the market is an increasing function of p, say, but decreases with increasing H according to the depreciation rate:

$$\dot{H} = b_0 + b_1 p - \delta H \tag{10.21}$$

where b_0, b_1 are positive constants. Equations (10.21) and (10.20) can again be written in matrix form

$$\begin{pmatrix} \dot{H} \\ \dot{p} \end{pmatrix} = \mathbf{A}\begin{pmatrix} H \\ p \end{pmatrix} + \begin{pmatrix} b_0 \\ -a_0/a_1 \end{pmatrix} \tag{10.22}$$

where

$$\mathbf{A} = \begin{pmatrix} -\delta & b_1 \\ 1/a_1 & r + \delta \end{pmatrix}$$

There is a unique equilibrium point, provided that

$$|\mathbf{A}| = -m/a_1 \neq 0$$

where

$$m = (r + \delta)\delta a_1 + b_1$$

It is found by equating the right-hand side of (10.22) to zero and solving for H and p:

$$\bar{H} = (a_0 b_1 + (r + \delta)a_1 b_0)/m$$
$$\bar{p} = (\delta a_0 - b_0)/m$$

If $b_0 \geqslant \delta a_0$, then $\bar{p} \leqslant 0$, giving a negative (or zero) stock price. This shows what happens (in a perfectly elastic market) if the flow of houses (b_0) is too large in relation to the depreciation of the total supply (δa_0). Thus we shall assume in our example that $b_0 < \delta a_0$. The characteristic equation of $\mathbf{A}$ is $\lambda^2 - r\lambda - m/a_1 = 0$. Since $r^2 + 4m/a_1 > 0$, the roots are real and distinct, with opposite sign, which means that the equilibrium is a saddle-point (that is, a 1-saddle in the terminology of section 10.2). If the roots are λ, $-\mu$, then each of H, p are linear combinations of $e^{\lambda t}$, $e^{-\mu t}$, but unless the coefficients of $e^{\lambda t}$ in both H and p are zero, we get a solution which diverges with time. Let us take H, p referred to $\bar{H}$, $\bar{p}$, so that $\bar{H} = \bar{p} = 0$. Then we have

$$H = Ae^{-\mu t}$$
$$p = Be^{-\mu t}$$

and by (10.21) $-A\mu e^{-\mu t} = b_1 Be^{-\mu t} - \delta Ae^{-\mu t}$, hence $A(\delta - \mu) = b_1 B$, so for some constant c we have

$$H = -b_1 ce^{-\mu t}, \quad p = (\mu - \delta)ce^{-\mu t},$$

where the sign has been chosen so that (with $c > 0$) the initial housing stock is less than the stock in equilibrium position. Now the initial value of p is $p_0 = \bar{p} + (\mu - \delta)c$; any other initial value will lead to a term in $e^{\lambda t}$ and so give a divergent solution. The situation is illustrated in Fig. 10.3.

The crucial question is whether there is any mechanism which will lead the market to hit on the 'right' initial price p_0. This question is far from completely resolved. Of course, it also has to be borne in mind that a number of simplifying assumptions were made in order to reach equation (10.22).

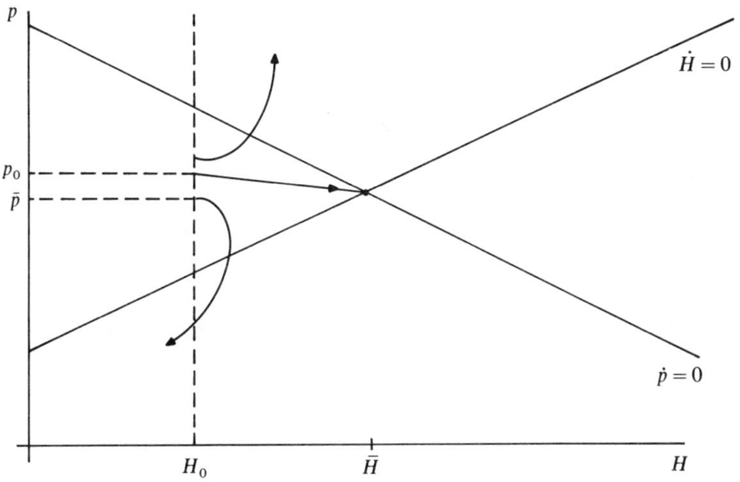

Fig. 10.3

10.5 HOMOGENEOUS SYSTEMS OF LINEAR DIFFERENTIAL EQUATIONS AND THE WRONSKIAN

Let us consider a homogeneous system of linear differential equations with variable coefficients. It will be convenient to use a dash rather than a dot to indicate the derivative, thus we shall write $x' = \mathrm{d}x/\mathrm{d}t$ and $x^{(n)} = \mathrm{d}^n x/\mathrm{d}t^n$. Our system in matrix form is

$$\mathbf{x}' = \mathbf{A}\mathbf{x} \tag{10.23}$$

where $\mathbf{A} = (a_{ij}(t))$ is an $n \times n$ matrix whose entries are functions of t. A solution of (10.23) is a vector $\mathbf{u}$ whose components are functions of t, such that $\mathbf{u}' = \mathbf{A}\mathbf{u}$. We shall assume as known the fact that for suitably well-behaved functions $a_{ij}(t)$ and any constant vector $\mathbf{c}$ the system (10.23) has a unique solution $\mathbf{u}$ such that $\mathbf{u}(0) = \mathbf{c}$. In particular, $\mathbf{u} = \mathbf{0}$ is the only solution vanishing for $t = 0$ (cf., for example, Hurewicz 1958). Under these assumptions we can give a precise description of the solution set of (10.23).

Theorem 10.1

The set of solutions of (10.23) is an n-dimensional vector space over **R**, *and a set of solutions* $\mathbf{u}_1, \ldots, \mathbf{u}_n$ *forms a basis for this space if and only if the determinant formed from their components is non-zero.*

A basis for the space of solutions is also called a **fundamental system of solutions**.

Proof

If is clear that the solutions of (10.23) form a vector space over **R**; we shall prove the result by finding a basis of n elements. Let $\mathbf{e}_1, \ldots, \mathbf{e}_n$ be any basis of $\mathbf{R}^n$, such as the standard basis, and let $\mathbf{u}_i$ be the unique solution of (10.23) satisfying $\mathbf{u}_i(0) = \mathbf{e}_i$. Then $\mathbf{u}_1, \ldots, \mathbf{u}_n$ are linearly independent over **R**, since they are linearly independent for $t = 0$. Given any solution $\mathbf{u}$ of (10.23), we can write $\mathbf{u}(0) = \sum a_i \mathbf{e}_i$, for some $a_i \in \mathbf{R}$. Then $\mathbf{u} - \sum a_i \mathbf{u}_i$ is a solution of (10.23) which reduces to 0 when $t = 0$, and so it is identically zero, hence $\mathbf{u} = \sum a_i \mathbf{u}_i$ and the $\mathbf{u}_i$ form indeed a basis of the solution space. Finally, given any n solutions $\mathbf{u}_i = (u_{i1}, \ldots, u_{in})^{\mathrm{T}}$, it is clear that they are linearly independent if and only if $\det(u_{ij}) \neq 0$ at any point t, and hence at every point t. This establishes the result.

We remark that if the matrix $\mathbf{B} = \int \mathbf{A}\,\mathrm{d}t$ has the property that $\mathbf{AB} = \mathbf{BA}$, then the solution of (10.23) can formally be written as $\mathbf{x} = \mathrm{e}^{\mathbf{B}}\mathbf{x}(0)$. This condition is satisfied, for example, when $\mathbf{A}$ is independent of t. The proof involves convergence questions which fall outside our framework (cf. Hurewicz 1958).

As already noted in section 10.4, a linear differential equation

$$x^{(n)} + a_1 x^{(n-1)} + \cdots + a_{n-1}x' + a_n x = 0 \tag{10.24}$$

can always be reduced to the form (10.23), and this still holds when the coefficients depend on t. We shall confine ourselves to **normalized** equations where the coefficient of $x^{(n)}$ is 1. The more general equation with coefficient a_0 of $x^{(n)}$ can be normalized provided that a_0 does not vanish for any t in the interval considered.

Let us carry out the reduction of (10.24). Thus we put $x_i' = x_{i+1}$ $(i = 1, \ldots, n-1)$, $x_n' = -a_1 x_n - a_2 x_{n-1} - \cdots - a_n x_1$; then (10.24) reduces to (10.23), where

$$\mathbf{A} = \begin{pmatrix} 0 & 1 & 0 & \cdots & 0 \\ 0 & 0 & 1 & \cdots & 0 \\ & \cdots & \cdots & \cdots & \\ 0 & 0 & 0 & 0 & 1 \\ -a_n & -a_{n-1} & \cdots & -a_2 & -a_1 \end{pmatrix}$$

This is just the companion matrix of the polynomial $p(\lambda) = \lambda^n + a_1\lambda^{n-1} + \cdots + a_n$. To each solution $u = u(t)$ of (10.24) there corresponds the solution $\varphi(u) = (u, u', \ldots, u^{(n-1)})$ of (10.23), and it is clear that the correspondence φ preserves linear dependence: if $u_1, \ldots, u_r$ are linearly dependent, say $\sum c_i u_i = 0$ ($c_i \in \mathbf{R}$, not all zero), then by differentiating we find $\sum c_i u_i' = 0, \ldots, \sum c_i u_i^{(n-1)} = 0$, hence $\sum c_i \varphi(u_i) = 0$. Conversely, if a family of vectors $\varphi(u)$ is linearly dependent, then the same is true of their first components. By applying Theorem 10.1 we conclude that (10.24) has a solution space of dimension n, and any n solutions $u_1, \ldots, u_n$ of (10.24) form a fundamental system of solutions precisely when the following determinant is non-zero for any (and hence every) value of t:

$$W(u_1, \ldots, u_n) = \begin{vmatrix} u_1 & u_2 & \cdots & u_n \\ u_1' & u_2' & \cdots & u_n' \\ \cdots & \cdots & \cdots & \\ u_1^{(n-1)} & u_2^{(n-1)} & \cdots & u_n^{(n-1)} \end{vmatrix} \neq 0 \qquad (10.25)$$

This determinant is called the **Wronskian** (after Josef-Maria Hoëne Wronski, 1778–1853).

The Wronskian can be defined more generally for any n (sufficiently differentiable) functions, not necessarily given as solutions of the same differential equation, but the vanishing of the Wronskian does not imply the linear dependence of the functions. For example, the functions t^3 and $|t^3|$ are linearly independent in the interval $[-1, 1]$, even though their Wronskian is zero:

$$\begin{vmatrix} t^3 & |t^3| \\ 3t^2 & 3t|t| \end{vmatrix} = 0$$

However, if $W \neq 0$, then the given functions are linearly independent. In fact, they form a fundamental system of solutions of a linear differential equation. Thus if $\mathbf{u}_1, \ldots, \mathbf{u}_n$ have a non-vanishing Wronskian, then

$$W(\mathbf{u}_1, \ldots, \mathbf{u}_n, \mathbf{x})/W(\mathbf{u}_1, \ldots, \mathbf{u}_n) = 0 \tag{10.26}$$

is a normalized differential equation of order n. It has $\mathbf{u}_1, \ldots, \mathbf{u}_n$ as solutions, because for $\mathbf{x} = \mathbf{u}_i$ the first determinant in (10.26) has two equal columns and so must vanish; since $W(\mathbf{u}_1, \ldots, \mathbf{u}_n) \neq 0$, it is a fundamental system of solutions.

10.6 INVERSION BY ITERATION

Although the methods for solving linear systems given in Chapters 2 and 4 are relatively straightforward, they become lengthy and time-consuming for large systems. In certain cases other methods are available and we shall consider one such method in this section.

Suppose that our system is of the form

$$\mathbf{x} = \mathbf{C}\mathbf{x} + \mathbf{b}, \tag{10.27}$$

where $\mathbf{C}$ is an $n \times n$ matrix and $\mathbf{b}$ is a column vector. If $\mathbf{u}(0)$ is an approximate solution, we can compute $\mathbf{u}(1) = \mathbf{C}\mathbf{u}(0) + \mathbf{b}$ and generally define

$$\mathbf{u}(r+1) = \mathbf{C}\mathbf{u}(r) + \mathbf{b} \tag{10.28}$$

Under suitable conditions the sequence $\mathbf{u}(0)$, $\mathbf{u}(1)$, $\mathbf{u}(2), \ldots$ will converge to a vector $\mathbf{u}$, in the sense that each component converges, thus if $\mathbf{u} = (u_1, \ldots, u_n)$, $\mathbf{u}(r) = (u_1(r), \ldots, u_n(r))$, then $u_i(r) \to u_i$ as $r \to \infty$. In that case $\mathbf{u}$ is a solution of (10.27), as we see by letting $r \to \infty$ in (10.28). Intuitively, it is plausible that we have convergence provided that $\mathbf{C}$ is not too large. Such a condition is given in the next result.

Theorem 10.2
Let $\mathbf{C} = (c_{ij})$ be an $n \times n$ matrix, where

$$|c_{ij}| \leqslant \gamma/n \tag{10.29}$$

for some constant γ satisfying $0 < \gamma < 1$. Then the system (10.27) has a unique solution $\mathbf{x} = \mathbf{u}$, obtained by taking any vector $\mathbf{u}(0)$ and defining $\mathbf{u}(r+1)$ recursively by (10.28). The sequence $\mathbf{u}(r)$ converges to a limit $\mathbf{u}$ which is the unique solution of (10.27).

Proof
Let $\mathbf{u}(0) = \mathbf{a}$; we claim that for all $r \geqslant 1$,

$$\mathbf{u}(r) = \mathbf{C}^r\mathbf{a} + (\mathbf{I} + \mathbf{C} + \mathbf{C}^2 + \cdots + \mathbf{C}^{r-1})\mathbf{b} \tag{10.30}$$

For this holds by hypothesis when $r = 1$. If it holds for $r = t \geqslant 1$, then

$$\begin{aligned}\mathbf{u}(t+1) = \mathbf{C}\mathbf{u}(t) + \mathbf{b} &= \mathbf{C}^{t+1}\mathbf{a} + (\mathbf{C} + \mathbf{C}^2 + \cdots + \mathbf{C}^t)\mathbf{b} + \mathbf{b} \\ &= \mathbf{C}^{t+1}\mathbf{a} + (\mathbf{I} + \mathbf{C} + \cdots + \mathbf{C}^t)\mathbf{b}.\end{aligned}$$

This proves that (10.30) holds for $r = t + 1$ and so by induction it holds for all r. In particular we have

$$\mathbf{u}(r+1) - \mathbf{u}(r) = \mathbf{C}^r\mathbf{p} \tag{10.31}$$

where $\mathbf{p} = \mathbf{Ca} - \mathbf{a} + \mathbf{b}$.

Our aim is to use (10.31) to show that the sequence $\mathbf{u}(r)$ converges, but first we need an estimate for the entries of $\mathbf{C}^r$. Let us write $\mathbf{C}^r = (c_{ij}(r))$; we claim that

$$|c_{ij}(r)| \leqslant \gamma^r/n \tag{10.32}$$

Clearly this holds for $r = 1$. If it holds for $r = t - 1$, then

$$|c_{ik}(t)| = \left|\sum c_{ij}c_{jk}(t-1)\right| \leqslant n(\gamma/n)\gamma^{t-1}/n = \gamma^t/n$$

where we have used (10.29) and the induction hypothesis. Thus (10.32) holds for $r = t$; hence by induction it holds for all t. Now by (10.31) we have

$$u_i(r+1) - u_i(r) = \sum c_{ij}(r)p_j$$

Hence on writing $c = \max\{p_1, \ldots, p_n\}$, we have

$$\begin{aligned}|u_i(r+1) - u_i(r)| &\leqslant n(\gamma^r/n)c \\ &\leqslant \gamma^r c\end{aligned}$$

It follows that

$$|u_i(r+s) - u_i(r)| \leqslant \sum_{\sigma=0}^{s-1} \gamma^{r+\sigma}c = \gamma^r(1 + \gamma + \cdots + \gamma^{s-1})c$$

hence for all $s > 0$,

$$|u_i(r+s) - u_i(r)| \leqslant \gamma^r c/(1-\gamma) \tag{10.33}$$

Since $\gamma^r/(1-\gamma) \to 0$ as $r \to \infty$, it follows that $u_i(r)$ converges as $r \to \infty$ for $i = 1, \ldots, n$. By Cauchy's convergence principle the convergent sequence $u_i(r)$ has a limit u_i and on writing $\mathbf{u} = (u_1, \ldots, u_n)$ we find that $\mathbf{u}(r)$ converges to $\mathbf{u}$. Letting $r \to \infty$ in (10.28), we obtain $\mathbf{u} = \mathbf{Cu} + \mathbf{b}$, so we have found a solution of (10.27).

To prove the uniqueness of the solution it is enough to show that when $\mathbf{b} = \mathbf{0}$, then (10.27) has only the solution $\mathbf{x} = \mathbf{0}$. Suppose that $\mathbf{v} = \mathbf{Cv}$, where $\mathbf{v} \neq \mathbf{0}$, and let v_j be the component of $\mathbf{v}$ of largest absolute value. Then $v_j = \sum c_{jk}v_k$, hence

$$|v_j| \leqslant n(\gamma/n)|v_j| = \gamma|v_j|$$

Since $\gamma < 1$, this is impossible unless $v_j = 0$, but then $\mathbf{v} = \mathbf{0}$, as we had to show. So the solution is unique and this completes the proof.

We remark that the proof provides a practical means of carrying out the calculation. For example, suppose that we are given

$$x_1 = 0.4x_1 - 0.3x_2 + 1$$
$$x_2 = 0.3x_1 + 0.1x_2 - 3$$

Then (10.29) is satisfied with $\gamma = 0.8$ and starting with $\mathbf{u}(0) = \mathbf{0}$, we have the following approximations:

$$\mathbf{u}(1) = \begin{pmatrix} 1 \\ -3 \end{pmatrix}$$

$$\mathbf{u}(2) = \begin{pmatrix} 1.3 \\ 0 \end{pmatrix} + \begin{pmatrix} 1 \\ -3 \end{pmatrix} = \begin{pmatrix} 2.3 \\ -3 \end{pmatrix}$$

$$\mathbf{u}(3) = \begin{pmatrix} 1.82 \\ 0.39 \end{pmatrix} + \begin{pmatrix} 1 \\ -3 \end{pmatrix} = \begin{pmatrix} 2.82 \\ -2.61 \end{pmatrix}$$

$$\mathbf{u}(4) = \begin{pmatrix} 1.911 \\ 0.585 \end{pmatrix} + \begin{pmatrix} 1 \\ -3 \end{pmatrix} = \begin{pmatrix} 2.911 \\ -2.415 \end{pmatrix}$$

$$\mathbf{u}(5) = \begin{pmatrix} 1.8889 \\ 0.6318 \end{pmatrix} + \begin{pmatrix} 1 \\ -3 \end{pmatrix} = \begin{pmatrix} 2.8889 \\ -2.3682 \end{pmatrix}$$

$$\mathbf{u}(6) = \begin{pmatrix} 1.866\,02 \\ 0.629\,85 \end{pmatrix} + \begin{pmatrix} 1 \\ -3 \end{pmatrix} = \begin{pmatrix} 2.866\,02 \\ -2.370\,15 \end{pmatrix}$$

Thus we obtain the approximate solution $\mathbf{u} = (2.866, \ -2.370)^T$, and an estimate for the error is provided by (10.33).

With the help of the Jordan normal form (8.11) we can obtain a more precise condition. Here it will be convenient to introduce the notion of convergence for a matrix. A sequence of $n \times n$ matrices $\mathbf{A}(r) = (a_{ij}(r))$ is said to **converge** to a matrix $\mathbf{A} = (a_{ij})$ if it converges component-wise, that is to say, if $a_{ij}(r) \to a_{ij}$ as $r \to \infty$.

Theorem 10.3

Let $\mathbf{C}$ *be an* $n \times n$ *matrix whose eigenvalues are all less than* 1 *in absolute value. Then the system*

$$\mathbf{x} = \mathbf{Cx} + \mathbf{b}$$

has a unique solution which can again be found by iteration as in (10.28).

Proof

We shall show that under the given conditions the matrix $\mathbf{I} - \mathbf{C}$ has an inverse, which is the sum of an infinite series

$$\mathbf{I} + \mathbf{C} + \mathbf{C}^2 + \cdots \qquad (10.34)$$

We begin by noting that $\mathbf{C}$ has the eigenvalue λ if and only if $\mathbf{I}-\mathbf{C}$ has the eigenvalue $1-\lambda$. For the condition for λ to be an eigenvalue of $\mathbf{C}$ is that $\mathbf{Cu}=\lambda\mathbf{u}$ for some $\mathbf{u}\neq\mathbf{0}$, and this is equivalent to the equation $(\mathbf{I}-\mathbf{C})\mathbf{u}=(1-\lambda)\mathbf{u}$. Since the eigenvalues of $\mathbf{C}$ are less than 1 in absolute value, $\mathbf{C}$ cannot have the eigenvalue 1, hence $\mathbf{I}-\mathbf{C}$ cannot have 0 as eigenvalue and it follows that $\mathbf{I}-\mathbf{C}$ is regular and so has an inverse. It only remains to show that the series (10.34) converges. For when this is so, then $\mathbf{C}^r\to\mathbf{0}$ and

$$(\mathbf{I}+\mathbf{C}+\mathbf{C}^2+\cdots+\mathbf{C}^{r-1})(\mathbf{I}-\mathbf{C})=\mathbf{I}-\mathbf{C}^r$$

and letting $r\to\infty$ we see that (10.34) is indeed the inverse of $\mathbf{I}-\mathbf{C}$.

Suppose first that $\mathbf{C}$ is a single Jordan block: $\mathbf{C}=\gamma\mathbf{I}+\mathbf{N}$, where $|\gamma|<1$ and $\mathbf{N}$ is a nilpotent matrix: $\mathbf{N}^n=\mathbf{0}$. We have

$$\mathbf{C}^r=(\gamma\mathbf{I}+\mathbf{N})^r=\gamma^r\mathbf{I}+\binom{r}{1}\gamma^{r-1}\mathbf{N}+\cdots+\binom{r}{n-1}\gamma^{r-n+1}\mathbf{N}^{n-1}$$

by the binomial theorem and the nilpotence of $\mathbf{N}$. To estimate the right-hand side, let m be an upper bound for the entries of $\mathbf{I},\mathbf{N},\mathbf{N}^2,\ldots,\mathbf{N}^{n-1}$, and $c=|\gamma|$. Then on writing $\mathbf{N}^r=(n_{ij}(r))$, we have $\binom{r}{i}\leqslant r^{n-1}$ for $i<n$ and so

$$\left|\gamma^r+\binom{r}{1}\gamma^{r-1}n_{ij}+\binom{r}{2}\gamma^{r-2}n_{ij}(2)+\cdots+\binom{r}{n-1}\gamma^{r-n+1}n_{ij}(n-1)\right|$$
$$\leqslant c^{r-n+1}r^{n-1}mn$$

It follows that

$$|c_{ij}(r)|\leqslant c^r r^{n-1}\frac{mn}{c^{n-1}}$$

It follows that the series (10.34) converges, by comparison with the convergent series $\sum c^r r^{n-1}$ for fixed n and $c<1$.

This establishes the result when $\mathbf{C}$ is a Jordan block; the same argument still applies when $\mathbf{C}$ is a diagonal sum of a number of Jordan blocks. In the general case we can write $\mathbf{C}=\mathbf{P}^{-1}\mathbf{BP}$, where $\mathbf{B}$ is a diagonal sum of Jordan blocks and has the same eigenvalues as $\mathbf{C}$. Hence $\sum\mathbf{B}^r$ converges, by what has been proved. Now if p is an upper bound for the entries of $\mathbf{P}=(p_{ij})$ and $\mathbf{P}^{-1}=(\bar{p}_{ij})$ and if we put $\mathbf{B}^r=(b_{ij}(r))$, then since $\mathbf{C}^r=\mathbf{P}^{-1}\mathbf{B}^r\mathbf{P}$, we have

$$|c_{ij}(r)|=\left|\sum\bar{p}_{ik}b_{kl}(r)p_{lj}\right|\leqslant np^2\max_{kl}\{|b_{kl}(r)|\}$$

and it follows that (10.34) converges by comparison with the convergent series $\sum\mathbf{B}^r$. This completes the proof.

It is of interest to note that in the proof we did not need the full Jordan form, but merely the reduction to a diagonal sum of terms $\gamma\mathbf{I}+\mathbf{N}$ with nilpotent **N**, which was relatively easy to accomplish; thus Lemma 2 of section 8.7 is not needed.

10.7 LINEAR DIFFERENCE EQUATIONS

The following problem was posed by Fibonacci (Leonardo of Pisa, son of Bonaccio, known as Fibonacci, *c.* 1175–*c.* 1250) in his *Liber abaci* in 1202. How many rabbits can be bred from a single pair in a given time if each pair gives birth to a new pair once a month from the second month onwards? Initially there is one pair; after one month there is still one pair, but a month later there are two pairs, and if at the end of n months there are u_n pairs, then after $n+1$ months there are in addition to the u_n pairs the offspring of the u_{n-1} pairs who were alive at $n-1$ months. Thus we have

$$u_{n+1} = u_n + u_{n-1} \tag{10.35}$$

Together with the initial data $u_0 = u_1 = 1$ this yields the sequence

$$1, 1, 2, 3, 5, 8, 13, 21, 34, \ldots$$

which is known as the **Fibonacci sequence**.

Equation (10.35) is an example of a difference equation and the general difference equation of order r can be written as

$$u(n) = a_1 u(n-1) + a_2 u(n-2) + \cdots + a_r u(n-r) \tag{10.36}$$

In order to solve it we write **T** for the translation operator

$$\mathbf{T}u(n) = u(n+1)$$

Then (10.36) can be written as

$$\mathbf{T}^r\mathbf{u} = a_1\mathbf{T}^{r-1}\mathbf{u} + \cdots + a_r\mathbf{u}$$

This leads to the equation

$$x^r - a_1 x^{r-1} - \cdots - a_r = 0$$

for the operator **T**, called the **auxiliary equation** for (10.36). We denote its roots by $\alpha_1 \ldots, \alpha_r$, so that

$$\mathbf{T}^r - a_1\mathbf{T}^{r-1} - \cdots - a_r = \prod_i (\mathbf{T} - \alpha_i)$$

The equation $(\mathbf{T} - \alpha)\mathbf{u} = \mathbf{0}$ has the solution $u(n) = \alpha^n u(0)$; so if $\alpha_1, \ldots, \alpha_r$ are all distinct, then we obtain the general solution of (10.36) as

$$u(n) = \sum_i \alpha_i^n p_i \tag{10.37}$$

where $p_1, \ldots, p_r$ are constants, determined by the initial conditions:

$$\sum_i \alpha_i^k p_i = u(k), \quad k = 0, 1, \ldots, r-1 \tag{10.38}$$

The determinant of this system is the Vandermonde determinant formed from $\alpha_1, \ldots, \alpha_r$. It is non-zero, because the αs were assumed to be distinct, so the equations (10.38) have a unique solution for any given initial conditions.

Let us see how this applies to the Fibonacci sequence. To solve (10.35) we have as auxiliary equation

$$x^2 - x - 1 = 0$$

Its roots are $\frac{1}{2}(1 \pm \sqrt{5})$, hence the solution is given by (10.37) in terms of p_1, p_2. To find their values, it is convenient to start the sequence with 0, taking $u(0) = 0$, $u(1) = 1$. Then the equations (10.38) become

$$p_1 + p_2 = 0$$
$$p_1(1 + \sqrt{5}) + p_2(1 - \sqrt{5}) = 2$$

Hence $p_1 = -p_2 = 1/\sqrt{5}$ and the solution is given by

$$u(n) = \frac{1}{\sqrt{5}}\left[\left(\frac{1+\sqrt{5}}{2}\right)^n - \left(\frac{1-\sqrt{5}}{2}\right)^n\right]$$

To deal with the case of equal roots, suppose that the auxiliary equation factorizes as $(x - \alpha)^r$. Let us write, for all n and r, where $r \leqslant n$,

$$n_r = n(n-1)\cdots(n-r+1)$$

Then we have

$$\begin{aligned}(n+1)_r - n_r &= (n+1)n(n-1)\cdots(n-r+2) - n(n-1)\cdots(n-r+1)\\ &= n(n-1)\cdots(n-r+2)r\\ &= rn_{r-1}\end{aligned}$$

Thus

$$(n+1)_r - n_r = rn_{r-1} \tag{10.39}$$

It follows that

$$(\mathbf{T} - \alpha)n_r\alpha^n = (n+1)_r\alpha^{n+1} - n_r\alpha^{n+1} = r\alpha n_{r-1}\alpha^n$$

Hence

$$(\mathbf{T} - \alpha)^r n_r \alpha^n = r!\alpha^r\alpha^n$$

and

$$(\mathbf{T} - \alpha)^r n_k \alpha^n = 0 \quad (k = 0, 1, \ldots, r-1)$$

Hence the general solution of the difference equation $(\mathbf{T} - \alpha)^r\mathbf{u} = \mathbf{0}$ is

$$u(n) = \alpha^n \sum_{k=0}^{r-1} n_k c_k$$

where $c_0, c_1, \ldots, c_{r-1}$ are arbitrary constants.

Now it is clear how the general case, with several possibly multiple roots, is to be treated. As an example we take the equation

$$u(n) = 3u(n-2) + 2u(n-3)$$

subject to $u(0) = u(1) = 0$, $u(2) = 9$. The auxiliary equation is

$$x^3 - 3x - 2 = 0$$

which factorizes as $(x-2)(x+1)^2$. Hence the general solution is

$$u(n) = p_1 2^n + p_2(-1)^n + np_3(-1)^n$$

with the initial conditions

$$\begin{aligned} p_1 + p_2 &= 0 \\ 2p_1 - p_2 - p_3 &= 0 \\ 4p_1 + p_2 + 2p_3 &= 9 \end{aligned}$$

Solving these equations, we find $p_1 = 1, p_2 = -1, p_3 = 3$, so the solution takes the form

$$u(n) = 2^n + (3n-1)(-1)^n$$

EXERCISES

10.1. Show that the function $\mathbf{u} = (x_1 + x_2 + x_3, x_1^2 + x_2^2 + x_3^2, x_1^3 + x_2^3 + x_3^3)$ has an inverse near any point (x_1, x_2, x_3) whose coordinates are all different.

10.2. Show that if $f(z)$ is a holomorphic function of z, then so is $\overline{f(\bar{z})}$.

10.3. Let $y_1, \ldots, y_r$ be smooth functions of $N > r$ variables $x_1, \ldots, x_N$ such that the $r \times N$ matrix $(\partial y_i/\partial x_j)$ has rank r at a given point $\mathbf{x} = \mathbf{p}$. Apply the inverse function theorem to show that after suitable renumbering of the xs, $x_1, \ldots, x_r$ can be expressed as functions of $x_{r+1}, \ldots, x_N$ and the ys. (This is the implicit function theorem.)

10.4. Find the critical points of the following functions and determine their nature. (i) $x_1x_2 + x_1x_3 + x_2x_3$. (ii) $x_1(x_2+1)(x_1x_2+1)$.

10.5. Find the normal form of the following function at its critical points, using the splitting lemma: $-x_1^2 - 2x_2^2 + 2x_4^2 - 2x_1x_2 - 2x_1x_3 - 2x_1x_4 - 2x_2x_3 - 4x_2x_4 - 2x_2x_5 + 2x_3x_4 - 2x_3x_5 - 6x_4x_5 + x_4^2x_5^2 - x_5^4$.

10.6. A region contains two animal populations, x_1, the predators, and x_2, the prey, which are related by the equations

$$\begin{aligned} \dot{x}_1 &= 4x_1 + x_2 \\ \dot{x}_2 &= -x_1 + 2x_2 \end{aligned}$$

where the unit of time is a year. Initially $x_1 = 500$, $x_2 = 100$; how long will it take for the prey to disappear?

10.7. Verify that the Wronskian W of the differential equation

$$y^{(n)} + a_1 y^{(n-1)} + \cdots + a_n y = 0$$

satisfies $W' = -a_1 W$. Deduce that $W = ce^{-b(x)}$, where $b(x) = \int a_1(x)\mathrm{d}x$ (Abel's formula).

10.8. Solve the system $\mathbf{x} = \mathbf{C}\mathbf{x} + \mathbf{b}$, where $\mathbf{b} = (1, -3, 5)^{\mathrm{T}}$, and

$$\mathbf{C} = \begin{pmatrix} -4 & -5 & -3 \\ 1 & 1 & 1 \\ 5 & 7 & 3 \end{pmatrix}$$

given that $\mathbf{C}$ is nilpotent.

10.9. If $u_1, u_2, \ldots$ denotes the Fibonacci sequence, show that $u_{n+1}/u_n \to \frac{1}{2}(1 + \sqrt{5})$, and find the first term in the sequence to exceed a million.

10.10. For any function $f(x)$ define the translation operator $\mathbf{T}$ as in section 10.7 by $\mathbf{T}f(x) = f(x+1)$ and define the differencing operator $\mathbf{D}$ by $\mathbf{D}f(x) = (\mathbf{T} - 1)f(x) = f(x+1) - f(x)$. Show that $f(n) = \sum \binom{n}{r} a_r$, where $a_r = \mathbf{D}^r f(0)$.

Answers to the exercises

Page v (Dedication): The great[7] grandparents are the ninth generation, representing the fraction $2 \times 2^{-9} = 0.0039$. the fraction indicated on p.v, 1.37%, is $3\frac{1}{2}$ times this amount. This discrepancy is explained by the fact that (because of cousin marriages) the couple were great[7] grandparents twice over and great[8] grandparents three times over, and $2 + \frac{3}{2} = 3\frac{1}{2}$.

CHAPTER 1

1.2. (ii) $(7,8,9) = 2(4,5,6) - (1,2,3)$. (iii) $(0,0,0,0) = \mathbf{0}$. (iv) $(4,0,3,2) = 3(2,1,-1,3) + (-2,-3,6,-7)$.

1.3. $(-15,25,10,-40) = -5/3(9,-15,-6,24)$.

1.4. Basis of $\mathbf{C}$: $1, \mathrm{i}$. $\mathbf{C}^3$: $\mathbf{e}_1, \mathbf{e}_2, \mathbf{e}_3, \mathrm{i}\mathbf{e}_1, \mathrm{i}\mathbf{e}_2, \mathrm{i}\mathbf{e}_3$.

1.5. The points represented by the vectors are not collinear with the origin. In 3 dimensions the three points represented by the vectors are coplanar with the origin.

1.7. Any two vectors except the second $((1,1,1,1))$.

1.8. Basis: $(0,1,0,0), (0,0,1,1)$. The vectors with $x_3 = 1$ do not include 0.

1.10. No: $x - x^2$ and x^2 are both of even degree, but their sum is not.

1.11. If $\lambda\mathbf{x} + \sum \mu_i \mathbf{y}_i = \mathbf{0}$, where $\mathbf{y}_i \in S$, and $\lambda \neq 0$, then we can divide by λ and obtain a dependence of $\mathbf{x}$ on the vectors $\mathbf{y}_i$ of S, contradicting the fact that $\mathbf{x} \notin U$. Hence $\lambda = 0$, and now all the μ_i vanish, by the linear independence of the $\mathbf{y}_i$.

1.12. If neither of U', U'' contains the other, take $\mathbf{x}$ in U' but not in U'' and $\mathbf{y}$ in U'' but not in U', then $\mathbf{z} = \mathbf{x} + \mathbf{y} \notin U' \cup U''$. For if $\mathbf{z} \in U'$, then $\mathbf{y} = \mathbf{z} - \mathbf{x} \in U'$, a contradiction; similarly $\mathbf{z} \notin U''$, so $U' \cup U''$ is not a subspace.

1.15. Take a basis $\mathbf{u}_1, \ldots, \mathbf{u}_r$ of U and adjoin vectors $\mathbf{u}_{r+1}, \ldots$ so as to have a linearly independent set. Since V is of finite dimension, say n, the process ends with $\mathbf{u}_n$ and $\mathbf{u}_1, \ldots, \mathbf{u}_n$ is then a basis of V. Let U' be the subspace spanned by $\mathbf{u}_{r+1}, \ldots, \mathbf{u}_n$. If $\mathbf{z} \in U \cap U'$, then $\mathbf{z} = \lambda_1\mathbf{u}_1 + \cdots + \lambda_r\mathbf{u}_r = \mu_{r+1}\mathbf{u}_{r+1} + \cdots + \mu_n\mathbf{u}_n$, hence $\sum \lambda_i\mathbf{u}_i - \sum \mu_j\mathbf{u}_j = \mathbf{0}$ and by linear independence $\lambda_i = \mu_j = 0$, so $U \cap U' = 0$. Any vector in V is a linear combination of $\mathbf{u}_1, \ldots, \mathbf{u}_n$, hence the sum of a vector in U and one in U'.

1.16. Define $\lambda\mathbf{u} = \mathbf{0}$ for all $\lambda, \mathbf{u}$; then V.1–8 hold, but not V.9.

1.17. $(1,0,1,0,\ldots)$ is in V_1 but not in V_2, $(1,1,1,\ldots)$ is in V_2 but not in V_3, $(1,1/2,1/3,\ldots)$ is in V_3 but not in V_4.

1.18. Given $V = V_0 \supset V_1 \supset \cdots \supset V_r = 0$, let $\mathbf{v}_i$ be in V_{i-1} but not in $\mathbf{V}_i$, then $\mathbf{v}_1, \ldots, \mathbf{v}_r$ are linearly independent, by Exercise 1.11. Define W_n as the subspace of V_4 of vectors with the first n components zero. Then each W_n is a subspace and $V_4 \supset W_1 \supset W_2 \supset \cdots$.

CHAPTER 2

2.1. (The solution is written as a row vector in each case.) (i) $(2, -1, 5)$. (ii) $(-3, 1, 0)$. (iii) $(4, -2, 1)$. (iv) $(26, 5, -11.4)$. (v) $(0, 0, 0)$. (vi) $(1.5, 1, -1, 2)$. (vii) $(5, -1, -2, 1)$. (viii) $(-208, -141, -9, 13)$.

2.4. Observe that the space spanned by the rows is unchanged by elementary operations.

2.6. Type γ alone is not enough to accomplish an interchange (because the latter, unlike γ, changes the determinant, cf. Chapter 5).

2.7. Subtracting the first row from the second, we get $(-1, -4)$. We now have to operate with this row, not the old second row.

2.8. By equating to zero the coefficients of $\lambda_0 f + \lambda_1 f' + \cdots + \lambda_{n-1} f^{(n-1)} = 0$ we obtain a triangular system with diagonal terms $(n-1)\ldots(n-r)a_0$, where a_0 is the coefficient of x^{n-1} in f.

2.9. Verify that a minimal spanning set is linearly independent by using a non-trivial dependence relation to find a smaller spanning set.

CHAPTER 3

3.2. (a) (i), (ii), (iii), (iv); (b) (i), (iii), (iv); (c) (iii).

3.3. $\mathbf{A}^2 - \mathbf{AB} - \mathbf{BA} + \mathbf{B}^2$.

3.4. $\begin{pmatrix} a & b \\ -a^2/b & -a \end{pmatrix}$ for all a, b subject to $b \neq 0$ and $\begin{pmatrix} 0 & 0 \\ c & 0 \end{pmatrix}$ for all c.

3.5. $\begin{pmatrix} -11 & -15 \\ 9 & -14 \end{pmatrix}$, $\mathbf{A}^2 - 3\mathbf{A} + 17\mathbf{I} = \mathbf{0}$.

3.7.

$$\mathbf{u}\mathbf{u}^{\mathrm{T}} = \begin{pmatrix} 9 & -6 & 3 & 12 \\ -6 & 4 & -2 & -8 \\ 3 & -2 & 1 & 4 \\ 12 & -8 & 4 & 16 \end{pmatrix}, \quad \mathbf{u}^{\mathrm{T}}\mathbf{u} = (30).$$

3.8. If $\mathbf{A} = (a_{ij})$, $\mathbf{B} = (b_{ij})$, then $\mathrm{tr}(\mathbf{AB}) = \sum_{ij} a_{ij} b_{ji} = \sum_{ij} b_{ji} a_{ij} = \mathrm{tr}(\mathbf{BA})$.

3.11. (i) $\begin{pmatrix} -2 & -1 & 3 \\ 1 & 0 & -1 \\ 1 & 2 & -2 \end{pmatrix}$; (ii) $\dfrac{1}{60}\begin{pmatrix} -14 & 55 & 80 \\ 10 & -35 & -40 \\ 2 & -25 & -20 \end{pmatrix}$;

(iii) $\frac{1}{6}\begin{pmatrix} 20 & -1 & -12 \\ -9 & 3 & 3 \\ 2 & -1 & 0 \end{pmatrix}$; (iv) $\begin{pmatrix} 5.5 & 9.5 & -5 \\ 1 & 2 & -1 \\ -2.5 & -4.5 & 2.4 \end{pmatrix}$;

(v) $\frac{1}{2}\begin{pmatrix} 7 & -2 & 17 \\ -6 & 2 & -16 \\ 7 & -2 & 19 \end{pmatrix}$; (vi) $\frac{1}{8}\begin{pmatrix} 26 & 1 & -17 & -30 \\ -72 & 0 & 40 & 64 \\ -4 & -2 & 10 & 20 \\ -36 & -2 & 26 & 44 \end{pmatrix}$;

(vii) $\frac{1}{6}\begin{pmatrix} 3 & 3 & 4 & -14 \\ 0 & 6 & 0 & 30 \\ 0 & 0 & -2 & 4 \\ 0 & 0 & 0 & 6 \end{pmatrix}$ (viii) $\frac{1}{4}\begin{pmatrix} 66 & -140 & -64 & 8 \\ 43 & -92 & -44 & 4 \\ 1 & -4 & -4 & 0 \\ -4 & 8 & 4 & 0 \end{pmatrix}$

3.14. $\mathbf{A} = (1\ \ 0) = \mathbf{B}^{\mathrm{T}}$; these matrices satisfy $\mathbf{AB} = \mathbf{I}_1$.

3.16. If $\mathbf{AE}_{ij} = \mathbf{E}_{ij}\mathbf{A}$ for all i, j, then $\mathbf{A} = \alpha\mathbf{I}$. If $\mathbf{AE}_{ij} = \mathbf{E}_{ji}\mathbf{A}$ for all i, j, then $\mathbf{A} = \mathbf{0}$.

3.17. Left multiplication by $\mathbf{B}_{ij}(a)$ adds a times row j to row i, right multiplication adds a times column i to column j. $\mathbf{B}_{12}(\lambda)\mathbf{B}_{21}(-\lambda^{-1})\mathbf{B}_{12}(\lambda) = \begin{pmatrix} 0 & \lambda \\ -\lambda^{-1} & 0 \end{pmatrix}$; hence $\mathbf{B}_{12}(\lambda)\mathbf{B}_{21}(-\lambda^{-1})\mathbf{B}_{12}(\lambda)(\mathbf{B}_{12}(1)\mathbf{B}_{21}(-1)\mathbf{B}_{12}(1))^3 = \mathrm{diag}(\lambda, \lambda^{-1})$.

3.19. Draw the graph with adjacency matrix $\mathbf{A}$ and take $\mathbf{B}$ to be its incidence matrix.

CHAPTER 4

4.2. Use Theorem 2.3.

4.3. (i) $\frac{1}{2}(-1, 1, 0) + \lambda(5, -3, 2)$. (ii) $\lambda(-14, 11, 16)$. (iii) No solution. (iv) $(2, -1, 1, 0) + \lambda(12, 1, -20, 7)$. (v) $(0, 1, -1, 1) + \lambda(1, -3, 0, 0)$, (vi) $(0, 0, 1, 1)$. (vii) $(1, 0, 0, 0) + \lambda(-1, 1, 1, 0) + \mu(0, 0, -2, 1)$.

4.4. (i) $t = 6$: $\frac{1}{7}(-10, 13, 0, 0) + \lambda(-10, 6, 7, 0) + \mu(1, -1, 0, 1)$. (ii) $t = 1$: $(0, 1, 0, 0) + \lambda(-2, 0, 1, 0) + \mu(-2, 1, 0, 1)$; $t = -\frac{2}{5}$: $\frac{1}{25}(21, -17, 0, 0) + \lambda(-2, 0, 1, 0) + \mu(-2, 1, 0, 1)$. (iii) $t = 5$: $\lambda(5, 1, -4)$.

4.5. (i) $\mathbf{x} = \frac{1}{7}(12, 5)$. (ii) $\mathbf{x} = (9, -14)$. (iii) No solution (two parallel lines).

4.6. 0 or 1.

4.7. If $r = m$, the system $\mathbf{Ax} = \mathbf{k}$ has a solution for all $\mathbf{k}$, so $\mathbf{B}$ exists satisfying $\mathbf{AB} = \mathbf{I}$; for $r = n$ apply the previous result to $\mathbf{A}^{\mathrm{T}}$.

4.8. 2, 3, 3.

4.9. 2 for $t = 0, -4$, otherwise 3; 1 for $t = 0$, 2 for $t = -3$, otherwise 3; 2 for all t.

4.10. If the system has at least one solution, the columns span $\mathbf{R}^m$, so $n \geqslant m$; if the system has at most one solution, the columns are linearly independent, so $n \leqslant m$.

4.13. (i) $(1,-1,2,4,3)$, $(2,5,-3,5,7)$; (ii) linearly independent.

N.B.: The solutions to questions 4.3 and 4.4 can take different forms, depending on the choice of the particular solution and the basis for the solution of the associated homogeneous system.

CHAPTER 5

5.1. (ii), (iv), (vi), (viii), (x), (xi): 1; (i), (iii), (v), (vii), (ix), (xii): -1.

5.2. (i) 1; (ii) -60; (iii) 12; (iv) 10; (v) 2; (vi) -8; (vii) -6; (viii) 4.

5.4. (a) (i) 2; (ii) 2; (iii) 2; (iv) 3; (v) 3; (vi) 4; (vii) 2. (b) (i) 2; (ii) 2; (iii) 2 if $t=5$, otherwise 3.

5.5. (i) -20; (ii) -1; (iii) 0; (iv) -16; (v) -1440.

5.6. (i) V; (ii) $abcdV$; (iii) $(a+b+c+d)V$; where $V=(d-c)(d-b)(d-a)\cdot(c-b)(c-a)(b-a)$. (iv) $(af-be+cd)^2$.

5.9. Take determinants in (5.15).

5.10. If an $n\times n$ matrix has an $r\times s$ block of zeros, it has the form

$$\begin{pmatrix}\mathbf{A} & \mathbf{0}\\ \mathbf{C} & \mathbf{B}\end{pmatrix}=\begin{pmatrix}\mathbf{A} & \mathbf{0}\\ \mathbf{C} & \mathbf{I}\end{pmatrix}\begin{pmatrix}\mathbf{I} & \mathbf{0}\\ \mathbf{0} & \mathbf{B}\end{pmatrix},$$

where the factors on the right are $n\times(2n-r-s)$ and $(2n-r-s)\times n$. So for $r+s>n$ the inner rank is less than n. If $r+s=n$, then the diagonal sum $\mathbf{A}\oplus\mathbf{B}$ is $n\times n$ and has an $r\times s$ block of zeros, and it has non-zero determinant when $\mathbf{A},\mathbf{B}$ are both regular.

5.11. The columns of the non-zero rth-order minor are linearly independent, but any other column is linearly dependent on them.

5.12. Add $\mathbf{A}$ times the last n rows to the first, then subtract the last n columns times $\mathbf{B}$ from the first.

CHAPTER 6

6.2. If $\mathbf{a},\mathbf{b},\mathbf{c}$ are vectors representing the sides, then $\mathbf{c}=\mathbf{a}-\mathbf{b}$. Now work out $\mathbf{c}\cdot\mathbf{c}=(\mathbf{a}-\mathbf{b})\cdot(\mathbf{a}-\mathbf{b})$.

6.3. Use (6.11) and (6.9).

6.6. $(7,3,5)$.

6.8. $\frac{7}{13}(4,3-1)$, $7\sqrt{2/13}$.

6.9. Let the vertices be A,B,C,D with position vectors $\mathbf{a},\mathbf{b},\mathbf{c},\mathbf{d}$, respectively. Then the mid-points of AB and CD are $\frac{1}{2}(\mathbf{a}+\mathbf{b})$ and $\frac{1}{2}(\mathbf{c}+\mathbf{d})$, respectively, and the mid-point of their join is $\frac{1}{4}(\mathbf{a}+\mathbf{b}+\mathbf{c}+\mathbf{d})$.

6.10. (i) $(1,1,1)+\lambda(1,0,0)$, $(1,1,1)+\lambda(0,1,0)$, $(1,1,1)+\lambda(0,0,1)$; (ii) $(1,1,1)+\lambda(0,1,1)$, $(1,1,1)+\lambda(1,0,1)$, $(1,1,1)+\lambda(1,1,0)$; (iii) $(1,1,0)+\lambda(0,-1,1)$, $(1,0,1)+\lambda(-1,1,0)$, $(0,1,1)+\lambda(1,0,-1)$.

6.11. (i) $(1,1,1)+\lambda(3,1,2)$; (ii) $(2,0,1)+\lambda(5,4,1)$; (iii) $(1,4,2)+\lambda(-1,1,1)$; (i) and (ii) 30°; (i) and (iii) 90°; (ii) and (iii) 90°.

6.12. $1/\sqrt{3}$ (angle 54.7356°), 1/3 (angle 70.5288°).

6.13. (i) $3x_1 - 10x_2 + 6x_3 = 1$; (ii) $7x_1 + 13x_2 + 14x_3 = 75$; (iii) $53x_1 + 5x_2 - 3x_3 = 54$. (i) and (ii) $(1,2,3) + \lambda(2,0,-1)$; (i) and (iii) $(1,2,3) + \lambda(0,3,5)$; (ii) and (iii) $(1,2,3) + \lambda(1,-7,6)$.

6.14. (i) $(1,2,-2)$, $(-1,-2,2)$; (ii) $\frac{1}{3}(-1,2,2)$, $\frac{1}{7}(6,-2,3)$.

6.15. $\lambda + \mu + \nu = 1$.

6.16. $\mathbf{a} + \mathbf{b} - 2\mathbf{c}$, $\mathbf{b} + \mathbf{c} - 2\mathbf{a}$, $\mathbf{c} + \mathbf{a} - 2\mathbf{b}$, $\mathbf{a} + \mathbf{b} + \mathbf{c}$.

6.21. $f_0 = 1/\sqrt{2}$, $f_1 = \sqrt{\frac{3}{2}}x$, $f_2 = \frac{1}{2}\sqrt{\frac{5}{2}}(3x^2 - 1)$, $f_3 = \frac{1}{2}\sqrt{\frac{7}{2}}(5x^3 - 3x)$, $f_4 = \frac{15}{8\sqrt{2}}(7x^4 - 6x^2 + \frac{3}{5})$.

CHAPTER 7

7.1. $\mathbf{x} = \mathbf{A}\mathbf{y}$, where $\mathbf{A} = \begin{pmatrix} 0 & 0 & 1 \\ 1 & 0 & 0 \\ 0 & 1 & 0 \end{pmatrix}$.

7.2. Reflexion in the plane through O perpendicular to $\mathbf{a}$.

7.3. $(-\mathbf{I})^2 = \mathbf{I}$, $|-\mathbf{I}| = (-1)^n$.

7.4. One eigenvalue of $\mathbf{A}$ is -1; if the other two are complex conjugate, then $|\mathbf{A}| = -1$ and $\mathbf{A}$ is improper. Similarly if the other two are α, α^{-1}; otherwise they are $1, -1, -1$, but the matrix $\text{diag}(1,-1,-1)$ represents a rotation of 180° about the 1-axis.

7.5. Define $\mathbf{I} + \mathbf{S}$ as $2(\mathbf{I} + \mathbf{A})^{-1}$ and verify that $\mathbf{S} + \mathbf{S}^{\mathrm{T}} = \mathbf{0}$, $\mathbf{A}(\mathbf{I} + \mathbf{S}) = \mathbf{I} - \mathbf{S}$.

7.12. AB and CD: $8/\sqrt{13}$; AC and BD: $\sqrt{5}$; AD and BC: $8/\sqrt{5}$.

7.13. Use the fact that a square matrix is regular (and hence invertible) if and only if its columns are linearly independent.

7.14. $\lambda a \mapsto \overline{\lambda a} = \bar{\lambda}\bar{a}$; this is a linear mapping only if $\bar{\lambda} = \lambda$, i.e. λ is real.

7.15. $\mathbf{T}^{-1}: f(x) \mapsto f(x-1)$.

7.16 Every polynomial is a derivative, but $\ker D$ is 1-dimensional, consisting of all constants. The definite integral, taken from 0 to x, is a linear mapping, injective but not surjective: the image does not include non-zero constants.

CHAPTER 8

8.1. For suitable $\mathbf{P}$, $\mathbf{P}^{-1}\mathbf{A}\mathbf{P}$ is upper triangular, for others it is lower triangular, so $\mathbf{A}$ must be both, that is, diagonal. The diagonal elements can be permuted by similarity transformation, so all are equal and $\mathbf{A} = \alpha\mathbf{I}$.

8.2. The only possible diagonal matrix is $\mathbf{0}$, but this is clearly ruled out.

8.3. $\text{diag}(1+i, 1-i)$, transforming matrix: $\begin{pmatrix} 1 & 1 \\ 1 & -1 \end{pmatrix}$.

8.4. $\text{diag}(-1,-4)$, impossible, $\text{diag}(0,1,2)$, $\text{diag}(3,2,2)$, via transforming matrices:

$$\begin{pmatrix}1 & 1\\ 2 & -1\end{pmatrix}, -, \begin{pmatrix}-2 & 1 & -3\\ 1 & 0 & 1\\ 1 & -2 & 2\end{pmatrix},$$

$$\begin{pmatrix}-2 & 1 & 2\\ 1 & 0 & -1\\ 1 & -2 & 0\end{pmatrix} \text{ or } \begin{pmatrix}-2 & 1 & -3\\ 1 & 0 & 1\\ 1 & -2 & 2\end{pmatrix}$$

8.5. Characteristic equation $x^2 - 2x + 1 = 0$, hence $\mathbf{A}^2 = 2\mathbf{A} - \mathbf{I}$. If $\mathbf{A}^n = n\mathbf{A} - (n-1)\mathbf{I}$, then $\mathbf{A}^{n+1} = n\mathbf{A}^2 - (n-1)\mathbf{A} = n(2\mathbf{A} - \mathbf{I}) - (n-1)\mathbf{A} = (n+1)\mathbf{A} - n\mathbf{I}$. Hence the result follows by induction on n.

8.6. $f_i(\alpha_j) = \delta_{ij}$, hence $\sum b_i f_i(\alpha_j) = b_j$.

8.7. (i) $\mathbf{P} = \mathbf{I}^{-1}\mathbf{PI}$. (ii) If $\mathbf{S}^{-1}\mathbf{PS} = \mathbf{Q}$, then $\mathbf{SQS}^{-1} = \mathbf{P}$. (iii) If $\mathbf{S}^{-1}\mathbf{PS} = \mathbf{Q}$ and $\mathbf{T}^{-1}\mathbf{QT} = \mathbf{R}$, then $(\mathbf{ST})^{-1}\mathbf{P}(\mathbf{ST}) = \mathbf{R}$.

8.8. If $f = \sum_0^n a_i x^i$, then $\mathbf{T}f = \sum_0^n a_{n-i} x^i$ and $\mathbf{T}^2 f = f$. Hence $\mathbf{T}^2 = \mathbf{I}$ and $\mathbf{T}$ has eigenvalues 1 and -1. The eigenspace for 1 has basis $x^n + 1, x^{n-1} + x, \ldots$; and that for -1 has basis $x^n - 1$, $x^{n-1} - x, \ldots$.

8.9. If $\mathbf{x}^{\mathrm{T}}\mathbf{Cx} = 0$, then $\sum c_{ij} x_i x_j = 0$, so the coefficient of $x_i x_j$, $c_{ij} + c_{ji}$, vanishes. Thus $\mathbf{C}^{\mathrm{T}} + \mathbf{C} = \mathbf{0}$, that is, $\mathbf{A}^{\mathrm{T}} - \mathbf{B}^{\mathrm{T}} + \mathbf{A} - \mathbf{B} = \mathbf{0}$. $\mathbf{A}$ need not equal $\mathbf{B}$, for example $\mathbf{A} = \mathbf{E}_{12}$, $\mathbf{B} = \mathbf{E}_{21}$, where $\mathbf{E}_{ij}$ is as in Exercise 3.15.

8.10. $\mathrm{tr}(\mathbf{ABC}) = \sum a_{ij} b_{jk} c_{ki} = \sum b_{jk} c_{ki} a_{ij} = \mathrm{tr}(\mathbf{BCA})$. If $\mathbf{A} = \mathbf{E}_{12}$, $\mathbf{B} = \mathbf{E}_{21}$, $\mathbf{C} = \mathbf{E}_{11}$, then $\mathbf{ABC} = \mathbf{E}_{11}$, $\mathbf{BAC} = \mathbf{0}$, so $\mathrm{tr}(\mathbf{ABC}) = 1$, $\mathrm{tr}(\mathbf{BAC}) = 0$.

8.11. $\mathbf{A} = (a_{ij})$, where $a_{ij} = \delta_{j,i+1}$ for $i = 1, \ldots, n-1$ and $j = 1, \ldots, n$, while the last row is $a_{nj} = a_j$.

8.13. $\mathbf{T} = (t_{ij})$, where $t_{ij} = \delta_{j,n+1-i}$ as in section 8.3.

8.15. $\mathbf{A}_1 \oplus \mathbf{A}_2$, where $\mathbf{A}_1 = 1^2 (= 1 \oplus 1)$ or $\mathbf{J}_2(1)$ and $\mathbf{A}_2 = 2^4$ or $\mathbf{J}_2(2) \oplus \mathbf{J}_2(2)$ or $\mathbf{J}_2(2) \oplus 2^2$ or $\mathbf{J}_3(2) \oplus 2$ or $\mathbf{J}_4(2)$.

8.16. $\mathbf{A}_1 \oplus \mathbf{A}_{-1}$, where (i) $\mathbf{A}_1 = \mathbf{J}_2(1)$ and $\mathbf{A}_{-1} = \mathbf{J}_2(-1)^2$ or $\mathbf{J}_2(-1) \oplus (-1)^2$, or (ii) $\mathbf{A}_1 = \mathbf{J}_2(1) \oplus 1$ and $\mathbf{A}_{-1} = \mathbf{J}_2(-1) \oplus (-1)$, or (iii) $\mathbf{A}_1 = \mathbf{J}_2(1)^2$ or $\mathbf{J}_2(1) \oplus 1^2$ and $\mathbf{A}_{-1} = \mathbf{J}_2(-1)$.

8.17. $\mathbf{J}_3(1) \oplus 1^3$ or $\mathbf{J}_3(1) \oplus \mathbf{J}_2(1) \oplus 1$ or $\mathbf{J}_3(1)^2$.

8.18. The expression is of degree n in x and vanishes for $x = \alpha_1, \ldots, \alpha_n$.

CHAPTER 9

9.1. (i) $7y_1^2 + 7y_2^2 - 7y_3^2$, where $\sqrt{5}y_1 = x_1 + 2x_2$, $\sqrt{70}y_2 = -6x_1 + 3x_2 + 5x_3$, $\sqrt{14}y_3 = 2x_1 - x_2 + 3x_3$; (ii) $2y_1^2 + 2y_2^2 + 8y_3^2$, where $\sqrt{3}y_1 = x_1 + x_2 + x_3$, $\sqrt{2}y_2 = x_1 - x_2$, $\sqrt{6}y_3 = x_1 + x_2 - 2x_3$; (iii) $y_1^2 - \frac{1}{2}y_2^2 - \frac{1}{2}y_3^2$ with the same transformation as in (ii) (in each case the transformation can take different forms, although the diagonal form itself is uniquely determined, up to a permutation of the variables).

9.2. $\mathbf{Ap} + \mathbf{a} = \mathbf{0}$ means that $\mathbf{a}$ is the centre.

9.4. (i) $x_1 + x_3 = \pm 1$; (ii) $2x_1 + x_2 - 2x_3 = \pm 6$; (iii) $4x_1 + 3x_2 + 3x_3 = \pm 2$.
9.5. (i) $1, -1, 2$; (ii) $1, -2, 4$; (iii) $1, 3, 10$.
9.6. (i) Two-sheeted hyperboloid; (ii) ellipsoid; (iii) one-sheeted hyperboloid; (iv) cone; (v) no surface.
9.7. **B** may be singular.
9.8. If $\mathbf{Au} = \alpha\mathbf{u}$, then $\bar{\mathbf{u}}^T\mathbf{A}^T = \bar{\mathbf{u}}^T\bar{\alpha}$, hence $\bar{\mathbf{u}}^T\mathbf{A}\mathbf{u} = \alpha\bar{\mathbf{u}}^T\mathbf{u} = -\bar{\alpha}\bar{\mathbf{u}}^T\mathbf{u}$, so $(\alpha + \bar{\alpha})\bar{\mathbf{u}}^T\mathbf{u} = 0$.
9.13. If $\mathbf{A}, \mathbf{B}$ can be simultaneously transformed to diagonal forms $\mathbf{A}', \mathbf{B}'$, respectively, then $\mathbf{A}' + \mathbf{B}'c$ is positive definite for some c, hence the same is true for **A**, **B**.

9.14. $\begin{pmatrix} 3 & -2 & 1 \\ -2 & 6 & -2 \\ 1 & -2 & 3 \end{pmatrix}$

9.15. $\frac{1}{3}\begin{pmatrix} 2 & 1 & 2 \\ 1 & 2 & -2 \\ -2 & 2 & 1 \end{pmatrix}\begin{pmatrix} 6 & 3 & 3 \\ 3 & 6 & 3 \\ 3 & 3 & 6 \end{pmatrix}$

9.16. $\frac{1}{3}\begin{pmatrix} 2 & 1 & 2 \\ 1 & 2 & -2 \\ -2 & 2 & 1 \end{pmatrix}\begin{pmatrix} 3 & 3 & 6 \\ 0 & 6 & 3 \\ 0 & 0 & 9 \end{pmatrix}$

9.17. Apply Theorem 9.6 and 9.7 to $\mathbf{A}^T$.
9.18. 13 for $\mathbf{x} = (3, 0, 4, 3)$.
9.19. $y = 3.06x + 0.90$.

CHAPTER 10

10.1. The Jacobian has determinant $6V(x_1, x_2, x_3)$.
10.4. (i) Morse 1-saddle at $\mathbf{x} = \mathbf{0}$; (ii) $(0, -1)$, $(1, -1)$ saddle-point.
10.5. $V = -y_1^2 - y_2^2 + y_3^2 + y_4^2 y_5^2$; $y_1 = x_1 + x_2 + x_3 + x_4$, $y_2 = x_2 + x_4 + x_5$, $y_3 = x_3 + 2x_4 - x_5$, $y_4 = x_4 - x_5$, $y_5 = x_5$.
10.6. 2 months.
10.8. $\mathbf{x} = (\mathbf{I} + \mathbf{C} + \mathbf{C}^2)\mathbf{b} = (1, -2, 2)^T$.
10.9. If $u = (1 + \sqrt{5})/2$, then $\sqrt{5}u(n) = u^n - (-u^{-1})^n$, $|u^{-1}| < 1$, hence $u(n+1)/u(n) \to u$. $u(31) = 1\,346\,269$.

Notation and symbols used

$\sum$	summation sign	5
$\mathbf{R}$	real numbers	8
$\mathbf{C}$	complex numbers	8
$\mathbf{R}^n, \mathbf{C}^n$	n-space of row vectors	8
$\in$	membership sign	8
$\lvert a \rvert$	absolute value	9
$\subseteq$	is contained in	11
$\subset$	is properly contained in	11
$\supseteq$	contains	11
$\supset$	contains properly	11
0	zero space	11
$\mathbf{e}_i$	standard basis of $\mathbf{R}^n$	13
$[a, b]$	interval $a \leqslant x \leqslant b$	14
$\cap$	intersection	18
$\cup$	union	18
$U + U'$	sum of subspaces	18
${}^n\mathbf{R}$	n-space of column vectors	20
^{T}	transpose	20
$\mathbf{A} \mid \mathbf{k}$	scheme in matrix form	24
$\dim V$	dimension of V	30
${}^m\mathbf{R}^n$	set of $m \times n$ matrices over $\mathbf{R}$	34
δ_{ij}	Kronecker δ	39
$\mathbf{I}, \mathbf{I}_r$	unit matrix (of order r)	39
$\mathbf{A} \oplus \mathbf{B}$, $\mathrm{diag}(\mathbf{A}, \mathbf{B})$	diagonal sum of matrices	41
$\mathbf{A}^{-1}$	inverse of $\mathbf{A}$	42
$\mathbf{E}_{ij}$	matrix unit	50
$\mathrm{rk}(\mathbf{A})$	rank of $\mathbf{A}$	53
$\min(m, n)$	the smaller of m, n	53
$(\mathbf{A}, \mathbf{k})$	augmented matrix	57
$n!$	n factorial	70
$\varepsilon(i_1 \cdots i_n)$	Kronecker ε function	71

$(i\ j)$	transposition of i and j	72
$\det \mathbf{A}, \lvert\mathbf{A}\rvert$	determinant of the matrix $\mathbf{A}$	73
$\operatorname{adj}(\mathbf{A})$	adjoint of $\mathbf{A}$	79
$\binom{n}{r}$	binomial coefficient	84
$\lvert\mathbf{a}\rvert$	length of the vector $\mathbf{a}$	90
$\mathbf{a}\cdot\mathbf{b}$	scalar product	91
$\mathbf{a}\wedge\mathbf{b}$	vector product	97
$\vec{l}$	oriented line	99
$\sim$	is approximately equal to	107
$[\mathbf{u}, \mathbf{v}, \mathbf{w}]$	vector triple product	125
$\operatorname{im} f$	image of the mapping f	134
$\ker f$	kernel of the mapping f	134
$\prod$	product sign	145
$V(x_1, \ldots, x_n)$	Vandermonde determinant	146
$\mathbf{J}_r(\alpha)$	Jordan matrix block	160
$\mathrm{D}f$	Jacobian matrix of a mapping f	190
$\mathrm{H}f$	Hessian matrix of a function f	191
$\dot{x}, x'$	derivative	192, 202
$W(u_1, \ldots, u_n)$	Wronskian	203

Bibliography

In addition to the books quoted in the text this list contains titles of books where the subjects discussed here are treated in more detail.

Ahlfors, L. V. (1979) *Complex Analysis*, 3rd edn, McGraw-Hill, New York.

Artin, E. (1957) *Geometric Algebra*, Interscience, New York.

Cohn, P. M. (1982) *Algebra*, 2nd edn, Vol. 1, J. Wiley & Sons, Chichester.

Cohn, P. M. (1989) *Algebra*, 2nd edn, Vol. 2, J. Wiley & Sons, Chichester.

Cohn, P. M. (1991) *Algebra*, 2nd edn, Vol. 3, J. Wiley & Sons, Chichester.

Coxeter, H. S. M. (1961) *Introduction to Geometry*, J. Wiley & Sons, New York.

Dantzig, G. B. (1963) *Linear Programming and Extensions*, Princeton University Press, Princeton, NJ.

Gelfond, A. O. (1971) *Calculus of Finite Differences*, Hindustan Publishing Co., Delhi.

Harary, F. (1969) *Graph Theory*, Addison-Wesley, Cambridge, MA.

Hurewicz, W. (1958) *Lectures on Ordinary Differential Equations*, MIT Press, Cambridge, MA.

Jacobson, N. (1985) *Basic Algebra I*, 2nd edn, Freeman, San Francisco.

Jacobson, N. (1989) *Basic Algebra II*, 2nd edn, Freeman, San Francisco.

Kolmogorov, A. and Fomine, S. (1974) *Éléments de la théorie des fonctions et de l'analyse fonctionnelle*, Mir, Moscow.

Lidl, R. and Pilz, G. (1984) *Applied Abstract Algebra*, Springer-Verlag, New York and Berlin.

Loomis, L. H. and Sternberg, S. (1968) *Advanced Calculus*, Addison-Wesley, Reading, MA.

Luenberger, D. (1973) *Introduction to Linear and Non-linear Programming*, Addison-Wesley, Reading, MA.

Poston, T. and Stewart, I. (1978) *Catastrophe Theory and its Applications*, Pitman, London.

Index